Fundamentals of Spin Exchange

Kev M. Salikhov

Fundamentals of Spin Exchange

Story of a Paradigm Shift

Springer

Kev M. Salikhov
Zavoisky Physical-Technical Institute
of Russian Academy of Sciences
Kazan, Russia

ISBN 978-3-030-26824-4 ISBN 978-3-030-26822-0 (eBook)
https://doi.org/10.1007/978-3-030-26822-0

This Springer imprint is published by the registered company Springer Nature Switzerland AG.
The registered company address is: Gewerbestrasse 11, 6330 Cham, Switzerland

"The important thing in science
is not so much to obtain new facts
as to discover new ways of thinking about them."

William Lawrence Bragg: Physicist
1957

Preface

This book is about the paradigm shift in spin exchange and its manifestations in Electron Paramagnetic Resonance (EPR). Since I entered the area of spin exchange research in the 1970s and until now the understanding of spin exchange has changed to the point that we now have to admit there is a definite shift in the paradigm of how we understand spin exchange, how we measure the rate of spin exchange using EPR methods, and how we can use the spin exchange.

The paradigm shift came about as a result of the following achievements.

Initially, when considering the elementary act of spin exchange in the course of bimolecular collisions, only a large exchange interaction was usually considered, while much smaller spin dependent interactions of individual paramagnetic particles were neglected. But these minor interactions can give major consequences for the spin exchange if the duration of collision of two particles is long enough. So in the new paradigm, we differentiate equivalent and nonequivalent spin exchange cases.

A novel description of the spin dynamics for dilute solutions of paramagnetic particles was formulated based on the independent collective modes. For example, based on this, the physical nature of the exchange narrowing effect of the EPR spectra in the case of fast spin exchange can be elucidated at a much deeper level than previously possible. It was shown that resonance lines of these collective modes in the case of slow spin exchange manifest an asymmetric shape, as they are sums of the Lorentzian symmetric absorption and asymmetric dispersion terms. This asymmetric shape of collective resonances forced the reconsideration of the algorithm of finding the rate of spin exchange from EPR data. In strong microwave fields, the new collective modes are formed which can be considered as spin polaritons.

To study bimolecular collisions of paramagnetic particles in solution using spin exchange, the effects of dipole-dipole interactions must be taken into account. I have theoretically predicted (see [3], Chap. 4) that dipole-dipole interaction also produces a dispersion term but with a sign opposite to that for Heisenberg exchange interaction. This fact had escaped the attention of workers in the field for a long time. It had also been overlooked in previous theories (see, e.g., [4], Chap. 8).

As we all know, the scientific community has a reasonable level of inertia when it comes to adopting new ideas. And this is a great thing, because it forces us to scrupulously scrutinize every fact and every observation to ensure that every little bit of knowledge is well researched and proven.

I hope that this book will help the reader to learn about the modern understanding of spin exchange and give enough of a foundation to continue research in this field. Because of the volume of confirmed concepts and solutions we now have, it allows me to confidently conclude that we are experiencing a paradigm shift in the field of spin exchange. This paradigm shift happened as a result of theoretical predictions and fundamental experiments.

In the late 1970s along with Yu.N. Molin and K.I. Zamaraev, we wrote the book *Spin Exchange: Principles and Applications in Chemistry and Biology* which was published by the Publishing House of the USSR Academy of Sciences [1] and was later released in English with minor revisions [2]. Over the last 40 years, the aforementioned book continues to be the most popular book on this subject.

My contribution to this book (Chap. 2) was a comprehensive overview of the state of theoretical achievements in this area. There were even a number of predictions which first appeared in this book which had previously never been published before.

As a scientist and theoretician, I feel fulfilled and fortunate because practically every theoretical prediction I made in this book and in some other publications [5–11] have now been proven experimentally [12–18].

The purpose of this book is to present the current state of the theory of spin exchange, the theory of paramagnetic relaxation in a liquid due to dipole-dipole interaction, and a theoretical analysis of the manifestations of spin exchange in EPR spectroscopy. In addition to the theoretical results presented in [1, 2, 3], new theoretical results [5–11] are presented here. They are confirmed by experimental data (see, e.g., [12–18]). All this led to a qualitatively new understanding of the role of spin exchange in the dynamics of electron spins in liquids and its manifestations in EPR spectroscopy. In fact, there was a change in the spin exchange paradigm in dilute solutions of paramagnetic particles, and especially its manifestations in EPR spectroscopy, in algorithms for measuring the rate of spin exchange using EPR spectroscopy.

Experimental data on spin exchange are an important part of this book: they provide confirmation of the theory of spin exchange and its manifestations in EPR spectroscopy. For experimenters, sections of the book in which the modern protocol for determining the spin exchange rate constant from the analysis of the shape of the EPR spectra which is described in detail will be of particular interest. The modern protocol is very different from the protocol that was used in previous years [1, 2].

At present, studies dedicated to spin exchange are expanding, which is largely facilitated by the development of methods for preparing spin-labeled samples (see, e.g., [19]). It is especially important to note that chemists have developed methods for targeted incorporation into biopolymer molecules of spin labels. As such, many other new promising areas of spin exchange have been developed. For example, it turns out that spin exchange plays an important role in the formation of dynamic polarization of nuclei. This makes it possible to significantly increase the intensity of the observed nuclear magnetic resonance signals without lowering the temperature or using very high inductions of a constant magnetic field (see, e.g., [20, 21]).

I consider myself very lucky to have been involved in studying the very interesting problem of spin exchange and over the years have greatly enjoyed my collaborative work with many colleagues around the world – all to whom I am very thankful. It was always especially stimulating to work alongside experimenters, as, for the paradigm shift to happen, it was crucial that theoretical predictions were confirmed by experiments.

I'd like to express my gratitude for the opportunities to work with colleagues in the Institute of Chemical Kinetics and Combustion of the Russian Academy of Sciences in Novosibirsk, the Zavoisky Physical-Technical Institute of the Russian Academy of Sciences in Kazan, Physics Department of the Free University of Berlin, as well as the Department of Physics and Astronomy, Center for Biological Physics, California State University at Northridge.

A special thank you to my daughter, Assia, who helped me to achieve greater clarity of communication about the nature of this paradigm shift and to Igor Axenov who has been helping me with beautiful presentation of images and illustrations for over 30 years.

Scientific research demands total concentration, and I consider my achievements to have been made possible due to the tremendous support of my wife, Zoya, who passed away when I was starting to write this book.

Kazan, Russia Kev M. Salikhov

References

1. Zamaraev, K.I., Molin, Y.N., Salikhov, K.M.: Spin exchange. Nauka, Sibirian branch, Novosibirsk (1977)
2. Molin, Y.N., Salikhov, K.M., Zamaraev, K.I.: Spin exchange. Principles and applications in chemistry and biology. Springer, Heidelberg/Berlin (1980)
3. Salikhov, K.M., Semenov, A.G., Tsvetkov, Y.D.: Electron spin echo and its application. Nauka, Sibirianbranch, Novosibirsk (1976)
4. Abragam, A.: The principles of nuclear magnetism. Clarendon Press, Oxford (1961)
5. Salikhov, K.M.: The contribution from exchange interaction to the line shifts in ESR spectra of paramagnetic particles in solutions. J. Magn. Reson. **63**, 271–279 (1985)
6. Galeev, R.T., Salikhov, K.M.: The theory of dipolar broadening of magnetic resonance lines in non-viscous liquids. Khimicheskaya Fisika. **15**, 48–64 (1996)
7. Salikhov, K.M.: Contributions of exchange and dipole–dipole interactions to the shape of EPR spectra of free radicals in diluted solutions. Appl. Magn. Reson. **38**, 237–256 (2010)

8. Salikhov, K.M., Mambetov, A.E., Bakirov, M.M., Galeev, R.T., Khairuzhdinov, I.T., Zaripov, R.B., Bales, B.: Spin exchange between charged paramagnetic particles in dilute solutions. Appl. Magn. Reson. **45**, 911–940 (2014)
9. Salikhov, K.M., Bakirov, M.M., Galeev, R.T.: Detailed analysis of manifestations of the spin coherence transfer in EPR spectra of ^{14}N nitroxide free radicals in non-viscous liquids. Appl. Magn. Reson. **47**, 1095–1122 (2016)
10. Salikhov, K.M.: Consistent paradigm of the spectra decomposition into independent resonance lines. Appl. Magn. Reson. **47**, 1207–1228 (2016)
11. Salikhov, K.M.: Peculiar features of the spectrum saturation effect when the spectral diffusion operates: system with two frequencies. Appl. Magn. Reson. **49**, 1417–1430 (2018)
12. Bales, B.L., Peric, M.: EPR line shifts and line shape changes due to spin exchange of nitroxide free radicals in liquids. J. Phys. Chem. B. **101**, 8707–8716 (1997)
13. Bales, B.L., Peric, M.: EPR line shifts and line shape changes due to spin exchange of nitroxide free radicals in liquids 2. Extension to high spin Exchange frequencies and inhomogeneously broadened spectra. J. Phys. Chem. A. **106**, 4846–4854 (2002)
14. Bales, B.L., Peric, M.: EPR line shifts and line shape changes due to spin exchange of nitroxide free radicals in liquids 3. Extension to five hyperfine lines. Additional line shifts due to re-encounters. J. Phys. Chem. A. **107**, 9086–9098 (2003)
15. Bales, B.L., Meyer, M., Smith, S., Peric, M.: EPR line shifts and line shape changes due to spin exchange of nitroxide free radicals in liquids 4. Test of a method to measure re-encounter rates in liquids employing ^{15}N and ^{14}N nitroxide spin probes. J. Phys. Chem. A. **112**, 2177–2181 (2008)
16. Bales, B.L., Meyer, M., Smith, S., Peric, M.: EPR line shifts and line shape changes due to spin exchange of nitroxide free radicals in liquids 6. Separating line broadening due to spin exchange and dipolar interactions. J. Phys. Chem. A. **113**, 4930–4940 (2009)
17. Bales, B.L., Cadman, K.M., Peric, M., Schwartz, R.N., Peric, M.: Experimental method to measure the effect of charge on bimolecular collision rates in electrolyte solutions. J. Phys. Chem. **115**, 10903–10910 (2011)
18. Bales, B.L., Bakirov, M.M., Galeev, R.T., Kirilyuk, I.A., Kokorin, A.I., Salikhov, K.M.: The current state of measuring bimolecular spin exchange rates by the EPR spectral manifestations of the exchange and dipole-dipole interactions in dilute solutions of nitroxide free radicals with proton hyperfine structure. Appl. Magn. Reson. **48**, 1399–1447 (2017)
19. Altenbach, C., Froncisz, W., Hemker, R., Mchaourab, H., Hubbell, W.L.: Accessibility of nitroxide side chains: absolute Heisenberg exchange rates from power saturation EPR. Biophys. J. **89**, 2103–2112 (2005)
20. Bates, R.D., Drozdoski, W.S.: Use of nitroxide spin labels in studies of solvent-solute interactions. J. Chem. Phys. **67**, 4038 (1977)
21. Tuerke, M.T., Bennati, M.: Comparison of overhauser DNP at 0.34 and 3.4 T with Fremy's salt. Appl. Magn. Reson. **43**, 129–138 (2012)

Contents

About the Author

Kev M. Salikhov is professor, full member of the Russian Academy of Sciences, Lenin Prize winner (USSR), gold medalist of the International EPR (ESR) Society, fellow of Wissenschaftskolleg zu Berlin, winner of Humboldt research award and eminent scientist of RIKEN (Japan).

He is the founder of *Applied Magnetic Resonance* journal and initiator of the international prize named after Zavoisky, which is annually awarded to scientists for their outstanding contribution to the development of magnetic resonance and its application.

Kazan-Akademgorodok (Novosibirsk)

Chapter 1
Introduction: Development of the Study of Spin Exchange from a Bird's Eye View

Abstract A phenomenological representation of spin exchange in bimolecular collisions of paramagnetic particles and manifestations of spin exchange in the form of EPR spectra are given. Two periods in the study of spin are distinguished: early (widespread) and modern. The main theoretical and experimental results that led to a change in the spin exchange paradigm are listed.

The change in the state of the electron spins of two paramagnetic particles caused by the exchange interaction between them during their bimolecular collisions is called spin exchange. This process makes a significant contribution to the relaxation of electron spins in solutions of paramagnetic particles, e.g., complexes of paramagnetic ions of elements of transition groups, free radicals, and triplet excitons. Spin exchange between paramagnetic particles occupies a special place among bimolecular processes, because it is an example of an elementary physicochemical process for which there are well-developed methods of detailed experimental research and theoretical analysis of the experimental data. Therefore, spin exchange can be used and it is used as a model bimolecular process.

Spin exchange is associated with remarkable transformations of the shape of the electron paramagnetic resonance (EPR) spectrum of solutions of paramagnetic particles when the rate of this bimolecular process changes. Therefore, EPR spectroscopy is an ideal tool for studying spin exchange. For example, at sufficiently fast spin exchange, the entire spectrum merges into one narrow homogeneously broadened line, which has a Lorentzian shape. This phenomenon is called the exchange narrowing of the spectra. The same as in any bimolecular process, the spin exchange rate can be controlled by changing the concentration of paramagnetic particles or the viscosity of solvents. The analysis of the transformations of the shape of the EPR spectrum, in principle, makes it possible to determine the spin exchange rate. However, the development of spin exchange studies shows that the protocol for analyzing the EPR spectra and determining the spin exchange rate changed as the theory of the elementary spin-exchange act and the theory of EPR spectra developed.

Two periods can be distinguished in the development of spin exchange studies: the early period (from 1950 to around 1990) and the modern period. The spin exchange paradigm, which was formulated at the early period and gained

K. M. Salikhov, *Fundamentals of Spin Exchange*,
https://doi.org/10.1007/978-3-030-26822-0_1

recognition, was thoroughly summarized, e.g., in monographs [1, 2]. In these books in the section of the theory of spin exchange, I formulated several theoretical predictions. For example, I showed that the EPR lines should have a mixed shape in the region of slow spin exchange: they should contain absorption and dispersion contributions. Therefore the resonance lines should be asymmetric. This asymmetry provides additional contributions to the broadening and shifts of position of maximum value of the resonance lines of spectra detected in EPR experiments. At that time these predictions had no experimental confirmation and therefore did not arise much interest. As a result, they were not taken into account either in the widely accepted paradigm of spin exchange or in the protocol for analyzing the shape of the EPR spectra in the presence of spin exchange.

The conventional theory of spin exchange was based on the assumption that during the collision of two paramagnetic particles, the exchange interaction completely determines the spin motion of unpaired electrons, and all spin-dependent interactions of isolated paramagnetic particles can be ignored. This situation can be called the case of equivalent spin exchange. However, the theory predicts that taking into account the spin-dependent interactions of isolated (individual) particles along with the exchange interaction between particles during their bimolecular collisions can significantly change the effect of the collision on the state of the spins. Minor interactions can produce major effects if they act for the time long enough. In this situation, which can be called nonequivalent spin exchange, an additional shift of the EPR resonant frequencies appears, and it is necessary to introduce different rates to describe the spin decoherence due to the spin exchange and the transfer of quantum coherence between spins in bimolecular collisions. The theory also predicted that, due to spin exchange, the shape of the individual resonance lines is asymmetric, i.e., each line is a sum of a symmetric Lorentzian absorption line and an asymmetric Lorentzian dispersion line.

Note also that, along with the exchange interaction, the paramagnetic relaxation of spins in solutions is caused by the spin-spin dipole-dipole interaction between paramagnetic particles. The contribution of the dipole-dipole interaction also depends on the concentration of spins. Therefore, to find the spin exchange rate due to the exchange interaction in the course of bimolecular collisions, it is necessary to separate the contributions of the exchange and dipole-dipole interactions to the paramagnetic relaxation, and, ultimately, to the shape of the observed EPR spectra. Early protocols for analyzing EPR spectra and extracting the spin exchange contribution (see [1–3]) did not take into account the spin coherence transfer due to the dipole-dipole interaction. Strictly speaking, they did not take into account the reaction of a partner spin to an action of a given spin due to the dipole-dipole interaction. Meanwhile, in the theory of pulsed EPR spectroscopy, I showed [4], see also [5] that the dipole–dipole interaction leads to the spin coherence transfer, but, surprisingly, the contributions of the exchange and dipole–dipole interactions to the spin coherence transfer do not add up, as it might be expected. On the contrary, they are subtracted.

The early paradigm of spin exchange, which did not take into account the above details, nevertheless gives a good idea of spin exchange on the phenomenological level. It also reflects some of the fundamentally important manifestations of spin exchange in the shape of EPR spectra: broadening of individual lines, shifts of positions of lines and exchange narrowing of the spectra. In the overwhelming number of studied systems, the early paradigm was a good guiding star and contributed to the development of the spin probe method and its application in chemistry and molecular biology. There were, of course, rare exceptions, in which the deviations from predictions on the basis of the early paradigm of equivalent spin exchange appeared in the interpretation of experiments,. For example, according to the then prevailing paradigm, the shift of the positions of lines, the broadening of the lines and the exchange narrowing should give the same spin exchange rate. In [6], when studying spin exchange between triplet excitons and interpreting data within the early protocol, different spin exchange rates were obtained from the data on the line position shift, line broadening, and exchange narrowing of the spectrum. Note that an alternative interpretation of this observation was suggested in [1, 7] based on the assumption that the nonequivalent spin exchange is implemented in the system [6]. In another interesting work [8], the authors paid attention to the fact that in the conditions of spin exchange the shape of the line is asymmetric. But these separate "cries" drowned in the powerful "chorus" of works, in which everything could be described within the conventional paradigm.

A new stage in the development of spin exchange studies began with a series of remarkable experiments, which was implemented by B. Bales et al. [3, 9–13]. They conducted systematic experimental studies of spin exchange between stable nitroxide radicals. By thorough experiments, they demonstrated an additional shift of the resonant frequencies, which was theoretically predicted [1, 2, 14] and is associated with the effect of spin-dependent interactions of isolated radicals on the spin dynamics when the exchange interaction between the radicals is switched on during their collision [10]. It was also proved convincingly that, in complete agreement with the theoretical prediction [1, 2] under conditions of relatively slow spin exchange, the lines of the EPR spectrum are asymmetric being the sum of absorption and dispersion lines [8–10, 15–17]. As a result, it was shown that the fraction of the dispersion contribution measured in the experiment makes it possible to determine the spin exchange rate [1, 2, 10, 18, 19]. It was shown that this method gives the most accurate values of the spin exchange rate, namely, the spin coherence transfer rate.

It was shown in [1, 2] in the second order perturbation theory that the spin coherence transfer between paramagnetic particles leads to the formation of independent collective modes of motion of the quantum coherence of spins. This collective mode approach was further developed comprehensively in [19, 20]. The main progress in [19, 20] compared with the results obtained in [1, 2] was that in [19, 20], the exact analytical solution of the problem was found for model situations. These exact solutions made it possible to better understand physics behind the exchange narrowing of the EPR spectra when the spin exchange is fast enough. Each collective mode has its own characteristic frequency and the resonance "width" at this frequency. In the case of slow spin exchange, spectral lines corresponding to these collective modes have a mixed shape: they consist of absorption and dispersion contributions.

This approach based on collective modes produces a basis for a completely new paradigm of manifestations of spin exchange in EPR spectroscopy. For example, the phenomenon of exchange narrowing of the EPR spectra at the sufficiently high spin exchange rate is explained as follows. We suppose that initially the EPR spectrum consists of N components, i.e., spins had N different frequencies. Given the coherence transfer between spins, the behavior of the coherence of the system as a whole can be reduced to N independent collective modes. Each mode has its own frequency and its own "oscillator" force. Under conditions of exchange narrowing, at least one of the collective modes has a resonant frequency, which is equal to the center of gravity frequency of the original EPR spectrum of the system, and only this mode contributes to the observed spectrum and it accumulates all the integral intensity of the spectrum. All other modes have negligibly small "oscillator" forces and are not excited by the alternating magnetic field. They are the so-called "dark" states. Note that this is a completely new look at the phenomenon of exchange narrowing of the spectrum. Traditionally, the exchange narrowing is interpreted as averaging of the Hamiltonian by any motion, as a result of which all spins have an average frequency. In fact, it turns out that even in the conditions of exchange narrowing of the spectrum, collective modes have different frequencies but only one of the collective modes is observed in the EPR experiment.

Even more interesting effects are expected if the EPR spectra are studied not in the linear response mode, but under the conditions of the saturation effect in intense alternating magnetic fields. The conventional paradigm is that the saturating microwave field broadens the lines of the EPR spectrum. In the presence of spin exchange in the intense microwave field, new collective states of the magnetization of the spins and the microwave field, like polaritons, are formed [20]. Spin exchange shifts the resonance frequencies of the spins to the center of gravity of the spectrum, while the intense alternating field tends to "push apart" the resonant frequencies of the collective modes [20].

These results of theoretical and experimental studies have led to a new level of the analysis of the manifestations of spin exchange in EPR spectroscopy. In fact, we deal with a new paradigm of spin exchange and its manifestations in EPR spectroscopy and a new protocol for determining the spin exchange rate by using EPR spectroscopy.

This book presents the current state of the theory of electron spin polarization transfer and spin decoherence in dilute solutions of paramagnetic particles caused by the spin-spin interaction between paramagnetic particles. The kinetic equations for single-particle spin density matrices are given with allowance for the Heisenberg exchange interaction in the course of bimolecular collisions of particles and also taking into account the spin-spin dipole-dipole interaction between particles. The results of calculating the spin exchange rate (effective spin exchange radius) for a number of model situations are summarized. The characteristic manifestations of spin exchange and spin-spin dipole-dipole interaction in the shape of electron paramagnetic resonance (EPR) spectra are described. The general solution of the problem for the shape of the EPR spectrum of paramagnetic particles in the presence of spin exchange is given. Modern algorithms for separating the contributions of the exchange and dipole-dipole interactions to the shape of the spectra and, ultimately, determining the bimolecular spin exchange rate from the analysis of the EPR spectra are presented. Special attention is paid to the formation of collective modes of the evolution of spin polarization, in particular quantum spin coherence, in dilute solutions of paramagnetic particles due to the exchange and dipole-dipole interactions. A new interpretation of the effect of exchange narrowing of the EPR spectra is presented. The experimental EPR data are referred which confirm the theory. It is demonstrated that the modern theory of spin exchange makes it possible to consistently describe changes in the shape of the EPR spectra caused by spin exchange and to determine the spin exchange rate from the analysis of the shape of the EPR spectra.

References

1. Zamaraev, K.I., Molin, Y.N., Salikhov, K.M.: Spin exchange. Nauka, Sibirian branch, Novosibirsk (1977)
2. Molin, Y.N., Salikhov, K.M., Zamaraev, K.I.: Spin exchange. Principles and applications in chemistry and biology. Springer, Heidelberg/Berlin (1980)
3. Bales, B.L.: Berliner, L.J., Reuben, J. (eds) Biological magnetic resonance, vol. 8, pp. 77. New York, Plenum Publishing Corporation (1989)
4. Salikhov, K.M., Semenov, A.G., Tsvetkov, Y.D.: Electron spin echo and its application. Nauka, Sibirianbranch, Novosibirsk (1976)
5. Galeev, R.T., Salikhov, K.M.: The theory of dipolar broadening of magnetic resonance lines in non-viscous liquids. Khim. Fizika. **15**, 48–64 (1996)
6. Jones, M.T., Chesnut, D.B.: Triplet spin exchange in some ion radical salts. J. Chem. Phys. **38**, 1311–1317 (1963)
7. Rumyantsev, E.E., Salikhov, K.M.: Spin exchange between triplet excitons. Opt. Spectroscopy. **35**, 570–573 (1973)
8. Chesnut, D.B., Philips, W.D.: EPR studies of spin correlation in some ion radical salts. J. Chem. Phys. **35**, 1002–1012 (1961)
9. Bales, B.L., Willett, D.: EPR investigation of the intermediate spin exchange regime. J. Chem. Phys. **80**, 2997–3004 (1984)
10. Bales, B.L., Peric, M.: EPR line shifts and line shape changes due to spin exchange of nitroxide free radicals in liquids. J. Phys. Chem. B. **101**, 8707–8716 (1997)
11. Bales, B.L., Peric, M.: EPR line shifts and line shape changes due to spin exchange of nitroxide free radicals in liquids 2. Extension to high spin Exchange frequencies and inhomogeneously broadened spectra. J. Phys. Chem. A. **106**, 4846–4854 (2002)
12. Bales, B.L., Peric, M., Dragutan, I.: EPR line shifts and line shape changes due to spin exchange of nitroxide free radicals in liquids 3. Extension to five hyperfine lines. Additional line shifts due to re-encounters. J. Phys. Chem. A. **107**, 9086–9098 (2003)
13. Bales, B.L., Meyer, M., Smith, S., Peric, M.: EPR line shifts and line shape changes due to spin exchange of nitroxide free radicals in liquids 4. Test of a method to measure re-encounter rates in liquids employing ^{15}N and ^{14}N nitroxide spin probes. J. Phys. Chem. A. **112**, 2177–2181 (2008)
14. Salikhov, K.M.: The contribution from exchange interaction to the line shifts in ESR spectra of paramagnetic particles in solutions. J. Magn. Res. **63**(271–279), (1985)

15. Kurban, M.R., Peric, M., Bales, B.L.: Nitroxide spin exchange due to re-encounter collisions in a series of n-alkanes. J. Chem. Phys. **129**, 064501 (2008)
16. Salikhov, K.M., Bakirov, M.M., Galeev, R.T.: Detailed analysis of manifestations of the spin coherence transfer in EPR spectra of ^{14}N nitroxide free radicals in non-viscous liquids. Appl. Magn. Reson. **47**, 1095–1122 (2016)
17. Bales, B.L., Bakirov, M.M., Galeev, R.T., Kirilyuk, I.A., Kokorin, A.I., Salikhov, K.M.: The Current State of Measuring Bimolecular Spin Exchange Rates by the EPR Spectral Manifestations of the Exchange and Dipole-dipole interactions in Dilute Solutions of Nitroxide Free Radicals with Proton Hyperfine Structure. Appl. Magn. Reson. **48**, 1399–1447 (2017)
18. Salikhov, K.M.: Contributions of exchange and dipole–dipole interactions to the shape of EPR spectra of free radicals in diluted solutions. Appl. Magn. Reson. **38**, 237–256 (2010)
19. Salikhov, K.M.: Consistent Paradigm of the Spectra Decomposition into Independent Resonance Lines. Appl. Magn. Reson. **47**, 1207–1228 (2016)
20. Salikhov, K.M.: Peculiar Features of the Spectrum Saturation Effect when the Spectral Diffusion Operates: System with Two Frequencies. Appl. Magn. Reson. **49**, 1417–1430 (2018)

Chapter 2
Theory of Spin Exchange in Dilute Solutions

Abstract Kinetic equations are presented that are used to describe the bimolecular process of changing the state of electron spins (spin exchange) of paramagnetic particles caused by their exchange interaction in collisions. The theoretically calculated effective radii and spin exchange rate constants are presented. The conditions for the realization of equivalent spin exchange are discussed in detail. Under conditions of nonequivalent spin exchange, a frequency shifts of the EPR transitions occur. The role of re-encounters of pairs of particles during meetings, the role of the extended nature of the exchange integral, the influence of the total spin of a paramagnetic particle, the anisotropic nature of the distribution of the spin density of paramagnetic particles and the charge of paramagnetic particles are discussed in detail.

Spin exchange in dilute solutions is a bimolecular process. The study of spin exchange and the measurement of the bimolecular spin exchange rate are of great importance from many points of view. Spin exchange can be used and it is used as a model bimolecular process. By measuring the spin exchange rate, it is possible to obtain information about the rate of diffusion bimolecular collisions. This information can be used in the study of any bimolecular reactions. This is particularly important when it comes to bimolecular processes, such as chemical reactions, in complex environments, in biological systems. Spin exchange data provide valuable information on the molecular mobility, effect of solvent and particle charge, etc. on the rate of diffusion bimolecular collisions.

Spin exchange itself is also of interest. It provides unique information about the exchange interaction in the course of collisions of paramagnetic particles, the overlap of wave functions of unpaired electrons of particles in collisions with each other. Spin exchange between spin probes and oxygen molecules makes it possible to measure the oxygen concentration, e.g., in biological objects in real time, etc.

In fact, spin exchange is an example of a bimolecular "spin chemical reaction". Similar to all bimolecular processes, spin exchange has two stages. One stage is the diffusion collision of two particles. It is expected and assumed that the frequency of diffusion bimolecular collisions is determined by the random walk of molecules, which does not depend on the spin state of unpaired electrons of paramagnetic

K. M. Salikhov, *Fundamentals of Spin Exchange*,
https://doi.org/10.1007/978-3-030-26822-0_2

particles. The description of diffusion collisions of particles is a problem of molecular physics.

Another stage of spin exchange is actually an elementary act of changing the spin state of two colliding particles, which is induced by the spin-dependent exchange interaction. The collision efficiency and the effective spin exchange radius in a collision are problems of quantum dynamics of a pair of colliding paramagnetic particles.

2.1 Bimolecular Spin Exchange Process

Bimolecular collisions in condensed media have significant differences from the situation in gases. This section presents the description of bimolecular collisions in condensed matter. The basic concepts and physical parameters that are used to describe bimolecular processes are given: the effective radius and the rate constant of the bimolecular process.

Bimolecular collisions determine the rate of many physical, chemical and biological processes, e.g., energy transfer between molecules, intermolecular charge transfer (electron, proton), recombination of electron-hole pairs or free radicals, etc. The most important characteristic of bimolecular processes is their rate. It depends on the local concentration of the particles involved in the bimolecular process, on the kinematics of the molecular motion, on the diffusion coefficient and on the interaction of the colliding particles. From the point of view of the use of spin exchange, it is important to note that the rate of bimolecular processes makes it possible to identify subtle details of collisions, e.g., the degree of availability of certain areas of polymer molecules for collisions with added molecules. As an actual example, collisions of substrate molecules with active catalytic centers of enzymes in biological systems can be noted.

Spin exchange is commonly referred to as the change in the state of electron spins of two colliding paramagnetic particles caused by the exchange interaction [1, 2]. The exchange interaction between two paramagnetic particles is due to the overlap of electron orbitals of these particles (the so-called direct exchange) and overlap of the electron orbitals due to the delocalization of electrons in the diamagnetic particles between the paramagnetic particles and the appearance of a nonzero spin polarization on the bridging atoms (molecules), (the so-called indirect exchange and superexchange). Quantitative calculations of exchange interaction parameters are complicated. But a qualitative idea about the nature of the dependence of the exchange interaction on the distance between the colliding paramagnetic particles in the solution can be obtained on the basis of the fact that the exchange interaction requires overlap of the electron wave functions of these particles. The overlap of the electron wave functions of two particles decreases exponentially with the distance between them. At distances on the order of van der Waals radii of particles, the exchange interaction in a sufficiently good approximation is proportional to the square of the overlap integral of the electron orbitals of particles [3]. This means that the exchange interaction is short-range. Therefore, in dilute solutions of paramagnetic particles spin exchange mainly occurs due to bimolecular collisions.

Schematically, bimolecular spin exchange can be represented as

$$A(\uparrow) + B(\downarrow) \rightleftarrows A(\downarrow) + B(\uparrow). \tag{2.1}$$

Here the arrows indicate the orientation of the electron spins of paramagnetic particles A and B before the collision and after the collision. It is well known that the exchange interaction leads to the formation of covalent bonds in molecules. However, for the implementation of spin exchange, the exchange interaction with much lower energy, which occurs when particles approach each other at distances on the order of the sum of their van der Waals radii, might be sufficient. In these situations, the exchange interaction is less than not only of the energy of chemical bonds, but also of the thermal energy kT at room temperature. Therefore, spin exchange occurs with little or no change in the structure and nature of the thermal motion of the colliding molecules. This circumstance greatly simplifies the theoretical consideration of bimolecular spin exchange.

The study of bimolecular spin exchange is of great interest from several points of view. The elementary act of spin exchange can be described in some detail theoretically. The methods of electron paramagnetic resonance (EPR) spectroscopy make it possible to experimentally investigate the spin exchange rate, its dependence on the concentration of paramagnetic particles, viscosity of a system, temperature, charge of particles, etc. Therefore, spin exchange is used as a model process for studying bimolecular collisions in complex systems, such as polymer molecule solutions, liquid inclusions in porous media, biological systems, etc. Spin exchange makes it possible to study collisions of two monomer units remote along the polymer chain. Suppose that spin labels are attached to a polymer molecule in two given places separated by a sufficiently large number of monomer units. In the course of random conformational transitions, the spin labels can approach so close ("collide") that an exchange interaction occurs between them and spin exchange can occur. As a result, analyzing the manifestations of spin exchange in the EPR spectra, it is possible to obtain unique information about the conformational transitions of polymer molecules. Spin exchange is also one of the important mechanisms of electron paramagnetic relaxation, including decoherence of electron spins.

The statistics of bimolecular collisions in condensed media, in particular in dilute solutions of paramagnetic particles, is fundamentally different from the situation in gases: in condensed media, repeated collisions (re-encounters) of the same partners in a "pair" are possible. In [4], it was experimentally shown for the first time that collisions of two partners in a pair occur by trains (see Fig. 2.1): the same partners repeatedly collide inside the train, the collision partner changes from one train to another.

The first collisions in trains (marked with bold lines in Fig. 2.1) obey the Poisson distribution with an average time τ_0 between trains (between the first collisions of a given molecule A with different molecules B). The statistics of re-encounters of the same partners inside the train (they are marked with thin lines in Fig. 2.1) are not

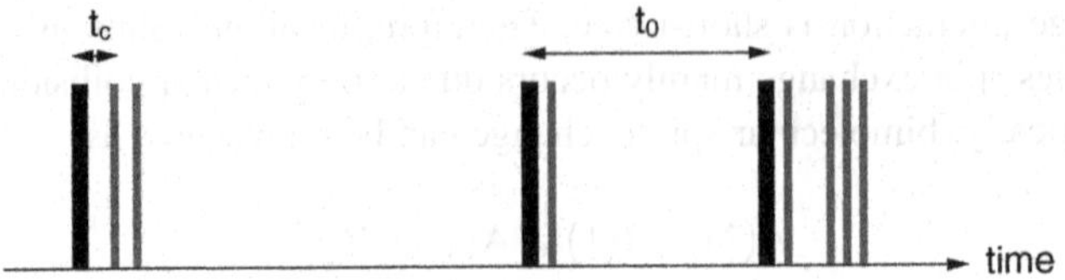

Fig. 2.1 Bimolecular collisions of particles in a condensed medium

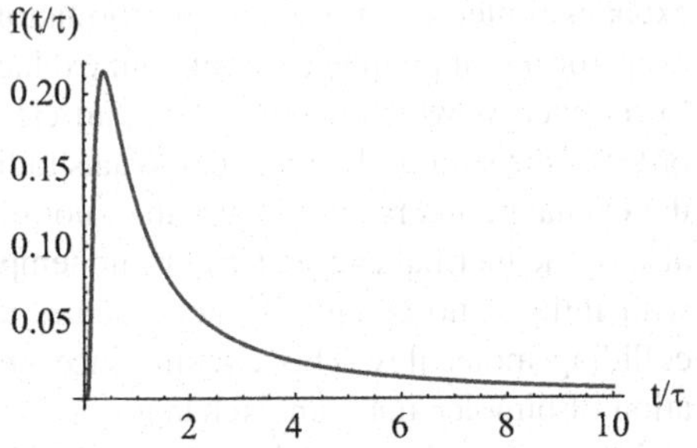

Fig. 2.2 Statistics of re-encounters of the same partners of a pair in a collision train, i.e., during one meeting. The curve is calculated for the parameters: $\lambda_D = r_0$, $p = 0.527$, $\tau = 10^{-12}$s

described by the Poisson distribution. For example, for the model of diffusion of molecules which are jumping with an average jump length in one elementary diffusion act equal to λ_D, the distribution of re-enconters over the time interval between re-encounters in a train of collisions is described by a function of the form [5] (see Fig. 2.2)

$$f(t) = m\left(1/t^{3/2}\right)\exp\left(-\pi m^2/\left(p^2 t\right)\right). \tag{2.2}$$

The parameters of this distribution are: p-probability of a re-encounter of a pair of particles. The parameter m is expressed in terms of the probability p and molecular kinetic parameters

$$m = (27/8\pi)^{1/2}(1-p)^2(r_0/\lambda_D)^2\tau^{1/2}. \tag{2.3}$$

Here τ is the average time interval between two subsequent diffusion jumps of molecules, r_0 is a distance of the closest approach of molecules in a pair.

Figure 2.2 shows the distribution function of a time interval between re-encounters in pairs of particles.

Considering repeated collisions of partners, in a condensed medium, it is possible to introduce a pair A . . . B of the colliding particles as a relatively long-lived intermediate state of two molecules (see, e.g., [6]). The frequency of the ***first*** collisions of molecule A with molecules B is the frequency of the formation of A . . . B pairs. Re-encounters occur within such a pair. When studying bimolecular collisions in a condensed medium, the concept of a meeting of two molecules is also used, which includes the first and all repeated collisions of an isolated pair of molecules. The statistics of these re-encounters, the role of spin dynamics in the time intervals between re-encounters are discussed in some detail in the books [6, 7] in connection with the problem of the influence of external magnetic fields on radical reactions, magnetic isotope effect and hyperpolarization of electron and nuclear spins during radical reactions. Spin dynamics in the time intervals between re-encounters also affects the efficiency of spin exchange during the meeting of two paramagnetic particles.

In a selected pair, different situations can be implemented: no re-encounter can occur, and several re-encounters can occur. In this sense, there might be an impression that it makes little sense to speak of some particular intermediate state: a pair of colliding particles. But the distribution of time between re-encounters statistically unambiguously characterizes the pair.

The probability of a re-encounter decreases with increasing length of a molecule jump in one elementary diffusion act. Therefore, the properties of the pairs in question depend not only on the nature of the partners of the pair, but also to a large extent on the solvent, the environment, and the kinematics of the particle motion. Note that in rarefied gases the probability of re-encounters is negligible, since the motion of molecules in a collision can be considered as a ballistic one.

An important parameter of these pairs is the characteristic lifetime. For example, for a pair of uncharged paramagnetic particles, e.g., in water with a viscosity of about 1 cP this time is about 10^{-10} s, in squalane with a viscosity of about 50 cP this time is about 10 ns. The lifetime of the pair increases significantly in environments with limited mobility. For example, the lifetime of pairs in the micelle can reach microseconds, this is a very long time in terms of the elementary act of spin exchange.

The frequency of diffusional bimolecular collisions in the condensed phase was first calculated by Smoluchowski [8]. He considered collisions of molecules of types A and B in dilute solutions, assuming that the random motion of molecules can be described by a continuous diffusion model and that molecules can be represented as solid balls. Under these assumptions, he obtained a well-known expression for the frequency of bimolecular collisions of any given type A molecule with type B molecules

$$\begin{gathered} Z_{A0} = 4\pi(r_A + r_B)(D_A + D_B)C_B = 4\pi r_0 D_{AB} C_B; \\ K_D \equiv 4\pi r_0 D_{AB} \end{gathered} \tag{2.4}$$

In this equation, r_A and r_B are the radii of balls representing the molecules, and D_A and D_B are the diffusion coefficients of molecules A and B, respectively. $D_{AB} = D_A + D_B$ is the coefficient of mutual diffusion of A and B molecules, and the distance of the closest approach of the balls $r_0 = r_A + r_B$ is called the collision radius.

Smoluchowski's formula (2.4) gives the frequency of the *first bimolecular collisions*. It can be noted that Smolukhowski's theory makes it possible to obtain the frequency of the first approach of pairs of molecules A and B not only to a distance r_0, but also at any distance $r > r_0 : Z_A(r) = 4\pi r D_{AB} C_B$. Therefore, it is possible to find the probability of a pair of molecules to be on the collision radius, if the entire ensemble of pairs starts from the initial distance r between the molecules

$$p(r) = r_0/r. \tag{2.5}$$

Suppose that $r_0 = 0.6$ nm and that after the first approach to the collision radius, the molecules-balls moved apart to $r = 1$ nm. Then they re-encounter on the collision radius with a fairly high probability $p = 0.6$. This shows that in the theory of bimolecular processes in condensed matter, including spin exchange, it is necessary to take into account the role of re-encounters at the meeting of two molecules.

For charged molecules (balls), an analog of formula (2.4) was obtained by Debye [9]. Debye showed that the frequency of the *first bimolecular collisions* is given by the formula similar to (2.4), but for charged particles the collision radius is not the sum of the radii of the balls, instead of it some effective radius is introduced. According to Debye, the frequency of the first bimolecular collisions and the effective collision radius of charged particles have the form

$$\begin{aligned} Z_{A0} &= 4\pi r_{eff} D_{AB} C_B, \\ r_{eff} &= f_D r_0, \\ f_D &= \left\{ r_0 \int_{r_0}^{\infty} \exp\left(\frac{U(r)}{kT}\right) \frac{dr}{r^2} \right\}^{-1}. \end{aligned} \tag{2.6}$$

For molecules with opposite charges $f_D > 1$, while for molecules with similar charges $f_D < 1$ (this issue is discussed in more detail in the Sects. 2.11 and 2.12).

Bimolecular processes are usually characterized by a value that does not depend on the concentration of molecules and is called the rate constant of the corresponding process,

$$K = 4\pi r_{eff} D_{AB}. \tag{2.7}$$

In this expression, the effective radius, r_{eff}, accumulates the ensemble average result of the interaction of two particles during one meeting. If there is no interaction, then $r_{eff} = 0$. If at the first approach of particles to the radius of their collision r_0 necessarily occurs the studied bimolecular process, then $r_{eff} = r_0$. If the interaction responsible for the studied bimolecular process is relatively weak, the effective radius may be smaller and even much smaller than the collision radius. In such a situation, the effective radius is often represented as a product of the collision radius on the probability (efficiency) of a single meeting of molecules

$$r_{eff} = p_{eff} r_0, \tag{2.8}$$

where the probability of the bimolecular process under consideration at one meeting, p_{eff}, takes values in the interval {0,1}. However, the interaction has an extended character, in general case, it is not zero also at distances between particles which are greater than the collision radius. Therefore, for a sufficiently strong interaction of a pair of encountered particles and sufficiently long "lifetime" of a pair of particles, the effective radius of the bimolecular process can be greater than the collision radius, i.e., it can be

$$r_{eff} > r_0. \tag{2.9}$$

In this case, it is no longer possible to use the effective radius representation as Eq. (2.8) because the probability cannot be greater than 1.

As for any bimolecular process, and for spin exchange, the rate constant can be represented as

$$Kex = 4\pi r_{ex} D_{AB}. \tag{2.10}$$

The effective exchange radius makes it possible to formulate a condition under which the solution can be considered sufficiently diluted to take into account only bimolecular collisions of paramagnetic particles and to ignore collisions of three or more particles. To do this, the inequality should be fulfilled

$$(4/3)\pi r_{EX}{}^3 C_B << 1. \tag{2.11}$$

The effective radius r_{ex} depends on the interaction between molecules, it accumulates the effect of interaction during the meeting of two molecules. Re-encounters affect the total time during which a given pair of spins is in the area of exchange interaction when they meet. Therefore, the effective spin exchange radius depends not only on the parameters of the exchange interaction, but also on the kinematics of the mutual motion of the interacting particles, e.g., for the continuous diffusion model, on the coefficient of mutual diffusion of colliding molecules.

2.2 The Exchange Interaction Between Two Paramagnetic Particles

Spin exchange is caused by the exchange interaction between the colliding paramagnetic particles. Therefore, it is advisable to briefly describe the physical nature of the exchange interaction. For the simplest situation, the exchange interaction is completely given by a single parameter, which is called the exchange integral. This section gives an estimate of the scale of the exchange integral on the collision radius r_0. It also provides examples of the elementary act of spin exchange in the collision of two paramagnetic particles, i.e., illustrations of changes in the state of electron spins as a result of exchange interaction in bimolecular collisions are given.

2.2.1 The Nature of Exchange Interaction

The so-called exchange interaction between paramagnetic particles is a very interesting and important manifestation of the indistinguishability of electrons and the presence of electron spin moment. Due to the spin moment, the electron has an additional (spin) degree of freedom. This ultimately leads to the fact that the energy of a multi-electron system depends on the spin of electrons, even in the absence of external magnetic fields. This dependence of the electron energy on the spin of electrons is called the exchange interaction.

To demonstrate how the spin dependence of electron terms appears, we consider the simplest system of two electrons. Suppose there are no external magnetic fields. In the nonrelativistic theory, the wave function of a system of two electrons $\psi(1,2)$ can be represented as a product of two functions [10]. One of these functions describes the distribution of electron density in the usual configuration space of two electrons: $\phi(\mathbf{r}_1,\mathbf{r}_2)$, where $\mathbf{r}_1,\mathbf{r}_2$ are the radius vectors of electrons with numbers 1 and 2. Another function describes the state of electrons in their spin space: $\chi(s_1,s_2)$, where s_1,s_2 are the spin degrees of freedom of the two electrons under consideration. Thus, the wave function of two electrons is represented as

$$\psi(1,2) = \varphi(\mathbf{r}_1,\mathbf{r}_2)\chi(s_1,s_2). \tag{2.12}$$

Electrons are indistinguishable. Therefore, when two electrons are permuted, the square of the module of the wave function (2.12) should not change. But for electrons with such a permutation, the wave function itself changes sign, i.e. $\psi(2,1) = -\psi(1,2)$, since electrons have a spin with a quantum number 1/2 and therefore are fermions. This property of the wave function of multi electron systems reflects the Pauli principle: two or more electrons can not be simultaneously in the same point of the configuration space and still have the same spin states.

For the situation described by Eq. (2.12), the anti-symmetry of the wave function $\psi(1,2)$ can be provided by two different options. In one case, the spatial distribution of electrons can be given by the symmetric function $\varphi_s(\mathbf{r}_1,\mathbf{r}_2) = \varphi_s(\mathbf{r}_2,\mathbf{r}_1)$, and the spin part of the wave function is antisymmetric, $\chi_a(s_1,s_2) = -\chi_a(s_2,s_1)$. In another case, the opposite is true: the spatial distribution of electrons is given by the antisymmetric function, and the symmetric wave function of spins. As a result, we obtain two types of electron states

$$\begin{aligned} \psi_1(1,2) &= \varphi_s(\mathbf{r}_1,\mathbf{r}_2)\chi_a(s_1,s_2); \\ \psi_2(1,2) &= \varphi_a(\mathbf{r}_1,\mathbf{r}_2)\chi_s(s_1,s_2);. \end{aligned} \tag{2.13}$$

Equations (2.13) show that the spatial distribution of electrons is correlated with the spin state of electrons. This means, e.g., that the energy of the Coulomb repulsion depends on the state of the spins of electrons. In the state $\psi_2(1,2)$ two electrons are forbidden to be in one point of space, since $\varphi_a(\mathbf{r},\mathbf{r}) = 0$, i.e. the state of spins completely excludes the variants of the greatest repulsion of electrons. And for the state $\psi_1(1,2)$, two electrons are allowed to be at the same time at the same point in space, that is, in this case the spin state of the electrons allowes variants of the greatest repulsion of electrons. It is obvious that in these states the energy of the Coulomb interaction should be different. At the same time, according to the virial theorem, the kinetic energies of electrons in the states of $\varphi_a(\mathbf{r}_1, \mathbf{r}_2)$и $\varphi_s(\mathbf{r}_1, \mathbf{r}_2)$ should differ as well. In these states, the energy of the Coulomb interaction of electrons with positively charged nuclei can also be different.

The antisymmetric and symmetric spin functions $\chi_a(s_1,s_2)$ and $\chi_s(s_1,s_2)$ (2.13) correspond to the states of two electrons with total spin S = 0 and S = 1, respectively. Thus, the example of a system of two electrons shows that the presence of the spin degree of freedom of electrons affects the movement of electrons in the coordinate space. So, the dependence of the coordinate part of the electron wave function on the total spin of electrons appears. As a result, the energy of a multi-electron system also depends on the total spin of electrons. This dependence of energy on the total spin of electrons is described as a result of some "interaction" (mutual correlation in the motion of electrons). This interaction is called "exchange interaction".

2.2.2 *Spin Hamiltonian of the Exchange Interaction*

During the meetings of paramagnetic particles in a condensed medium, at distances on the order of van der Waals radii, the dependence of the electron terms on the total electron spin **S** of two paramagnetic particles can be formally described using some operator that depends on spin operators. The simplest form of such a spin operator, the spin Hamiltonian, is the scalar product of two vector operators $\mathbf{S}_1$ and $\mathbf{S}_2$. Formally, the electron energy terms of relatively weakly interacting paramagnetic particles can be described with the help of the spin Hamiltonian of the so-called exchange interaction in the form [3, 10, 11].

$$V_{ex} = \hbar J(\mathbf{r}_{12})\,(\mathbf{S}_1\,\mathbf{S}_2), \tag{2.14}$$

where $\mathbf{r}_{12}$ is the radius vector connecting the positions of the spins $\mathbf{S}_1$ and $\mathbf{S}_2$, and the parameter J (r) quantitatively characterizes the splitting of electron terms, which correspond to different values of the total spin S of electrons. The eigenstates of the spin Hamiltonian (2.14) are multiplets with a total spin S, which can take the values $S_1 + S_2$, $S_1 + S_2 - 1, \ldots$ $|S_1 - S_2|$, so that eigenstates of (2.14) correctly reflect the multiplicity of electron terms of the pair of paramagnetic particles. The degree of degeneration of electron terms is 2S + 1.

The spin Hamiltonian of the exchange interaction in the form (2.14) contains a single parameter-exchange integral J. For example, for a system with two electrons, the total electron spin S can take only two values 1 and 0. The state with S = 1 is threefold degenerate, since three values of the projection of the spin on the quantization axis are possible, +1, 0, −1, and such electron terms are called triplets. A state with S = 0 has no degeneracy due to the state of spins, and such electron terms are called singlets. For the spin Hamiltonian (2.14), the energy for a triplet electron term is $E_T = (1/4)\,\hbar J(r_{12})$, and for a singlet term it is $E_S = -(3/4)\hbar J\,(r_{12})$. It follows that the exchange integral in a two-electron system is equal to the difference of the energies of adiabatic electron terms, E_T-E_S,

$$E_T - E_S = \hbar J(\mathbf{r}_{12}). \qquad (2.15)$$

When $J\,(r_{12}) > 0$, the singlet term is located below the triplet term, i.e. the exchange interaction of the antiferromagnetic type is implemented. When $J\,(r_{12}) < 0$, the singlet term is located above the triplet term, i.e. the ferromagnetic type of the exchange interaction is implemented.

If the spin of at least one of the interacting partners is 1/2, for example, $S_1 > 1/2$, and $S_2 = ½$, then only two multiplets are possible, corresponding to the total spin $S_1 + 1/2$ and $S_1 - 1/2$. In this case, also one parameter, the exchange integral, is sufficient for the effective Hamiltonian (2.14) to correctly specify the multiplicity of electron terms and the splitting of the multiplets. To this end, the exchange integral should be set equal to

$$\hbar J(\mathbf{r}_{12}) = \{E(S = S_1 + 1/2) - E(S = S_1 - 1/2)\}/(S_1 + 1/2). \qquad (2.16)$$

When particles with spins greater than 1/2 interact, the number of multiplets is greater than 2 and, in the general case, one parameter (one exchange integral) may not be enough to describe the relative position of all multiplets. However, in the case of the relatively weak interaction between paramagnetic particles that we are considering, the situation is significantly simplified and the spin Hamiltonian (2.14) can satisfactorily describe the spin dependence of the exchange interaction energy of two paramagnetic particles for such high-spin particles [3, 11].

2.2.3 *Exchange Integral*

For the simplest situation, we give an expression for the exchange integral. This will clearly demonstrate some important aspects of the exchange interaction and will also give an idea of why the concepts of exchange interaction, the exchange integral, were introduced.

Let us analyze the exchange interaction of two atoms, each of which has one electron. Let *a* and *b* denote the electron orbitals of atoms A and B. At large distances between A and B, when the interatomic interaction is still relatively small, but electron clouds *a* and *b* already overlap, a fairly good approximation is the representation of the wave function of the system as a product of orbitals *a* and *b* (the valence bond approximation formulated by Heitler and London [10]). Note that for very large distances between atoms, the Heitler-London approximation incorrectly describes the dependence of the exchange integral on the distance between atoms

[12]. However, at medium distances (on the order of the van der Waals radii), the results obtained according to the asymptotic theory [12] and within the Heitler-London approximation almost coincide [13]. The properly symmetrized and normalized wave functions describing the spatial distribution of electrons for the singlet and triplet terms of the system under consideration are [10].

$$\begin{aligned} \phi_s &= (a(1)b(2) + a(2)b(1))/\left(2 + 2|S_{ab}|^2\right)^{1/2}, \\ \phi_a &= (a(1)b(2) - a(2)b(1))/\left(2 - 2|S_{ab}|^2\right)^{1/2}. \end{aligned} \tag{2.17}$$

The parameter S_{ab} is called the overlap integral of the orbitals a and b and is

$$S_{ab} = \int a * b\,dv. \tag{2.18}$$

Here integration is carried throughout the volume. Note that two orbitals a, b are called orthogonal if $S_{ab} = 0$.

According to Eq. (2.15), in the system under consideration, the exchange integral is equal to the energy difference in the states with wave functions ϕ_a and ϕ_s (see (2.17)). The electron Hamiltonian of the system is

$$\mathbf{H}_e = \mathbf{H}_1 + \mathbf{H}_2 + e^2/r_{12}, \tag{2.19}$$

where $\mathbf{H}_k = \mathbf{T}_k + \mathbf{V}_k$ is the sum of the operators of the kinetic (**T**) and potential (**V**) energy of the kth electron, and **V** is the Coulomb interaction of the electron with the nuclei of both atoms, $\mathbf{V}_k = \mathbf{V}_{kA} + \mathbf{V}_{kB}$. The last term in (2.19) characterizes the Coulomb repulsion of electrons. Calculating the energy of the electron term using the formula $E = \int \phi * (r_1, r_2)\,\mathbf{H}_e\,\phi\,(r_1, r_2)\,dv\,(r_1)\,dv\,(r_2)$ and the wave functions (2.17), for the difference of the energies of the triplet and singlet terms, we obtain the next result

$$\begin{aligned} J_{AB} = &-(2/\hbar)\{S*_{ab}\int a * V_A b\,dv + S_{ab}\int b * V_A a\,dv \\ &-|S_{ab}|^2(\int a * V_B a\,dv + \int b * V_A b\,dv) \\ &-|S_{ab}|^2 \iint a * (1) b * (2)(e^2/r_{12}) a(1) b(2) dv_1 dv_2 \\ &+\iint a * (1) b * (2)(e^2/r_{12}) b(1) a(2) dv_1 dv_2\}/(1 - |S_{ab}|^4). \end{aligned} \tag{2.20}$$

The physical nature of the multiplet term splitting is especially clearly detected if using the wave functions (2.17) to determine the pattern of the electron distribution in the singlet (ϕ_s) and triplet (ϕ_a) states. To do this, we introduce one- and two-electron densities P_1 (r) and P_2 (r_1, r_2).

The two-electron density P_2 (r_1, r_2) gives a probability to find one electron in the volume element dv ($\mathbf{r}_1$), provided that the second electron is in the volume element dv ($\mathbf{r}_2$), and it is equal to the square of the modulus of the wave functions (2.17). For the singlet and triplet states, we obtain, respectively,

$$\begin{aligned} {}^1\mathrm{P}_2(\mathbf{r}_1,\mathbf{r}_2) &= (\mathrm{q}(\mathbf{r}_1,\mathbf{r}_2)+\mathrm{q}_\mathrm{e}(\mathbf{r}_1,\mathbf{r}_2))/\left(1+|\mathrm{S}_{ab}|^2\right), \\ {}^3\mathrm{P}_2(\mathbf{r}_1,\mathbf{r}_2) &= (\mathrm{q}(\mathbf{r}_1,\mathbf{r}_2)-\mathrm{q}_\mathrm{e}(\mathbf{r}_1,\mathbf{r}_2))/\left(1-|\mathrm{S}_{ab}|^2\right), \end{aligned} \tag{2.21}$$

where

$$\begin{aligned} \mathrm{q}(\mathbf{r}_1,\mathbf{r}_2) &= \left(|a(\mathbf{r}_1)|^2|b(\mathbf{r}_2)|^2+|a(\mathbf{r}_2)|^2|b(\mathbf{r}_1)|^2\right)/2, \\ \mathrm{q}_\mathrm{e}(\mathbf{r}_1,\mathbf{r}_2) &= Re\{a*(\mathbf{r}_1)b(\mathbf{r}_1)b*(\mathbf{r}_2)a(\mathbf{r}_2)\}. \end{aligned}$$

The one-electron density P_1 (r) gives a probability to find an electron in the volume element dv (r), provided that the position of the second electron is arbitrary, it equals to $P_1(r) = \int P_2(r, r_2)\, dv\, (r_2)$. Using (2.21), we obtain for singlet and triplet terms, respectively,

$$\begin{aligned} {}^1\mathrm{P}_1(\mathbf{r}) &= (\mathrm{q}'(\mathbf{r})+\mathrm{q}'_\mathrm{e}(\mathbf{r}))/\left(1+|\mathrm{S}_{ab}|^2\right), \\ {}^3\mathrm{P}_1(\mathbf{r}) &= (\mathrm{q}'(\mathbf{r})-\mathrm{q}'_\mathrm{e}(\mathbf{r}))/\left(1-|\mathrm{S}_{ab}|^2\right), \end{aligned} \tag{2.22}$$

where

$$\begin{aligned} \mathrm{q}'(\mathbf{r}) &= \left(|a(\mathbf{r})|^2+|b(\mathbf{r})|^2\right)/2, \\ \mathrm{q}'_\mathrm{e}(\mathbf{r}) &= Re\{a*(\mathbf{r})b(\mathbf{r})\,\mathrm{S}_{ba}\}. \end{aligned}$$

It follows from Eqs. (2.21 and 2.22) that in the general case of arbitrary atomic orbitals *a* and *b*, the one-electron and two-electron densities are different for the singlet and triplet spin states. The situation becomes simpler for orthogonal orbitals *a* and *b*, when $S_{ab} = 0$. Then the one-electron densities in the singlet spin state and the triplet spin state coincide, ${}^1P_1(r) = {}^3P_1(r)$ (see 2.22), but the two-electron densities do not coincide (see 2.21)

$$ {}^1\mathrm{P}_2(\mathbf{r}_1,\mathbf{r}_2)-{}^3\mathrm{P}_2(\mathbf{r}_1,\mathbf{r}_2) = 2\,Re\{a*(\mathbf{r}_1)b(\mathbf{r}_1)b*(\mathbf{r}_2)a(\mathbf{r}_2)\}. \tag{2.23}$$

In the considered case of orthogonal orbitals a and b, the singlet-triplet term splitting is solely the result of the difference in the two-electron distribution in these states, which is described by Eq. (2.23). Indeed, at $S_{ab} = 0$ from (2.20) we get

$$J_{AB} = -(2/\hbar)\iint a*(r_1)b(r_1)(e^2/r_{12})b*(r_2)a(r_2)dv(r_1)dv(r_2). \quad (2.24)$$

Comparison of Eqs. (2.23) and (2.24) shows that for orthogonal orbitals the exchange integral is determined by the Coulomb interaction of two electrons distributed in space with densities $\rho(1) = a*(r_1)\,b(r_1)$ and $\rho(2) = a(r_2)\,b*(r_2)$. These electron densities appear only within quantum mechanics. Indeed, they mean that the same electron simultaneously occupies *a* and *b* orbitals of two atoms A and B. It looks as if the electron quickly "changed" the occupied orbital. Perhaps this type of association has led to the introduction of the terms exchange integral (J_{AB}) and exchange interaction.

Equation (2.24) for the exchange integral makes interesting observations possible. The exchange integral of the form (2.24) shows that the differential overlap of the electron orbitals of unpaired electrons plays a crucial role in the exchange interaction. Indeed, the exchange integral (2.24) is zero if there is no differential overlap of the orbitals *a* and *b*, i.e. $a*(r)\,b(r) = 0$ at any point in space.

Equation (2.24) predicts that for orthogonal orbitals a and b, the exchange integral has the negative sign [6]. So, in this case, the energy of the triplet term is lower than that of the singlet term (see 2.15), and the exchange interaction of the ferromagnetic type is implemented. The ferromagnetic type of the exchange interaction is implemented, e.g., between electrons occupying different orbitals of the same atom, which leads to the well-known Hund's rule, according to which for a given electronic configuration the ground term of the atom is the state with the largest possible spin multiplicity.

In the case of non-orthogonal orbitals, when $S_{ab} \neq 0$, the singlet and triplet states have different one- and two-electron densities P_1 (r) and P_2 (r_1, r_2) (2.21 and 2.22). Therefore, along with the inter-electron Coulomb repulsion effect, the electrostatic interaction (Coulomb attraction) of electrons with nuclei (the first four terms in (2.20) and the kinetic energy of electrons can make a significant contribution to the exchange integral! As a result, the sign of the exchange integral can become positive. It is this antiferromagnetic nature of the exchange interaction that is typical for interatomic interactions. Note that in most cases the covalent chemical bond is created by two electrons in the singlet spin state, i.e., when a covalent chemical bond is formed, the exchange interaction of the antiferromagnetic type, as a rule, is manifested. This correlates with the fact that electrons on different atoms can occupy non-orthogonal orbitals to each other.

For non-orthogonal orbitals, all terms in (2.20) are on the same order of magnitude, roughly speaking, they are all proportional to the square of the overlap integral [3].

$$J_{AB} \tilde{} |Sab|^2. \quad (2.25)$$

Since the overlap integral decays exponentially with increasing distance r between atoms, it follows from Eq. (2.25) that the exchange integral is also an exponentially decreasing function

$$J_{AB}(r) = J_{AB}(r_0)exp(-\varkappa(r - r_0)), \tag{2.26}$$

where J_{AB} (r_0) is the exchange integral at the distance of the collision radius, and the parameter $\varkappa$ characterizes the declining slope of the exchange integral with increasing distance between the interacting atoms. For estimates, in the zeroth approximation for many atoms, $1/\varkappa \approx 0.03$–0.04 nm [14].

2.2.4 *Exchange Interaction of Multi Electron Systems. Electron Delocalization and Spin Polarization*

In the study of spin exchange, the exchange interaction of paramagnetic particles with not one, but with many electrons is of the great interest. For such systems in the Hartree-Fock approximation, the exchange integral can be represented as a sum [15]

$$J_{AB} = \Sigma J_{AB}(a, b)/(n_A n_B), \tag{2.27}$$

in which the terms describe the contributions of all pairs of orbitals *a* and *b* of interacting particles A and B, while n_A, n_B are the number of unpaired electrons of A and B. In [16], in the approximation (2.27), the exchange interaction for several pairs of atoms was calculated. It turned out that at distances on the order of the sum of the van der Waals atomic radii, J_{AB} decreases with increasing distance r between atoms even faster than it follows from Eq. (2.25). According to [16], in the range of distances on the order of the sum of van der Waals radii, the relation is

$$J_{AB} = J_0|S_{ab}|^n, \tag{2.28}$$

in which the index n is somewhat greater than 2. The results obtained in [16] are summarized in Table 2.1.

In polyatomic paramagnetic particles — free radicals, complexes of paramagnetic ions, triplet excited molecules — unpaired electrons occupy molecular rather than atomic orbitals, i.e., unpaired electrons are distributed along the orbitals of all atoms of the molecule. This effect is called the electron delocalization. The delocalization of unpaired electrons plays an important role in the exchange interaction of paramagnetic

Table 2.1 Parameters of the approximation formula (2.28)

Pairs of atoms	J_0 10^{-16} rad/s	n
H-H	3.53	2.22
Li-Li	0.94	2.33
B-B	2.40	2.30
C-C	5.78	2.36
N-N	9.46	2.42
O-O	10.92	2.49
F-F	12.96	2.54

particles, since in most cases the exchange interaction is determined by the overlap of the orbitals of only those atoms of paramagnetic particles A and B that are in direct contact. Therefore, the scale of the exchange interaction will crucially depend on the degree of delocalization of unpaired electrons on the orbitals of the atoms in contact.

Let us demonstrate the role of electron delocalization in the exchange interaction by the example of the interaction between atom A and the paramagnetic complex. Let φ_a be the orbital of the unpaired electron of atom A, and let the unpaired electron of the complex be distributed between the orbital $\varphi_m$of the central ion of the complex and the orbital φ_b of the ligand, i.e., occupies the molecular orbital of the complex $\varphi = c_1\varphi_m + c_2\varphi_b$ (see Fig. 2.3).

In real systems, the unpaired electron is predominantly concentrated on the central ion of the complex, so that $|c_1|^2 \gg |c_2|^2$. Hence, it would seem that the main contribution should be made to the exchange interaction by the direct overlap of the orbitals φ_m of the central ion of the complex with the orbital φ_a of the atomic-A partner. However, the overlap of φ_a with φ_m is substantially less than the overlap of φ_a with the orbital of φ_b (see Fig. 2.3). Therefore, in the exchange interaction, the delocalization of the unpaired electron of the complex from the central ion to the ligand and the overlapping of the orbitals of the atom A with the orbital of the atom of the ligand usually comes to the fore.

Along with the effect of the delocalization of unpaired electrons in the exchange interaction of many-electron paramagnetic particles, the spin polarization effect may play an important role. As a result of the interaction of unpaired electrons with electrons of the filled molecular orbitals, the molecular spin-orbitals of the filled shells are split, as the unpaired electron perturbs differently the state of the electrons in orbitals occupied by electrons with the same spatial, but different (α and β) spin wave functions. As a result, the local spin polarization appears on the atomic orbitals of paramagnetic particles, even in filled electron shells. In this case, the total electron spin polarization in the filled shells is zero, but on individual atoms the spin polarization creates a nonzero spin density, and this spin polarization is positive on some atoms, and the polarization is negative on others. Thus, in many-electron paramagnetic particles, the spin density on individual atoms is the sum of the contribution of the delocalization of unpaired electrons and the contribution of the spin polarization of electrons on the spin-orbitals of the inner filled shells.

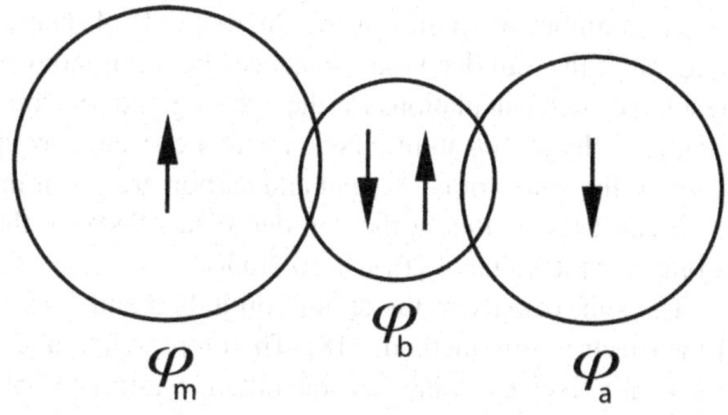

Fig. 2.3 Exchange interaction via a diamagnetic ligand. There is no direct overlap between φ_a and φ_m orbitals. However both orbitals overlap with the orbital φ_b of the bridging diamagnetic molecule

In such a situation, the exchange interaction of paramagnetic particles depends not only on the overlap of the orbitals of the unpaired electrons themselves, but also on the overlap of the orbitals of the filled shells of the particles perturbed by the unpaired electrons.

The interaction that occurs as a result of the overlap of the orbitals of paramagnetic particles (ϕ_a and ϕ_m in the above example) is called the direct exchange interaction. The interaction due to the delocalization of the unpaired electron to the orbitals of the diamagnetic ligand or the spin polarization of the diamagnetic bridge between paramagnetic particles is called the indirect exchange interaction (or superexchange interaction).

2.2.5 Semiempirical Estimates of the Exchange Integral

Consistent theoretical calculations of the exchange integral for polyatomic many-electron paramagnetic particles are quite a difficult task. Therefore, a semiempirical estimate of the exchange integral is sometimes used using the relation proposed by Owen [17]. According to [17], the decisive contribution to the exchange interaction of two paramagnetic particles A and B makes the overlap of the orbitals *a* and *b*, which are in direct contact in the course of the bimolecular collisions. The rationale for this assumption is that it is the indirect exchange that is often the main mechanism of exchange interaction in magnetic dielectrics [1–3], and the qualitative reasoning that was given above. Within this approximation, the exchange interaction of particles A and B is different from zero, when the orbitals *a* and *b* of the contacting atoms of particles A and B have a nonzero spin density. Based on these considerations, Owen introduced the exchange integral in the form [17]

$$J_{AB} = J_{ab}(r_0)\, \rho_a \rho_b / (n_A n_B), \tag{2.29}$$

The following notations are introduced in this expression: J_{ab} (r_0) is the exchange integral between atoms with orbitals *a* and *b* at a distance of the collision radius, i.e., at a distance equal to the sum of the van der Waals radii of these atoms; ρ_a and ρ_b are the spin densities on the atomic orbitals of a and b; n_A and n_B are the number of unpaired electrons in A and B.

In a number of studies, using the methods of quantum chemistry, the exchange integral at the van der Waals distances between atoms was estimated [16–18]. The results of such calculations for the exchange interaction of unpaired electrons at 1 s-orbitals of hydrogen atoms (1s-1s interaction) and 2pσ-2pσ interactions in the case of pairs of fluorine atoms, oxygen and carbon are given in Table 2.2.

It can be seen that at the van der Waals distance the exchange integral for two atoms has the values $J_{ab}(r_0) \approx 10^{14}$rad/s.

The spin density at the atomic orbitals *a* and *b* can be directly measured using EPR spectroscopy methods [18]. Therefore, relation (2.29) is a useful working tool for qualitative reasoning and quantitative estimates of the exchange integral. For

Table 2.2 Exchange integral J (r_0) for pairs of atoms at van der Waals distances

Pairs of atoms	Collision radius, r_0, nm	$J(r_0)$, rad/s
H. . . .H	0.24	$3.5\ 10^{14}$
O. . . .O	0.28	$1.1\ 10^{14}$
C. . . .C	0.37	$0.6\ 10^{14}$
F. . . .F	0.28	$0.4\ 10^{14}$

example, paramagnetic complexes collide with their ligands, and on ligands can be delocalized to several tens of percent of the spin density of the unpaired electron. This means that in real systems on ligand atoms in contact with a collision, the product of spin densities $\rho_a\rho_b$ can be of the order of 0.1–0.01.

The above arguments make it possible to estimate the magnitude of the exchange integral over the collision radius of paramagnetic particles. If the direct exchange interaction is implemented and $\rho_a = 1$, $\rho_b = 1$, then $\rho_a\rho_b J_{ab}(r_0) \sim 10^{14}$rad/sec. If indirect exchange interaction is implemented and $\rho_a\rho_b$ is about 0.1–0.01, then

$$J(r_0) \approx \rho_a\rho_b J_{ab}(r_0)\tilde{}10^{12} - 10^{13}\text{rad/sec}.$$

According to (2.29), the J_{AB} parameter of the spin Hamiltonian of the exchange interaction has the factor $1/(n_A n_B)$. The appearance of this multiplier can be interpreted as follows. In the model under discussion, the exchange interaction is determined by the spin density on the orbitals *a* and *b* and the overlap of the orbitals *a* and *b* of the two atoms that are in contact. Therefore, the scale of the exchange interaction of A and B can be estimated as $E_{ex} = \hbar J_{ab}(r_0)\rho_a\rho_b$. The scale of the same exchange interaction of paramagnetic particles A and B, which is described by the spin Hamiltonian $\hbar J_{AB}\mathbf{S}_A\mathbf{S}_B$ (1.14), can be estimated as the total splitting of all multiplets with total spin $S = S_A + S_B, S_A + S_B - 1, \ldots |S_A\text{-}S_B|$. This total splitting is $\Delta E_{ex} = \hbar J_{AB}(2S_A + 1)S_B$ when $S_B \leq S_A$. Comparing these two estimates of the scale of the exchange interaction of two paramagnetic particles and taking into account that $n_A = 2S_A$, $n_B = 2S_B$, we obtain (2.29).

Nowadays one can calculate the exchange integrals using the density functional theory (see, e.g., [19]).

2.3 Dynamics of Spins Caused by the Exchange Interaction

This section presents examples of how the state of electron spins can change under the influence of an exchange interaction in a bimolecular collision.

2.3.1 Spin Density Matrix. Quantum Spin Coherence

The state of the isolated spins is fully given by the wave function. For example, an arbitrary spin state of a single electron is given by the function

$$\chi = c_1\alpha + c_2\beta, |c_1|^2 + |c_2|^2 = 1, \tag{2.30}$$

where α and β are the eigenfunctions of the operator of the projection of the spin moment on the quantization axis, which corresponds to the eigenvalues of the spin projection on the quantization axis, equal to +1/2 and – 1/2. The wave function fully describes the state of the spin. With its help, one can calculate any spin-dependent physical quantity **Q**

$$< Q >=< \chi \mid Q \mid \chi >= |c_1|^2 Q_{\alpha\alpha} + c_1 c_2 * Q_{\alpha\beta} + c_1 * c_2 Q_{\beta\alpha} + |c_2|^2 Q_{\beta\beta}. \tag{2.31}$$

This shows that, along with the wave function (2.30), a full description of the spin state also provides a matrix composed of the coefficients c_1 and c_2 of the wave function,

$$\rho = \begin{Bmatrix} \rho_{\alpha\alpha} & \rho_{\alpha\beta} \\ \rho_{\beta\alpha} & \rho_{\beta\beta} \end{Bmatrix}; \tag{2.32}$$

$$\rho_{\alpha\alpha}, = |c_1|^2, \rho_{\alpha\beta} = c_1 * c_2, \rho_{\beta\alpha} = c_1 c_2 *, \rho_{\beta\beta} = |c_2|^2.$$

A matrix composed in this way is called a density matrix. Its diagonal elements $\rho_{\alpha\alpha} = |c_1|^2$ and $\rho_{\beta\beta} = |c_2|^2$ give the population of the spin in the states α and β, respectively. The off-diagonal elements of the density matrix $\rho_{\alpha\beta} = c_1 * c_2$, $\rho_{\beta\alpha} = \rho_{\alpha\beta}*$ define the quantum coherence of the spin state.

It is straightforward to illustrate in the simplest example, the spin of a single electron, that all the elements of the spin density matrix S = 1/2 are measurable physical quantities.

Note that the experimentally measured values of the observable **Q** can be calculated using the equation (see 2.31)

$$Q = \mathrm{Tr}\{\rho Q\}, \tag{2.33}$$

where Tr denotes the trace of the product of the density matrix and the matrix representation of the operator **Q** of the observable quantity. The matrix representation of the operators of spin projections on the X, Y, Z axes in the basis of eigenfunctions of the operator S_z has the form

$$S_x = \begin{Bmatrix} 0 & 1/2 \\ 1/2 & 0 \end{Bmatrix}; S_y = \begin{Bmatrix} 0 & -i/2 \\ i/2 & 0 \end{Bmatrix}; S_z = \begin{Bmatrix} 1/2 & 0 \\ 0 & -1/2 \end{Bmatrix} \tag{2.34}$$

From Eqs. (2.32, 2.33 and 2.34) we have

$$\begin{aligned} S_X &= Re(\rho_{\alpha\beta}), \\ S_Y &= -Im\{\rho_{\alpha\beta}\}, \\ S_Z &= (\rho_{\alpha\alpha} - \rho_{\beta\beta})/2. \end{aligned} \tag{2.35}$$

We represent the complex number $\rho_{\alpha\beta}$ in the form $\rho_{\alpha\beta} = |\rho_{\alpha\beta}| \exp(-i\phi)$, then the transverse components of the spin moment can be represented as

$$\begin{aligned} S_X &= |\rho_{\alpha\beta}| \cos(\phi), \\ S_Y &= |\rho_{\alpha\beta}| \sin(\phi). \end{aligned} \tag{2.36}$$

From Eqs. (2.35 and 2.36), it can be seen that the difference of the diagonal elements of the density matrix determines the longitudinal polarization of the spin (along the quantization axis of the spin Z), and the off-diagonal elements determine the projection of the spin on the plane perpendicular to the quantization axis of the spins. The phase of the nondiagonal element of the density matrix $\rho_{\alpha\beta}$ is equal to the angle ϕ, which forms with the X axis the average value of the spin projection on the XY plane.

Bearing in mind the normalization condition for the density matrix $\rho_{\alpha\alpha} + \rho_{\beta\beta} = 1$, we find that all the elements of the density matrix are expressed through from observables

$$\begin{aligned} \rho_{\alpha\beta} &= S_X - i\, S_Y; \\ \rho_{\alpha\alpha} &= 1/2 + S_Z; \\ \rho_{\beta\beta} &= 1/2 - S_Z. \end{aligned} \tag{2.37}$$

Direct methods for measuring the spin coherence $\rho_{\alpha\beta}$ is provided by EPR spectroscopy. Indeed, in EPR spectroscopy in alternating fields perpendicular to a constant magnetic field, the projection of the spin magnetic moment in the X, Y plane $\{S_X, S_Y\}$ is recorded.

Note that for systems with two or more unpaired electrons, the experimental data on the projections of spin dipole moments are not enough to fully describe the spin state, as is the case for the ensemble of systems with one unpaired electron considered above. To measure the parameters of quantum spin coherence for many-electron systems, it is necessary to apply multipulse protocols of the experiment [20, 21].

In EPR spectroscopy experiments, the characteristics of an ensemble of spins, rather than a single spin, are measured. Therefore, for comparison with the experiment, Eqs. 2.32, 2.33 and 2.35) should be averaged over the ensemble. The advantages of the density matrix formalism in comparison with the wave function are revealed at this stage of statistical averaging. The density matrix, all the elements of the density matrix, in principle, are the quantities which can be measured in the experiment. Therefore, it is possible to apply the averaging of the density matrix of a quantum system over an ensemble of systems. The wave function is not an observable quantity and cannot be averaged over an ensemble of particles. The density matrix formalism makes it possible to consider the behavior of open systems, to obtain kinetic equations taking into account relaxation processes, in this case, processes of electron spin relaxation. In the next chapter, examples will be given of the implementation of the potential of the density matrix formalism in studying the spin coherence behavior of an ensemble of unpaired electrons in paramagnets.

2.3.2 *Elementary Act of Spin Exchange. Spin Dynamics of a Pair of Paramagnetic Particles*

The change in the spin state of the colliding particles, caused by their exchange interaction, is essentially reduced to the exchange of spin states between partners. This follows from the fact that the exchange interaction in the form of $\mathbf{V}_{ex} = \hbar J(\mathbf{r}_{12})(\mathbf{S}_1\mathbf{S}_2)$ (2.14) preserves the total electron spin moment of both partners. Indeed, the operators of all projections of the total spin moment commute with the spin Hamiltonian (1.14):

$$\begin{aligned} &\mathbf{S}_\mu = \mathbf{S}_{A\mu} + \mathbf{S}_{B\mu}, \mu = x, y, z; \\ &\left[\mathbf{S}_\mu, \mathbf{V}_{ex}\right] = 0. \end{aligned} \tag{2.38}$$

From the equation of motion for spin projection operators

$$\partial \mathrm{S}\mu/\partial \mathrm{t} = -(\mathrm{i}/\hbar)\,[\mathbf{S}_\mu, \mathbf{V}_{ex}], \tag{2.39}$$

and taking into account Eq. (2.38), we find that the exchange interaction preserves the projections of the total spin, $S\mu = \text{const}$, and the projections of the spin moments of the partners change in the opposite way

$$\partial\, \mathbf{S}_{A\mu}/\partial t = -\partial\, \mathbf{S}_{B\mu}/\partial t. \tag{2.40}$$

Thus, in the case when the exchange interaction (2.14) is the only spin-dependent term in the spin Hamiltonian of a pair of paramagnetic particles, the meeting of paramagnetic particles in solution leads to the exchange of spin states, i.e., to spin exchange.

The manifestation of the exchange interaction during the meeting is not as simple as it is given by (2.39 and 2.40), when we take into account the influence of other spin-dependent terms in the Hamiltonians of individual paramagnetic particles: Zeeman interaction of spins of unpaired electrons with an external constant magnetic field, spin-orbit interaction, hyperfine interaction with magnetic nuclei, zero field splitting, and spin-lattice interaction. In such situations, when, along with exchange interaction, other spin-dependent interactions operate, the result of the meeting of two paramagnetic particles is not necessarily reduced to equivalent spin exchange, as predicted by (2.40). These issues will be discussed below in this Chapter of the book.

In order to illustrate the manifestation of the exchange interaction in the collision of two paramagnetic particles and the changes in the spin states during their collisions, we consider the simplest model examples.

Usually, the spin exchange process is studied by EPR spectroscopy, i.e., under conditions when the spins are placed in an external constant magnetic field. Therefore, along with the exchange interaction, we include the Zeeman interaction with the external magnetic field in our consideration. We choose the direction of the external magnetic field as a Z axis of the laboratory coordinate system. Then the spin Hamiltonian of a pair of paramagnetic particles encountered in the solution can be described using the spin Hamiltonian

$$\mathbf{H} = \hbar\omega_A \mathbf{S}_{Az} + \hbar\omega_B \mathbf{S}_{Bz} + \hbar J_{AB}(\mathbf{S}_A \mathbf{S}_B), \tag{2.41}$$

where ω_A and ω_B are the EPR frequencies of individual paramagnetic particles. The difference between these frequencies can be due, e.g., to different g factors of paramagnetic particles A and B.

The change in the state of the two considered spins is determined by the equation of motion of the density matrix

$$\partial\rho/\partial t = -(i/\hbar)[\mathbf{H}, \rho]. \tag{2.42}$$

The solution of this equation is

$$\rho(t) = \exp(-i\mathbf{H}t/\hbar)\,\rho(0)\exp(i\mathbf{H}t/\hbar). \tag{2.43}$$

Using Eq. (2.43), the changes in the populations and the coherence of the spins under the action of the spin Hamiltonian (2.41) are calculated below for several model initial states of the spins of the pair. In parallel, the changes in the average values of the projections of the spin moments are calculated using Eqs. (2.33 and 2.43).

Exchange of Longitudinal Polarization of Spins Consider the situation when both spins at the initial moment are in their eigenstates of the projections of the spin moments on the Z axis: $\alpha_A\beta_B$. This means that initially the density matrix of the considered pair of spins has a single nonzero element $\rho(0) = \delta_{\alpha\beta,\alpha\beta}$. In this state, there is no quantum coherence, and therefore the transverse projections of the spin moment of each of the electrons of the pair are zero: $S_{Ax} = 0$; $S_{Bx} = 0$; $S_{Ay} = 0$; $S_{By} = 0$. Of course, the total transverse projections of the spin moment are also zero, $S_x = 0$, $S_y = 0$.

For the considered initial state, the longitudinal projections of the spin moments of electrons are $S_{Az} = 1/2$ and $S_{Bz} = -1/2$, and the total longitudinal projection is zero, $S_z = 0$. As a result of the motion of the spins under the action of the spin

Hamiltonian (2.41), we obtain the following result. Four elements of the density matrix are nonzero:

$$\rho_{\alpha\beta,\alpha\beta}, \rho_{\alpha\beta,\beta\alpha}, \rho_{\beta\alpha,\alpha\beta}, \rho_{\beta\alpha,\beta\alpha}.$$

For illustration, we give explicit results for the case when Zeeman frequencies are equal, $\omega_A = \omega_B$:

$$\begin{aligned} \rho_{\alpha\beta,\alpha\beta} &= 1 - sin^2(Jt/2), \\ \rho_{\alpha\beta,\beta\alpha} &= (i/2) sin\,(Jt), \\ \rho_{\beta\alpha,\alpha\beta} &= \rho*_{\alpha\beta,\beta\alpha}, \\ \rho_{\beta\alpha,\beta\alpha} &= 1 - \rho_{\alpha\beta,\alpha\beta} = sin^2(Jt/2). \end{aligned} \quad (2.44)$$

It can be seen that the population of the two states αβ and βα changes. This causes the longitudinal projections of the spins to change and take on values

$$S_{Az} = 1/2 - sin^2(Jt/2), S_{Bz} = -1/2 + sin^2(Jt/2). \quad (2.45)$$

The last equation shows that with probability

$$p = sin^2(Jt/2) \quad (2.46)$$

there is a mutual flip-flop of spins $\alpha_A\beta_B$ ς $\beta_A\alpha_B$ (see (2.1)). Equation (2.45) shows that the exchange interaction carries out the equivalent exchange of longitudinal polarization of spins. Indeed, taking into account (2.46) the longitudinal polarizations of spins, which are given by (2.45), can be represented as the sum of two terms

$$\begin{aligned} S_{Az} &= (1/2)(1 - p) + (-1/2)p, \\ S_{Bz} &= (1/2)p + (-1/2)(1 - p). \end{aligned} \quad (2.47)$$

Equation (2.47) shows that the exchange of longitudinal polarization of spin, the mutual transfer of longitudinal polarization, occurs with the preservation of the sign of the transfered polarization and without the loss of the polarization. In this case, the total polarization of both spins remains zero.

It has already been noted above that in the example under consideration there is no quantum coherence in the initial state. Over time, the quantum coherence $\rho_{\alpha\beta,\beta\alpha}$ = (*i*/2) sin (Jt) (2.44) is formed. But this coherence does not contribute to the transverse projections of spin moments, which remain zero. This coherence is called zero-quantum coherence, and its manifestations can be observed in multi-pulse experiments (see [20, 21]).

Spin Coherence Transfer Consider another situation, when at the initial moment one of the spins is in the eigenstate of the projection of the spin moment on the Z axis, and the second spin is in the eigenstate of the projection of the spin moment on the X axis: $\alpha_A(\alpha_B + \beta_B)/\sqrt{2}$. This means that at the initial moment the density matrix of the spin pair under consideration has four nonzero matrix elements

$$\rho(0)_{\alpha\alpha,\,\alpha\alpha} = 1/\sqrt{2}, \rho(0)_{\alpha\alpha,\,\alpha\beta} = 1/\sqrt{2}, \rho(0)_{\alpha\beta,\,\alpha\alpha} = 1/\sqrt{2}, \rho(0)_{\alpha\beta,\,\alpha\beta} = 1/\sqrt{2}. \tag{2.48}$$

In this state

$$S_{Ax} = 0; S_{Ay} = 0; S_{Az} = 1/2; S_{Bx} = 1/2; S_{By} = 0; S_{Bz} = 0. \tag{2.49}$$

As a result of the spin motion under the action of spin Hamiltonian (2.41), we obtain the following result. Nine matrix elements are nonzero: $\rho_{\alpha\alpha,\,\alpha\alpha}$, $\rho_{\alpha\alpha,\,\alpha\beta}$, $\rho_{\alpha\alpha,\,\beta\alpha}$, $\rho_{\alpha\beta,\,\alpha\alpha}$, $\rho_{\alpha\beta,\,\alpha\beta}$, $\rho_{\alpha\beta,\,\beta\alpha}$, $\rho_{\beta\alpha,\,\alpha\alpha}$, $\rho_{\beta\alpha,\,\alpha\beta}$, $\rho_{\beta\alpha,\,\beta\alpha}$. For illustration, we present explicit results only for the case when Zeeman frequencies of the spins are equal, $\omega_A = \omega_B$:

$$\rho(t) = \begin{Bmatrix} 1/2 & (1+e^{-iJt})/4 & (1-e^{-iJt})/4 & 0 \\ (1+e^{iJt})/4 & cos^2(Jt/2)/2 & isin(Jt)/4 & 0 \\ (1-e^{iJt})/4 & -isin(Jt)/4 & sin^2(Jt/2)/2 & 0 \\ 0 & 0 & 0 & 0 \end{Bmatrix}. \tag{2.50}$$

It can be seen that the population of the two states $\alpha\beta$ and $\beta\alpha$ changes. This causes the longitudinal projections of the spins to change and take on values

$$S_{Az} = (1/2)\left(1 - sin^2(Jt/2)\right), S_{Bz} = (1/2)\,sin^2(Jt/2). \tag{2.51}$$

Equation (2.51) shows that the initial longitudinal polarization of spin A is transferred to spin B with the probability (2.46), but the transfer of longitudinal polarization from spin B to spin A does not occur, since spin B initially had no longitudinal polarization. It is interesting to note that in this example there is transfer of longitudinal polarization, but not equivalent exchange of longitudinal polarizations, as it was for the other initial state of spins (cf. 2.45 and 2.51). The probabilities of spin exchange (2.45) and spin polarization transfer (2.51) are the same and are given by Eq. (2.46).

As a result of spin dynamics under the action of spin Hamiltonian (2.41), the same as in the previous example, the zero-quantum coherence $\rho_{\alpha\beta,\,\beta\alpha}$ is formed. But in contrast to the previous example, here there are also coherences, which contribute to the transverse projections of spin moments A and B. We have

$$S_{Ax} = (1/2)\,sin^2(Jt/2), S_{Ay} = -(1/4)\,sin(Jt) = -(1/2)sin^2(Jt/2)ctg(Jt/2),$$
$$S_{Bx} = (1/2)\left(1 - sin^2(Jt/2)\right), S_{By} = (1/4)sin\,(Jt) = (1/2)sin^2(Jt/2)ctg(Jt/2). \tag{2.52}$$

It can be seen that due to the exchange interaction, the coherence transfer from spin B to spin A occurs, and the probability of the coherence transfer is determined by the probability of spin exchange (2.46).

Spin Coherence Exchange If both spins are in the quantum coherent state before the exchange interaction is switched on, the exchange interaction induces the coherence exchange between the spins. Suppose that at the time of switching on of the exchange interaction the average values of the spin projections were:

$$S_{Ax} = 1/2, S_{Ay} = 0, S_{Az} = 0,$$
$$S_{Bx} = -1/2, S_{By} = 0, S_{Bz} = 0.$$

After switching on of the exchange interaction we have

$$S_{Ax} = (1/2)cos(Jt) \equiv (1/2)\,cos^2(Jt/2) - (1/2)sin^2(Jt/2), S_{Ay} = 0, S_{Az} = 0,$$
$$S_{Bx} = -(1/2)cos(Jt) \equiv -(1/2)sin^2(Jt/2) + (1/2)\,cos^2(Jt/2), S_{By} = 0, S_{Bz} = 0.$$

It follows that there is equivalent spin exchange, the exchange of x-projections, between spins (cf. 2.47).

2.4 Modified Bloch Equations Taking into Account Spin Exchange for Paramagnetic Particles with Spin S = 1/2

An important aspect of the theory of spin exchange is the description of the manifestations of spin exchange in EPR spectroscopy. The first theoretical interpretation of the changes in the shape of the spectrum observed in EPR experiments was formulated in [22]. Considering the change in the spin state of paramagnetic particles in bimolecular collisions as a certain "chemical reaction", phenomenological kinetic equations for the magnetization vectors of colliding particles were formulated in [22]. To describe the contribution of the exchange interaction in bimolecular collisions to the equation of motion of magnetization of paramagnetic particles, it was proposed by analogy with the equations of formal kinetics of reversible chemical reactions to add a term of the form [22]

$$(\partial \mathbf{M}_A/\partial t)_{ex} = -K_{ex}C_B\mathbf{M}_A + K_{ex}C_A\mathbf{M}_B;$$
$$(\partial \mathbf{M}_B/\partial t)_{ex} = K_{ex}C_B\mathbf{M}_A - K_{ex}C_A\mathbf{M}_B. \quad (2.53)$$

These equations are applicable when the equivalent spin exchange is implemented. In this situation, it is assumed that only the Heisenberg exchange interaction, which preserves the total spin of the colliding pair of particles, can be taken into account when describing the spin motion in the region of exchange interaction. Note that in real systems, the spin exchange in a bimolecular collision may cease to be equivalent due to the "interference" of the exchange interaction with other interactions. Therefore, the rate constants in (2.53) may not coincide. Moreover, they may be complex quantities. Examples of nonequivalent spin exchange will be discussed below. Moreover, Eq. (2.53) should be modified even in the case of equivalent spin exchange if paramagnetic particles in the collision have different electron spins (see below, e.g., 2.137 and 2.138).

In the majority of cases, spin exchange is studied with the stable nitroxide radicals, S = 1/2, and, as a rule, for nitroxide radicals the conditions of equivalent spin exchange are fulfilled [1, 2]. Therefore, in the study of spin exchange between free nitroxide radicals, phenomenological equations such as chemical kinetics Eq. (2.53) are widely used in the interpretation of spin exchange experiments.

In order to theoretically calculate the signals observed in EPR experiments from the system of paramagnetic particles, it is necessary to have kinetic equations of motion of the system magnetization in the presence of external constant and alternating magnetic fields. Suppose that a constant magnetic field with induction B_0 acts on the paramagnetic particles along the z axis of the laboratory coordinate system, and a circularly polarized alternating magnetic field with frequency ω and amplitude B_1 acts on the perpendicular plane. Thus, the magnetic field has projections $B_{1x} = B_1\cos(\omega t)$, $B_{1y} = B_1\sin(\omega t)$, $B_z = B_0$. The motion of the magnetization vector **M** in the external field is determined by the equation

$$\partial \mathbf{M}/\partial t = -\gamma_e[\mathbf{M} \times \mathbf{B}(t)]. \tag{2.54}$$

Here γ_e gyromagnetic ratio for electron. It is convenient to analyze the motion of the magnetization vector in a coordinate system rotating around the z axis with a frequency of ω. In the rotating coordinate system, the magnetization vector performs a Larmor precession in an effective magnetic field with the $\mathbf{B}_{эфф}\{B_1, 0, B_0 - \omega/\gamma_e\}$ components. To describe the process of establishing thermodynamic equilibrium, it is necessary to add terms into the equation of motion for magnetization that describe the process of paramagnetic relaxation. For an ensemble of independent paramagnetic particles in solutions, the behavior of spin magnetization in the external magnetic fields can be described using phenomenological equations, which are well known as Bloch equations (see [23])

$$\begin{aligned}
\partial M_x/\partial t &= -\gamma_e[\mathbf{M} \times \mathbf{B}(t)]_x - M_x/T_2; \\
\partial M_y/\partial t &= -\gamma_e[\mathbf{M} \times \mathbf{B}(t)]_y - M_y/T_2; \\
\partial M_z/\partial t &= -\gamma_e[\mathbf{M} \times \mathbf{B}(t)]_z - (M_z - M_0)/T_1.
\end{aligned} \tag{2.55}$$

In these equations, in the right-hand side, the terms are added that phenomenologically describe the tendency of the system to thermodynamic equilibrium, if the initial magnetization is nonequilibrium. The relaxation of the transverse magnetization components is given by the characteristic relaxation time T_2, and the longitudinal component comes into equilibrium with the time T_1. In magnetic resonance spectroscopy, the time T_1 of longitudinal relaxation is more often called the spin-lattice relaxation time, since there is an exchange of energy between the spins and the environment (thermostat). The transverse relaxation time T_2 is called the phase relaxation time or of the spin decoherence time. Often T_2 is also called the spin-spin relaxation time. This suggestion is due to the fact that transverse relaxation may have a contribution related to the interaction between spins, e.g., exchange and dipole-dipole. It has already been noted that spin exchange can be studied in detail using EPR spectroscopy methods. To this end, in [22] the Bloch equations were modified to take into account spin exchange (2.55).

Consider a dilute solution containing paramagnetic particles A and B with equal spins and have concentrations C_A and C_B, respectively. Consider a situation in which the Zeeman frequencies of paricles A and B are different, ω_A и ω_B, respectively, to make sense to discuss the effect of spin exchange, since the exchange interaction between particles with the same frequencies in the EPR is not manifested. For such a model situation in a rotating coordinate system we have [22, 24, 25].

$$\begin{aligned}
\partial M_{Ax}/\partial t &= -(\omega_A-\omega)M_{Ay} - M_{Ax}/T_{2A} - K_{ex}C_BM_{Ax} + K_{ex}C_AM_{Bx};\\
\partial M_{Ay}/\partial t &= (\omega_A-\omega)M_{Ax} + \omega_1M_{Az} - M_{Ay}/T_{2A} - K_{ex}C_BM_{Ay} + K_{ex}C_AM_{By};\\
\partial M_{Az}/\partial t &= -\omega_1M_{Ay} - (M_{Az}-M_{A0})/T_{1A} - K_{ex}C_BM_{Az} + K_{ex}C_AM_{Bz};\\
\partial M_{Bx}/\partial t &= -(\omega_B-\omega)M_{By} - M_{Bx}/T_{2B} + K_{ex}C_BM_{Ax} - K_{ex}C_AM_{Bx};\\
\partial M_{By}/\partial t &= (\omega_B-\omega)M_{Bx} + \omega_1M_{Bz} - M_{By}/T_{2B} + K_{ex}C_BM_{Ay} - K_{ex}C_AM_{By};\\
\partial M_{Bz}/\partial t &= -\omega_1M_{By} - (M_{Bz}-M_{B0})/T_{1B} + K_{ex}C_BM_{Az} - K_{ex}C_AM_{Bz}.
\end{aligned} \tag{2.56}$$

Here the Rabi nutation frequency is introduced

$$\omega_1 = \gamma_eB_1. \tag{2.57}$$

We assume that g-factors of spectroscopic splitting for unpaired electrons A and B are different, therefore the frequencies $\omega_A = (g_A/g_e)\gamma_eB_0$ и $\omega_B = (g_B/g_e)\gamma_eB_0$ differ. By solving the kinetic Eq. (2.56) for the corresponding experimental protocols in EPR spectroscopy and adjusting the unknown magnetic resonance parameters to the experimental data, it is possible to find, in particular, the spin exchange rate constant K_{ex}.

2.5 On the Potential of Steady State EPR Spectroscopy in the Study of Spin Exchange

The results obtained by many scientists, since the 1950s, on finding (experimental evaluation) the frequency of bimolecular collisions in liquids using the concentration dependence of the shape of the EPR spectra of spin probes, indicate a large potential of EPR spectroscopy in solving this problem. This approach is successfully applied for studying of complex environments, such as electrolytes, porous structures filled with fluid, heterogeneous environment, etc. (see, e.g., [1, 2, 26]).

To illustrate the potential of EPR in the study of spin exchange, we theoretically calculate the shape of the absorption EPR spectrum in steady-state conditions. To do this, we need to find a steady-state solution of (2.56). The spectrum of steady-state absorption observed in the EPR experiment is proportional to the product of the amplitude of the alternating field on the y-component of magnetization (the magnetization component that is perpendicular to the field $\mathbf{B}_1$ in the rotating coordinate system [23]):

$$I_{epr} = I_0\left(M_{Ay} + M_{By}\right). \tag{2.58}$$

Calculations are particularly simplified if we use alternating magnetic fields with a sufficiently small amplitude B_1, i.e., in the linear response limit of the system to the microwave field. In such a situation, the change of the z-component of magnetization in (2.56) can be neglected. As a result, for the calculation of the EPR spectrum shape in the steady-state mode, the problem is reduced to solving a system of algebraic equations only for the transverse component of the magnetization

$$\begin{aligned}
0 &= -\left(\omega_A - \omega\right)M_{Ay} - M_{Ax}/T_{2A} - K_{ex}C_B M_{Ax} + K_{ex}C_A M_{Bx};\\
0 &= \left(\omega_A - \omega\right)M_{Ax} + \omega_1 M_{A0} - M_{Ay}/T_{2A} - K_{ex}C_B M_{Ay} + K_{ex}C_A M_{By};\\
0 &= -\left(\omega_B - \omega\right)M_{By} - M_{Bx}/T_{2B} + K_{ex}C_B M_{Ax} - K_{ex}C_A M_{Bx};\\
0 &= \left(\omega_B - \omega\right)M_{Bx} + \omega_1 M_{B0} - M_{By}/T_{2B} + K_{ex}C_B M_{Ay} - K_{ex}C_A M_{By}.
\end{aligned}$$

Having solved this system of equations and using expression (2.58), the form of the steady-state EPR spectrum of the model system under linear response conditions was calculated. Calculations were carried out for the situation when $C_A = C_B = C$. Figure 2.4 shows the calculated EPR spectra. They differ in the spin exchange rate $K_{ex}C$. It is possible to see how much the shape of the EPR spectrum changes with the increase in the spin exchange rate.

The upper left figure gives the spectrum of the model system of paramagnetic particles with two EPR frequencies, the spectrum lines are separated by 15 G, the spin exchange is considered off. When the spin exchange rate is nonzero and increases, the spectrum lines broaden and shift to the center (see spectra in the upper row in the center and right). With the further increase in the spin exchange rate, the two lines merge into one broad line (see spectra in the lower row on the left and in the center). And then the phenomenon of exchange narrowing of the spectrum occurs (see spectrum in the lower right corner).

If we look more closely at the shape of the spectrum at different values of the coherence transfer rate V, we can see another interesting feature. At intermediate coherence transfer rates (see spectra in the upper row of Fig. 2.4, in the center and right) two well-resolved lines are observed in the spectrum, but each line is not

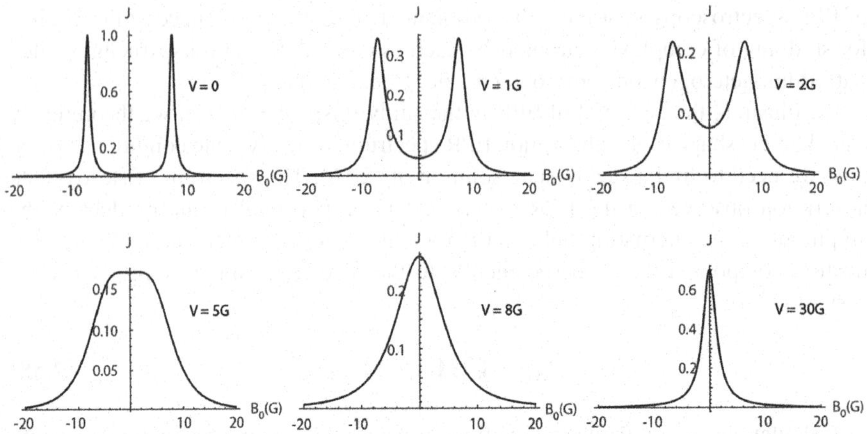

Fig. 2.4 Illustration of changes in the shape of the EPR spectrum with the increase in the spin exchange rate $V = K_{ex}C$. In EPR spectroscopy, the angular frequency is usually set in units of the magnetic field. Therefore, for converting a spin exchange rate, given in gauss, it should be multiplied by the value of $1.76\ 10^7$ to obtain its value in the frequency units rad/s. (Adapted with permission from [29])

symmetric (!) For a long time, the asymmetry of individual lines in the spectrum was not detected in experiment and therefore was not used in the interpretation of experimental data, although theoretically this effect was predicted more than 40 years ago in the book [1, 2]. This effect will be further discussed in detail (see Sect. 5.4).

Thus, the results show that spin exchange can have a very strong and very characteristic effect on the shape of the EPR spectrum. The spin exchange rate can be varied by changing the solvent, temperature (diffusion coefficient of molecules) and/or concentration of paramagnetic particles. The characteristic transformations of the EPR spectrum with increasing spin exchange rate are the broadening of the resonance lines of a spectrum, the asymmetric (mixed) shape of the individual lines in the spectrum, the shift of the lines, merging of all resonance lines into one broad line and further narrowing of the spectrum can serve as a criterion that the coherence transfer between spins with different resonance frequencies operates.

Qualitatively, transformations of the EPR spectrum with the change of the spin exchange rate are interpreted by two considerations. One consideration stems from the uncertainty principle. Spin exchange changes the state of the spin. This means that the lifetime of the spin in a given state is reduced, and this leads to the "uncertainty" of spin energy levels and to the broadening of lines in the spectrum. Such reasoning is used to explain the broadening of the spectrum lines in the region of relatively slow spin exchange (see spectra in the upper row in Fig. 2.4). Another consideration relates to the effect of averaging of the interaction by rapid random molecular motion. It is assumed that something like this takes place in the conditions of the exchange narrowing of the EPR spectrum. However, the situation is much more complicated and interesting. A consistent theoretical approach to the interpretation of spectroscopic manifestations of spin exchange was formulated in [1, 2] and further developed in [27–32] and will be presented below, e.g. in Sect. (5.4).

2.6 Kinetic Equations for Magnetization of Solutions of Free Radicals with the Arbitrary Hyperfine Structure of EPR Spectra Taking into Account Equivalent Spin Exchange

It has already been noted above that the greatest number of spin exchange studies have been conducted for stable nitroxide radicals (see [1, 2, 26]. They have one unpaired electron, spin $S = 1/2$. Both theoretical estimates and experimental data on spin exchange show that in the case of nitroxide radicals, the exchange integral is large enough and the situation of equivalent spin exchange, moreover, strong exchange, is implemented.

In real systems, free radicals have many magnetic nuclei. And there is a need to take into account the hyperfine interaction of unpaired electrons with all magnetic nuclei. For this purpose, consider the ensemble of free radicals. Due to the hyperfine interaction with magnetic nuclei, radicals have a set of resonance frequencies for the electron spin, which manifests itself in the EPR spectrum as a hyperfine structure. We divide the whole ensemble of radicals into isochromatic sub-ensembles, each of which gives a certain hyperfine component of the EPR spectrum. Let us denote the resonance frequency and magnetization of the k-th sub-ensemble by ω_k and $\mathbf{M}_k$, respectively. The contribution of spin exchange between two particles with different resonance frequencies to the kinetic equation has been described in detail above (see, e.g., 2.56). Summing up the contribution of all bimolecular collisions of particles, we obtain kinetic equations for the k-th sub-ensemble magnetizations. For nitroxide radicals, conditions of equivalent spin exchange are expected to be met. For this case, we have kinetic equations for magnetization (see also [27–34])

$$\begin{aligned} \partial M_{k-}/\partial t &= -\, i(\omega_k + \delta_k)M_{k-} - M_{k-}/T_{2k} - K_{ex}CM_{k-} + \varphi_k K_{ex} C \Sigma M_{n-}, \\ \partial M_{kz}/\partial t &= -\, (M_{kz} - M_{k0})/T_{1k} - K_{ex}CM_{kz} + \varphi_k K_{ex} C \Sigma M_{nz}. \end{aligned} \tag{2.59}$$

In fact, these are modified Bloch equations for partial magnetizations $\mathbf{M}_k$, T_{1k}, T_{2k} are the longitudinal and transverse paramagnetic relaxation times of spins, respectively. Here, in Bloch equations, an additional shifts δ_k of the resonance frequencies of spins ω_k are added, which might arise due to spin dynamics in the course of bimolecular meeting of two radicals [27]. This frequency shift is discussed below, e.g., see Sect. 2.8.2.

2.7 Theoretical Calculations of the Spin Exchange Rate Constant

2.7.1 Phenomenological Description of Spin Exchange

As already noted, the first theories of spin exchange kinetics were constructed by analogy with bimolecular chemical reactions [22, 24]. In formal chemical kinetics, equations are written for the concentrations of molecules. Spin exchange includes changes in populations of spin levels of molecules, and the description of changes in populations of spin levels using the equations of formal chemical kinetics is justified. The populations of the levels do not give a complete description of the quantum state of spins, it is necessary to have kinetic equations for the quantum coherence of the state of spins. The modern theory of spin exchange is based on quantum kinetic equations for single-particle spin density matrices.

However, phenomenological approaches played a positive role in the development of study of spin exchange, they made it possible to give qualitatively correct explanation of the features of the manifestations of spin exchange in the shape of EPR spectra [1, 2]. This greatly stimulated the development of experimental research. Currently, the use of spin probes to study the molecular mobility of molecules and bimolecular collisions in complex media, in the study of the role of particle charge for their bimolecular collisions in polar media, etc. has been well developed [1, 2]. From the EPR experiment, it is possible to find the rate and the rate constant for the bimolecular process.

It was shown in Sect. 2.1 that phenomenologically the rate constant of any bimolecular process, including spin exchange, can be expressed by several parameters. One parameter characterizes the mobility of molecules, it is the D_{AB} coefficient of mutual diffusion of the two molecules involved in the process. Another important parameter is the effective r_{ex} radius for the process in question. Based on the theory of Smolukhovski [8] and Debye [9] for the frequency of bimolecular collisions of molecules in solutions, the spin exchange rate constant can be represented as

$$K_{ex} = 4\pi r_{ex} D_{AB}. \tag{2.60}$$

The effective exchange radius r_{ex} (2.60) accumulates some averaged result of the motion of electron spins induced by the exchange interaction at the meeting of two paramagnetic particles. This result depends not only on the exchange interaction at the moments of the closest approach of particles, i.e., at the moments of collision of particles at the collision radius r_0, but also in the time intervals between repeated collisions of particles at the collision radius. The result of the exchange interaction depends on the duration of the meeting of paramagnetic particles. Increasing the mobility of molecules reduces the meeting time of colliding particles, reduces the total interval of time during which the exchange interaction can change the state of the spins of colliding particles. So, increasing the mobility of molecules reduces the efficiency of spin exchange. Therefore, the spin exchange rate constant (2.60) increases with the incease in the mutual diffusion coefficient of molecules $D = D_{AB}$, according to Smolukhovski equation. At the same time, the effective radius r_{ex} of the spin exchange may decrease with the growth of D. The deviation of the dependence of the spin exchange rate constant on the mutual diffusion coefficient $K_{ex}(D)$ on the linear one in the region of relatively large D can serve as an experimental criterion for ineffective (weak) spin exchange.

The Smoluchowski formula (2.1 and 2.60) for the rate constant of the bimolecular process is so simple due to the fact that it treated each molecule as a ball. The theories below assume that paramagnetic particles (spin probes) can be considered as balls whose radius is determined by the structure of paramagnetic particles and van der Waals radii of atoms.

Thus, in the theory of spin exchange in dilute solutions the main task is to calculate the effective spin exchange radius.

2.7.2 *Early Theories of Spin Exchange*

In a number of papers [22, 24, 25] the spin exchange rate contant between free radicals (spin probes with spin $S = 1/2$) was published. The simplest possible model was considered. It is assumed that paramagnetic particles can be represented by spheres and that when two particles meet, only one collision occurs, re-encounters are not taken into account. It is assumed that the exchange interaction $\mathbf{V} = \hbar J \mathbf{S}_1 \mathbf{S}_2$ is switched on suddenly only during a collision and also suddenly switched off. A flip-flop of spins of the colliding particles is considered as the elementary act of spin exchange.

In the described model of the sudden collisions, the spin exchange rate constant is equal to (2.60), in which the effective spin exchange radius is $r_{ex} = p_{ex} r_0$ (see 2.8). Here r_0 is the collision radius, i.e., the distance between the radicals at their closest approach, p_{ex} is the probability of spin exchange during the residence time at the collision radius in a single collision of a pair of molecules in solution. Within such a model, the effective radius r_{ex} of the spin exchange between particles should always be smaller than the collision radius of the molecules, since the probability cannot be greater than unity, it should be $p_{ex} \leq 1$.

The exchange interaction (2.14) causes a mutual flip-flop of spins when $S = 1/2$ with the probability $p(t) = \sin^2(Jt/2)$, (see 2.46).

It is assumed that the duration of collisions t in an ensemble of pairs has a Poisson distribution

$$\varphi(t) = (1/\tau c)\, exp\, (-t/\tau c). \tag{2.61}$$

Averaging over the ensemble, we obtain an average probability (2.46) of a mutual flip-flop of spins

$$p_{ex} = \int p(t)\varphi(t)dt = (1/2)J^2\tau_c{}^2/(1+J^2\tau_c{}^2). \tag{2.62}$$

As a result, for the spin exchange rate constant, we obtain

$$K_{ex} = (1/2)\left[J^2\tau_c{}^2/\left(1+J^2\tau_c{}^2\right)\right]K_D. \tag{2.63}$$

Substituting the constant K_D of the diffusion collision rate of molecules found by Smoluchowski [8], we get

$$K_{ex} = 2\pi r_0 D_{AB} J^2\tau_c{}^2/\left(1+J^2\tau_c{}^2\right) \tag{2.64}$$

Here, J is the exchange integral at the collision radius r_0 in units of rad/s, and τ_c is the average residence time of a pair of colliding particles at the collision radius. Within the model of sudden collisions, we have $p_{ex} \leq 1/2$. This means that within the model of the sudden switching on of the exchange interaction at the collision radius r_0, the spin exchange rate constant would always have to be no more than half of the rate constant of bimolecular diffusion collisions.

Although the considered model has found a wide application in interpreting experimental data on spin exchange between spin probes, it should be borne in mind that it describes the elementary act of bimolecular spin exchange in the simplest possible way. Equation (2.62) gives a lower estimate of the efficiency of spin exchange for the following reasons:

Firstly, this equation does not take into account re-encounters of two spins of a pair of paramagnetic particles. Remaining within the sudden switching on of the exchange interaction, it is possible to try to take into account repeated collisions if we assume that in (2.62) the time τ_c is equal to the total residence time of a pair of particles in the area of exchange interaction taking into account all the first and repeated collisions in the area of exchange interaction. To find this total time of collision of two particles during one meeting, we define the area of exchange interaction. Suppose that the exchange interaction suddenly turns on when the distance between two particles is in a narrow interval $\{r_0, r_0 + a\}$, $a << r_0$. Within the model of continuous diffusion of molecules, the total time that, during one meeting, two particles spend in the region of configuration space between spheres with radii r_0 and $r_0 + a$, i.e., in the field of exchange interaction, is

$$\tau_c = r_0 a/D_{AB}. \tag{2.65}$$

It can be noted that the use of the total time (2.65) of finding two particles in the exchange interaction region for calculating the efficiency p_{ex} (2.62) overestimates it, since it does not take into account the spin decoherence, which can occur in the time intervals between repeated collisions of particles at collision radius. But even taking into account decoherence of spins in the intervals between repeated collisions when considering repeated collisions, the total time τ_c in (2.65) should be greater than the average duration τ_c of a single collision of two particles. Suppose that when two molecules meet, n collisions occur on average (the first one and n-1 re-encounters). Without taking into account spin decoherence in the intervals between repeated collisions, the effective duration of all collisions in (2.62) should be $\tau_{eff} = \tau_c n$. It was shown (in [34], see also below) that $\tau_{eff} = \tau_c(n)^{1/2}$, if in the interval between repeated collisions the spin coherence is completely lost. From this we can conclude that $\tau_c n > \tau_{eff} > \tau_c$. (here n is the average number of collisions per one meeting of two particles).

Secondly, a sudden collision model (2.62) does not take into account the extended nature of the exchange interaction. In the situation under consideration, the intermolecular exchange interaction of paramagnetic particles, namely, the exchange integral can be considered as proportional to the square of the module of the overlap integral of electron orbitals occupied by unpaired electrons of interacting molecules, so that the exchange integral is an exponentially decreasing function

$$J_{AB}(r) = J_{AB}(r_0)\,exp\,(-\kappa(r - r_0)), \tag{2.66}$$

where J_{AB} (r_0) is the exchange integral at the collision radius, and the parameter κ characterizes the steepness of the decline of the exchange integral with increasing distance between the interacting atoms. For interatomic interactions, typical values are $1/\kappa \approx 0.03 \div 0.04$ nm [14]. This value can be used to estimate the “thickness” a of the exchange interaction region, i.e., $a \approx 1/\kappa \approx 0.03 \div 0.4$ nm. For free radicals, the exchange integral at a distance of the collision radius is estimated as $J_{AB}(r_0)$ ~$10^{12} \div 10^{13}$rad/s [1, 2]. Using these estimates, we can verify that the condition J (r_0) $\tau_c > > 1$ is expected for free radicals. Then, according to (2.62), $p_{ex} = 1/2$, and the spin exchange rate constant is equal to half of the rate constant of bimolecular collisions, $K_D = 4\pi\, r_0 D_{AB}$. However, the fulfillment of condition J (r_0) $\tau_c > > 1$ means that even for $r > r_0$, the exchange integral can be quite large and when calculating the spin exchange rate, one should take into account the extended nature of the exchange integral. Therefore, the true spin exchange rate contant can become equal or even greater than the rate constant of the first collisions of molecules K_D (2.6). The theoretical results for the spin exchange rate constant taking into account the extended nature of the exchange integral will be given below in Sect. 2.10.

2.7.3 *Kinetic Equations for the Spin Density Matrix in a Model of Sudden Collisions*

For a detailed description of spin exchange and its manifestations in the experiment, it is necessary to have the corresponding kinetic equations for spin density matrices with account for bimolecular collisions. Within the sudden collision model, the consistent kinetic equations for the description of spin exchange processes in dilute solutions were formulated in [35].

Consider a solution of paramagnetic particles A and B. According to [35], the spins are divided into two subsystems: isolated paramagnetic particles and pairs of particles with the switched on exchange interaction. These two subsystems are in dynamic equilibrium: with a rate K_DC_B (C_B concentration of spins), the separated paramagnetic particle A leaves the subsystem of isolated spins and forms pair AB, and with a rate of τ_c^{-1}, the pair break up into two isolated spins. Particle solution is considered sufficiently dilute to neglect triple collisions of paramagnetic particles. Before writing out the kinetic equations, we introduce the following notation: $\mathbf{H}_0(k)$ -spin Hamiltonian of an isolated k-th paramagnetic particle, $\mathbf{V}_{ex}$-spin Hamiltonian of exchange interaction in a pair of AB (see (2.14)), ρ-density matrix of the spin, $\rho^{(2)}$ -spin density matrix of the pair AB. It is assumed that at the moment of pair formation, its density matrix is equal to the direct product of the spin density matricies of the collision partners.

Within the assumptions made for spin exchange between identical particles, the kinetic equations for single-particle and two-particle spin density can be written as [35].

$$\begin{aligned} \partial\rho(1)/\partial t &= -i\hbar^{-1}[H_0(1), \rho(1)] - K_D C[\rho(1) - \mathrm{Tr}_{1'}\rho^{(2)}(1,1')]_{1'\to 1}; \\ \partial\rho^{(2)}(1,1')/\partial t &= -i\hbar^{-1}[H_0(1) + H_0(1'), \rho^{(2)}(1,1')] - (1/t_c)[\rho^{(2)}(1,1') - \rho(1)\times\rho(1')]. \end{aligned} \tag{2.67}$$

Here Tr denotes the convolution of the spin density matrix with respect to the variables of the partner spin in a collision. Using these equations, a number of questions in the theory of spin exchange between free radicals, paramagnetic complexies and triplet excitons was analyzed (see, e.g., [36–39]).

In the case of spin exchange between different particles, Eq. (2.67) are written as [35, 40].

$$\begin{aligned} \partial\rho_A/\partial t &= -i\hbar^{-1}[H_A, \rho_A] - K_D C_B[\rho_A - \mathrm{Tr}_B\rho^{(2)}{}_{AB}]; \\ \partial\rho_B/\partial t &= -i\hbar^{-1}[H_B, \rho_B] - K_D C_A[\rho_B - \mathrm{Tr}_A\rho^{(2)}{}_{AB}]; \\ \partial\rho^{(2)}{}_{AB}/\partial t &= -i\hbar^{-1}[H_A + H_B + V, \rho^{(2)}{}_{AB}] - (1/t_c)[\rho^{(2)}{}_{AB} - \rho_A \times \rho_B]. \end{aligned} \tag{2.68}$$

In dilute solutions, the average collision duration τ_c is small compared with the time interval between collisions, i.e., $K_DC\tau_c \ll 1$. Under this condition the system of Eq. (2.68) allows a substantial simplification [40]. We solve the equation for the pair density matrix $\rho^{(2)}{}_{AB}$, and find

$$\begin{aligned} \rho^{(2)}{}_{AB}(t) &= \exp(-t/t_c)\exp(-i\hbar^{-1}Ht)\rho^{(2)}{}_{AB}(0)\exp(i\hbar^{-1}Ht) \\ &\quad + \int_0^t (d\tau/\tau_c)\exp(-\tau/\tau_c)\exp(-iH\tau/\hbar)\rho^A(t-\tau)\rho^B(t-\tau)\exp(iH\tau/\hbar). \end{aligned} \tag{2.69}$$

In this equation, $\rho^{(2)}{}_{AB}(0)$ is the initial two-particle density matrix, $\mathbf{H} = \mathbf{H}_A + \mathbf{H}_B + \mathbf{V}_{ex} = \mathbf{H}_0 + \mathbf{V}_{ex}$. Collisions cause changes in single-particle spin matrices with a characteristic free-motion time between collisions which is estimated as being around $1/(K_DC)$. The integrand in (2.69) is nonzero only for $\tau < \tau_c$. Therefore, up to small values on the order of $K_DC\tau_c \ll 1$ in the integrand, it is possible to ignore the changes in single-particle spin density matrices ρ_A, ρ_B due to collisions and at times $\tau < \tau_c$ accept that

$$\rho_{A(B)}(t - \tau) \approx \exp(i\hbar^{-1}\mathbf{H}_0\tau)\ \rho_{A(B)}(t)\exp(-i\hbar^{-1}\mathbf{H}_0\tau).$$

Substituting this expression into (2.69) and considering that for times $t\sim 1/(K_DC) >> \tau_c$, $\rho^{(2)}{}_{AB}(t)$ ceases to depend on the initial state, we get

$$\rho^{(2)}{}_{AB}(t) \approx \int_0^{\infty} exp(-\tau/\tau_c)\mathbf{S}(\tau)\rho_A(t) \times \rho_B(t)\mathbf{S}^{-1}(\tau)d\tau/\tau_c.$$

The collision matrix is introduced here

$$\mathbf{S}(\tau) = \exp(-i\hbar^{-1}\mathbf{H}\tau)\exp(i\hbar^{-1}\mathbf{H}_0\tau), \tag{2.70}$$

where $\mathbf{H}_0 = \mathbf{H}_A + \mathbf{H}_B$; $\mathbf{H} = \mathbf{H}_0 + \mathbf{V}_{ex}$.

Using these expressions, the kinetic equations proposed in [35] can be represented as (see, for example, [1, 2, 40], Eq. (2.131) below)

$$\begin{aligned}\partial\rho_A/\partial t &= -i\hbar^{-1}[\mathbf{H}_A, \rho_A] - K_DC_B(\rho_A - \mathrm{Tr}_B < \mathbf{S}(\tau)\rho_A\rho_B\mathbf{S}^{-1}(\tau) > ;\\ \partial\rho_B/\partial t &= -i\hbar^{-1}[\mathbf{H}_B, \rho_B] - K_DC_A(\rho_B - \mathrm{Tr}_A < \mathbf{S}(\tau)\rho_A\rho_B\mathbf{S}^{-1}(\tau) > .\end{aligned} \tag{2.71}$$

Here $\rho_{A(B)} = \mathrm{Tr}_{B(A)}\rho_A\rho_B$. The spin Hamiltonian of the exchange interaction is given by (2.14). In Eq. (2.71), $\mathrm{Tr}_{A(B)}$ means convolution with respect to spin variables of A or B, respectively. Averaging <. . . > means averaging over the distribution of τ, which is given by Eq. (2.61), see also Eq. (2.62).

When writing the kinetic Eq. (2.71), it is taken into account that there is a hierarchy of processes over time. There are collisions of two molecules in a pair state and meetings with new molecules. The characteristic time of one meeting, τ_c, is significantly less than the time of diffusion "free run", which can be estimated as $1/(K_DC_{A(B)})$. The kinetic Eq. (2.71) describe the changes of single-particle spin density matrices on a time scale $1/(K_DC_{A(B)})$. Therefore, in (2.71) one can carry out averaging over a faster process on the scale of τ_c. In fact, in (2.71), in all terms, single-particle density matrices ρ_A and ρ_B are the mean values on the collision time scale. But the change of the single-particle matrix on the τ_c scale due to collisions can be neglected. Therefore, in (2.71) the averaging is indicated only of the term describing the spin dynamics during the collision.

In Sect. 2.9 are presented the kinetic equations taking into account the extended nature of the exchange interaction and the arbitrary kinematics of the mutual motion of two molecules. In the limit, which corresponds to the model of the sudden collisions, the general theory reproduces the theory based on the model of the sudden collisions presented here.

Equations (2.71) can be written in a more compact form if we introduce a superoperator of collision efficiency $\mathbf{T} = \langle \mathbf{S}(\tau)\mathbf{S}^{-1}(\tau)\rangle$. Elements of $\mathbf{T}$ are expressed in terms of $\mathbf{S}$ matrix elements as

$$T_{mn,\,kl} = \int \exp(-\tau/\tau_c) S_{mk}(\tau)\, S_{ln}{}^{-1}(\tau)\, d\tau/\tau_c. \tag{2.72}$$

With the help of the superoperator $\mathbf{T}$, Eq. (2.71) take the form

$$\begin{aligned} \partial\rho_A/\partial t &= -i\hbar^{-1}[H_A, \rho_A] - K_D C_B(\rho_A - Tr_B T \rho_A \rho_B); \\ \partial\rho_B/\partial t &= -i\hbar^{-1}[\mathbf{H}_B, \rho_B] - K_D C_A(\rho_B - Tr_A T \rho_A \rho_B). \end{aligned} \tag{2.73}$$

The collision efficiency superoperator satisfies the equation [40].

$$[\mathbf{T}, \mathbf{Q}] = \mathbf{\Omega T} - (\mathbf{T} - \mathbf{E})/\tau_c. \tag{2.74}$$

$$Q_{mn,\,kl} = -i\hbar^{-1}(H_{0mk}\delta_{nl} - H_{0ln}\delta_{mk}),$$

$$\Omega_{mn,\,kl} = -i\hbar^{-1}(V_{mk}\delta_{nl} - V_{ln}\delta_{mk}),$$

E-unit operator.

For typical conditions of EPR experiments, spin exchange is studied in systems close to equilibrium. Let us introduce matrices σ, which describe small deviations of the spin system from thermodynamic equilibrium

$$\rho_A = \rho_{0A} + \sigma_A, \rho_B = \rho_{0B} + \sigma_B. \tag{2.75}$$

Substituting (2.75) into (2.73) and leaving only the linear in σ_A and σ_B terms, we obtain linear equations that describe the kinetics of spin exchange under conditions when the system is not by far deviated from equilibrium.

Within the model of sudden collisions, in principle, it is possible to take into account not only the exchange, but also other spin-dependent interactions of isolated paramagnetic particles, i.e., $\mathbf{H}_A$, $\mathbf{H}_B$ spin Hamiltonians in (2.70) can include the Zeeman interaction with the external magnetic field, hyperfine interaction with magnetic nuclei, splitting of the energy levels in the zero magnetic field, if a paramagnetic particle contains more than one unpaired electron, and also the spin relaxation of isolated paramagnetic particles. To do this, it is enough to add the terms describing the relaxation of spins to the kinetic equations [36, 38]. Within this model, the effect of rotational diffusion of paramagnetic particles on spin exchange was also considered in the case of the anisotropic distribution of the spin density of paramagnetic particles [37].

2.8 Calculations of the Spin Exchange Rate Constant in the Approximation of the Sudden Collisions

In the approximation of the sudden collisions, the rate constant of the bimolecular spin exchange in a consistent manner, based on the kinetic Eqs. (2.73, 2.74 and 2.75), was calculated for several cases.

2.8.1 *Spin Exchange Between Particles with Spin S = 1/2*

There are many examples of paramagnetic particles with one unpaired electron, S = 1/2: atoms (e.g., H, Li, K), complexes of transition element ions (e.g., $(Fe(CN)_6)^{3-}$ $(Co(H2O)_6)^{2+}$), defects in solids, etc. It is especially possible to distinguish free radicals, which often play a crucial role in chemical reactions, as active intermediate states. But among the free radicals there are fairly stable, chemically inactive ones. Examples include nitroxide radicals, which are widely used as spin labels and probes. Apparently, it is precisely stable radicals that have found the greatest application in the study of spin exchange and its application for the study of translational mobility and bimolecular collisions of molecules in complex media.

Within the phenomenological model [22], the equivalent spin exchange rate constant in collisions of paramagnetic particles with spins S = 1/2 was already given in Sect. 2.1. This section presents the results obtained for particles with spin S = 1/2 using kinetic Eqs. (2.73, 2.74 and 2.75). It is not assumed that the conditions of equivalent spin exchange are satisfied, when at the moment of collision only the exchange interaction (2.14) is taken into account. There will be presented results concerning the equivalent as well as nonequivalent spin exchange cases.

Consider the model situation. We assume that the spin Hamiltonians of particles A and B (see 2.70) contain only the Zeeman interaction with a constant magnetic field B_0 and that the spins have different spectroscopic g-factors, and therefore different Larmor (resonant in the EPR experiment) frequencies:

$$\mathbf{H}_A = g_A\beta B_0 \mathbf{S}_{Az} = \hbar\omega_A \mathbf{S}_{Az}, \mathbf{H}_B = g_B\beta B_0 \mathbf{S}_{Bz} = \hbar\omega_B \mathbf{S}_{Bz}. \tag{2.76}$$

Here, β is the Bohr magneton, ω_A, ω_B are Larmor frequencies of spins. The frequency difference of the spins can also be caused by the hyperfine interaction of unpaired electrons with magnetic nuclei. In this case, the entire ensemble of paramagnetic particles can be divided into sub-ensembles, in each of which the nuclear spins are in a certain configuration. Each sub-ensemble gives a certain hyperfine component in the EPR spectrum with the corresponding resonant frequency. If we consider the collision of two particles that belong to different sub-ensembles of paramagnetic particles, then the problem reduces to a model situation of collisions of two spins with different resonant frequencies. Of course, in the case of the hyperfine structure of the EPR spectra, it is necessary to take into account the collisions of particles of the selected sub-ensemble with particles from other sub-ensembles with different resonant frequencies. Here we consider in detail the effect of the frequency difference of the colliding pair of spins on the elementary act of a change in the spin state caused by the exchange interaction. For illustration, it is only important that for one reason or another the frequencies of the colliding spins are different. We assume that at the moment of collision the exchange interaction $\mathbf{V}\mathbf{ex} = \hbar\, \mathbf{J}\mathbf{S}_A\mathbf{S}_B$ is switched on suddenly.

The collision matrix **S** (2.70) in the representation $|m_A, m_B>$, where $|m_A>$, $|m_B>$ are eigenstates of the operators $\mathbf{S}_{Az}$, $\mathbf{S}_{Bz}$, has the following nonzero matrix elements

$$\begin{aligned}
&< 1/2, 1/2 \mid \mathbf{S}(\tau) \mid 1/2, 1/2 >= < -1/2, -1/2 \mid \mathbf{S}(\tau) \mid -1/2, -1/2 >= exp(-iJ\tau);\\
&< 1/2, -1/2 \mid \mathbf{S}(\tau) \mid 1/2, -1/2 >= exp(i\delta\tau/2)\left[b^2 exp(-iE_b\tau) + c^2 exp(-iE_c\tau)\right);\\
&< -1/2, 1/2 \mid \mathbf{S}(\tau) \mid -1/2, 1/2 >= exp(-i\delta\tau/2)\left[c^2 exp(-iE_b\tau) + b^2 exp(-iE_c\tau)\right);\\
&< 1/2, -1/2 \mid \mathbf{S}(\tau) \mid -1/2, 1/2 >= < -1/2, 1/2 \mid \mathbf{S}(\tau) \mid 1/2, -1/2 > exp(-i\delta\tau) =\\
&\qquad = exp(-i\delta\tau/2)\left[exp(-iE_b\tau) - exp(-iE_c\tau)\right).
\end{aligned} \tag{2.77}$$

Here are introduced notations:

$$\begin{aligned}
&\delta = \omega_A - \omega_B;\\
&E_b = -J/4 + R/2;\; E_c = -J/4 - R/2;\\
&R = \left(J^2 + \delta^2\right)^{(1/2)};\\
&tg(\varphi) = J/\delta; b = cos(\varphi/2); c = sin(\varphi/2);\\
&-\pi/2 \le \varphi \le \pi/2.
\end{aligned}$$

Using Eq. (2.77), it is possible to calculate the component of the collision efficiency operator and thereby find the necessary kinetic coefficients. The result is

$$\begin{aligned}
(\partial/\partial t)\rho^A{}_{1/2,1/2} &= -K_e C_B\left(\rho^A{}_{1/2,1/2}\rho^B{}_{-1/2,-1/2} - \rho^A{}_{-1/2,-1/2}\rho^B{}_{1/2,1/2}\right) +\\
&\quad + 2\,Re\left\{K_1 \rho^A{}_{1/2,-1/2}\rho^B{}_{-1/2,1/2}\right\};\\
(\partial/\partial t)\rho^A{}_{1/2,-1/2} &= -i\omega_A\rho^A{}_{1/2,-1/2} - K_D C_B\left(1 - p_1\rho^B{}_{1/2,1/2} - p_2\rho^B{}_{-1/2,-1/2}\right)\rho^A{}_{1/2,-1/2} +\\
&\quad + K_D C_B\left(p'_1\rho^A{}_{1/2,1/2} + p'_2\rho^A{}_{-1/2,-1/2}\right)\rho^B{}_{1/2,-1/2}.
\end{aligned} \tag{2.78}$$

Here

$$
\begin{aligned}
&\mathrm{Ke} = 2\mathrm{b}^2\mathrm{c}^2(1 - <cos(\mathrm{Rt})>)\mathrm{K_D};\\
&\mathrm{p}_1 = <exp(-iJ\tau/4 + i\delta\tau/2)\left(\mathrm{c}^2 exp(i\mathrm{E_b}\tau) + \mathrm{b}^2 exp(\mathrm{iE_c}\tau)\right)>;\\
&\mathrm{p}_2 = <exp(iJ\tau/4 + i\delta\tau/2)\left(\mathrm{b}^2 exp(-i\mathrm{E_b}\tau) + \mathrm{c}^2 exp(-i\mathrm{E_c}\tau)\right)>;\\
&\mathrm{p}_1' = <exp(-iJ\tau/4 - i\delta\tau/2)(exp(i\mathrm{E_b}\tau) - exp(i\mathrm{E_c}\tau))\mathrm{bc}>;\\
&\mathrm{p}_2' = <exp(iJ\tau/4 - i\delta\tau/2)(exp(-i\mathrm{E_b}\tau) - exp(-i\mathrm{E_c}\tau))\mathrm{bc}>;\\
&\mathrm{K}_1 = \mathrm{K_D C_B bc} <exp(i\delta\tau)\left(\mathrm{b}^2 - \mathrm{c}^2 - \mathrm{b}^2 exp(-iR\tau) + \mathrm{c}^2 exp(iR\tau)\right)>.
\end{aligned}
$$

Similar equations also hold for the density matrix of a spin partner in collision, ρ_B.

Thus, within the model of sudden switching on and switching off of the exchange interaction during a collision, kinetic equations are given that describe the behavior of the diagonal and off-diagonal elements of the spin density matrix of particles A ($S_A = 1/2$) and B ($S_B = 1/2$) taking into account bimolecular collisions A + B, but neglecting re-encounters of A and B molecules in a course of their meeting and forming a pair state. Note that the off-diagonal density matrix elements describe quantum coherence. It is the off-diagonal elements of the density matrix that determine the behavior of transverse components of the magnetization of the spins, which are directly measured in the EPR experiments.

This example makes it possible to demonstrate that, in the general case, the manifestations of exchange interactions during bimolecular collisions cannot be described by phenomenological equations such as the equations of chemical kinetics, as suggested in [22]. In Eq. (2.78), the kinetic coefficients in the general case are complex numbers, and not real as is the case in the kinetics of chemical reactions.

Incidentally, this suggests that there may be situations in which the equations of the kinetics of chemical reactions should be generalized, taking into account the effects of quantum coherence of the state of reagents.

Below it will be shown that the spin kinetics taking into account the exchange interaction during bimolecular collisions can be described using equations of the type of chemical kinetics equations only when at the time of the collision all the spin interactions can be neglected, except for the exchange interaction. In this case, the exchange interaction performs equivalent exchange of spin states of colliding partners.

Kinetic equations are greatly simplified if the systems are considered whose state is slightly different from the equilibrium state. Such a situation takes place in EPR experiments on the study of spin exchange (2.75). Substituting (2.75) into the kinetic Eq. (2.78) and leaving only terms linear in σ_A and σ_B, we obtain linear equations that describe the kinetics of spin exchange under conditions when the system is not too far deviated from equilibrium. For the convenience of writing the resulting equations, eigenfunctions of the operator S_z, $|+1/2>$ and $|-1/2>$, we denote as α and β states, respectively:

$$
\begin{aligned}
&\partial\sigma_{A\alpha\alpha}/\partial t = -\mathrm{K_{ex}C_B}\left(\sigma_{A\alpha\alpha}\rho_{0B\beta\beta} - \sigma_{A\beta\beta}\rho_{0B\alpha\alpha}\right) + \mathrm{K_{ex}C_B}\left(\sigma_{B\alpha\alpha}\rho_{0A\beta\beta} - \sigma_{B\beta\beta}\rho_{0A\alpha\alpha}\right);\\
&\partial\sigma_{A\alpha\beta}/\partial t = -i\omega_A\sigma_{A\alpha\beta} - \mathrm{W1}\sigma_{A\alpha\beta} + \mathrm{W2}\sigma_{B\alpha\beta}.
\end{aligned}
\tag{2.79}
$$

Here the following notations are introduced:

$$\begin{aligned}
&\mathrm{Kex} = p_{ex} K_D;\\
&p_{ex} = (J^2/R^2) < sin^2(Rt/2) >= (1/2)\left(J^2\tau_c{}^2/\left(1 + R^2\tau_c{}^2\right)\right);\\
&W1 = K_D C_B\{1- < exp(i(\omega_A - \omega_B)t/2)\ (cos(Rt/2) - i((\omega_A - \omega_B)/R) sin(Rt/2)\ )\\
&\qquad (cos(Jt/2) - isin(Jt/2)(\rho_{0B\alpha\alpha} - \rho_{0B\beta\beta})) >\};\\
&W2 = K_D C_B < (J/R)\ exp(-i(\omega_A - \omega_B)t/2)\ sin(Rt/2)\ (sin(Jt/2)+\\
&\qquad i\,cos(Jt/2)(\rho_{0A\alpha\alpha} - \rho_{0A\beta\beta})) > ;\\
&R = \left(J^2 + (\omega_A - \omega_B)^2\right)^{1/2}.
\end{aligned} \tag{2.80}$$

Similar equations for spins B can be obtained by replacing indices A by B and B by A in (2.79 and 2.80). These equations make several observations possible. The rate constant K_{ex} of mutual flip-flops of two spins caused by the exchange interaction depends on the difference in the resonance frequencies of the spins

$$K_{ex} = (1/2)\left(J^2\tau_c{}^2/\left(1 + \left(J^2 + (\omega_A - \omega_B)^2\right)\tau_c{}^2\right)\right)K_D. \tag{2.81}$$

In the limiting situation, when $J^2 > > (\omega_A - \omega_B)^2$, $R \approx J$, and $(\omega_A - \omega_B)^2)\tau_c{}^2) < 1$ the spin exchange rate constant K_{ex} coincides with expression (2.46) for equivalent spin exchange obtained in [24, 25]. If the difference of resonant frequencies and/or the average duration of the collision of a pair of spins is sufficiently large and the condition $(\omega_A - \omega_B)^2)\tau_c{}^2) > 1$ is satisfied, then with the increase in the exchange integral the rate constant of mutual flip-flop of two spins tends not to 1/2, but behaves as

$$p_{ex} \approx (1/2)J^2/\left(J^2 + (\omega_A - \omega_B)^2\right). \tag{2.82}$$

In Eq. (2.80) <…> means averaging over the distribution of the duration of collisions (2.61). It can be seen from Eq. (2.80) that the kinetic coefficients W1, W2 are generally complex quantities. The real part of W1 describes the decoherence rate, or the spin coherence relaxation rate caused by the exchange interaction in bimolecular collisions of paramagnetic particles. The imaginary part of W1 gives the frequency shift of the spin ω_A. The W2 characterizes the coherence transfer to the selected spin, in this case A, from the collision partner, B Eq. (2.80).

To highlight the results of the spin dynamics during the collision of particles, we consider two limiting situations: (1) there is almost equivalent spin exchange, when $|J| > |(\omega_A - \omega_B)|$ and $(\omega_A - \omega_B)t_c < < 1$, and (2) nonequivalent spin exchange case when $(\omega_A\text{-}\omega_B)\tau_c > 1$ and the exchange integral has an arbitrary value. Equivalent exchange is precisely implemented only when the difference of resonant frequencies during the collision of particles can be completely neglected. To do this, at the time of collision it should be considered $(\omega_A\text{-}\omega_B) = 0$.

In the case of equivalent spin exchange, the expressions for W1 and W2 (2.80) are greatly simplified:

$$\begin{aligned}
&R = J;\\
&W1 = K_D C_B < sin^2(Jt/2) > -i < (sin(Jt)/2)\left(\rho_{0B\alpha\alpha} - \rho_{0B\beta\beta}\right) > =\\
&\quad K_D C_B\left((1/2)\left(J^2\tau_c{}^2/\left(1 + J^2\tau_c{}^2\right)\right) - i\left(J\tau_c/\left(2\left(1 + J^2\tau_c{}^2\right)\right)\right)\left(\rho_{0B\alpha\alpha} - \rho_{0B\beta\beta}\right)\right);\\
&W2 = K_D C_B < sin^2(Jt/2) > +i < cos(Jt/2)\left(\rho_{0A\alpha\alpha} - \rho_{0A\beta\beta}\right) > =\\
&\quad K_D C_B\left((1/2)\left(J^2\tau_c{}^2/\left(1 + J^2\tau_c{}^2\right)\right) + i\left(2/\left(4 + J^2\tau_c{}^2\right)\right)\left(\rho_{0A\alpha\alpha} - \rho_{0A\beta\beta}\right)\right).
\end{aligned} \tag{2.83}$$

It can be seen that the decoherence rate, as expected, in the case of an equivalent spin exchange given by a well-known expression

$$Re\{W1\} = (1/2)\left(J^2\tau_c{}^2/\left(1 + J^2\tau_c{}^2\right)K_D C_B.\right.$$

But it follows from Eqs. (2.79 and 2.83) that in this case of equivalent exchange there is a frequency shift caused by the exchange interaction

$$\omega'_A = \omega_A - K_D C_B\left(J\tau_c/\left(2\left(1 + J^2\tau_c{}^2\right)\right)\right)\left(\rho_{0B\beta\beta} - \rho_{0B\alpha\alpha}\right). \tag{2.84}$$

This frequency shift can be called the paramagnetic shift. The frequency shift for B spins is obtained from (2.84) by permutation of indices A and B. Assuming that $(\rho_{0\beta\beta} - \rho_{0\alpha\alpha}) > 0$ in equilibrium, we find that the sign of the paramagnetic frequency shift depends on the sign of the exchange integral. For the spin Hamiltonian in the form (2.14) the ferromagnetic type of the exchange interaction corresponds to $J < 0$, and the antiferromagnetic type to $J > 0$. Frequency shifts of both particles A and B have the same sign. This shift occurs as a result of the fact that the spins, due to the predominant orientation in one direction in presence of constant B_0 field, produce a certain average nonzero local magnetic field (such as the Weiss field in magnetic materials) at the location of the paramagnetic particles. Under normal conditions of the EPR experiments in liquids (T~300K, the induction of the external magnetic field B_0~3000 Gs, X – band) the equilibrium dipole polarization of the electron spins $(\rho_{0B\beta\beta} - \rho_{0B\alpha\alpha}) \sim 10^{-3}$. Therefore, the frequency shift of the type (2.84) in EPR experiments, as a rule, should be small enough and it can be neglected in experiments in which there is no special polarization of electron spins. Note that this paramagnetic frequency shift was experimentally detected in [41] when studying spin exchange between Rb atoms with electron spins optically polarized in gases. The ratio of the resonance frequency shift to the the exchange line broadening was 0.025. Similar measurements in the system of electrons+ cesium atoms were performed in [42]. In the conditions of optical pumping, the spin polarization, $\rho_{1/}$

$_{2,1/2}$-$\rho_{-1/2,-1/2}$, is significantly greater than the equilibrium value, and therefore this paramagnetic frequency shift is manifested in the experiment.

If the equilibrium dipole polarization of spins is negligible, then in the considered situation of equivalent spin exchange we have

$$\begin{aligned} W1 &= W2 = K_{ex} C_B = p_{ex} K_D C_B, \\ p_{ex} &= (1/2)\left(J^2\tau_c{}^2/\left(1 + J^2\tau_c{}^2\right)\right). \end{aligned} \tag{2.85}$$

This result coincides with the result of early theories of spin exchange [22, 24, 25] based on the analogy of bimolecular spin exchange with the bimolecular chemical reactions. The coincidence of the results of the consistent theory (2.81, 2.82 and 2.83) and phenomenological theory [22, 24, 25] occurs when the result of the collision can be reduced only to the equivalent exchange of spin states, i.e., to the mutual flip-flop of spins.

In the general case of spin exchange, the consistent theory yields a result that does not coincide with the phenomenological theory. To highlight the features of spin exchange in general case, suppose that the dipole polarization of spins, and hence the paramagnetic shift of lines, can be neglected. Assuming $\rho_{0\alpha\alpha}$-$\rho_{0\beta\beta}$=0, from (2.80), we obtain

$$\begin{aligned} W1 &= p_{Aex} K_D C_B; \\ p_{Aex} &= \{1 - < exp(i(\omega_A - \omega_B)t/2)\ (cos(Rt/2) - i((\omega_A - \omega_B)/R) sin(Rt/2))\, cos(Jt/2) >\} \\ &= 1 - (1/4)\{(1 - ((\omega_A - \omega_B)/R)\ [< exp(ix_1) > + < exp(ix_2) >] \\ &\quad + (1 + ((\omega_A - \omega_B)/R)[< exp(ix_3) > + < exp(ix_4) >]\}; \\ W2 &= p_{Bex} K_D C_B; \\ p_{Bex} &= < (J/R)\, exp(-i(\omega_A - \omega_B)t/2)\, sin(Rt/2)\, sin(Jt/2) > \\ &= -(J/(4R))\,[- < exp(-ix_1) > + < exp(-ix_2) > + < exp(-ix_3) > - < exp(-ix_4) >]. \end{aligned} \tag{2.86}$$

Here the notations are introduced

$$\begin{aligned} x_1 &= (R - J + (\omega_A - \omega_B))\tau_c/2, \\ x_2 &= (R + J + (\omega_A - \omega_B))\tau_c/2, \\ x_3 &= (-R - J + (\omega_A - \omega_B))\tau_c/2, \\ x_4 &= (-R + J + (\omega_A - \omega_B))\tau_c/2, \\ &< exp(ix) >= (1 + ix)/\left(1 + x^2\right). \end{aligned}$$

The case $|J| \gg |\omega_A$-$\omega_B|$ is of practical interest. Note that in this case, if the condition of nonequivalent spin exchange $(\omega_A$-$\omega_B)\tau_c > 1$ is implemented, we obtain the situation of the strong exchange $|J|\tau_c \gg 1$.

So, if the condition $|J| \gg |\omega_A - \omega_B|$ is fulfilled, we have

$$R \approx J, x_1 \approx x_4 \approx (\omega_A - \omega_B)\tau_c/2, x_2 \approx J\tau_c + (\omega_A - \omega_B)\tau_c/2, x_3 \approx -J\tau_c + (\omega_A - \omega_B)\tau_c/2.$$

Substituting these values into (2.86), we obtain

$$p_{Aex} \approx 1 - (1 + i(\omega_A - \omega_B)\tau_c/2)/\left(2\left(1 + ((\omega_A - \omega_B)t_c/2)^2\right)\right) - (1/4)(1 - (\omega_A - \omega_B)/J)(1 + i(J + (\omega_A - \omega_B)/2)\tau_c)/\left(1 + (J + (\omega_A - \omega_B)/2)^2\tau_c^2 - (1/4)(1 + (\omega_A - \omega_B)/J)(1 + i(-J + (\omega_A - \omega_B)/2)\tau_c)/\left(1 + (-J + (\omega_A - \omega_B)/2)^2\tau_c^2\right).$$

If the condition $(\omega_A - \omega_B)\tau_c > 1$ is fullfilled,

$$p_{Aex} \approx 1 - i/((\omega_A - \omega_B)\tau_c). \quad (2.87)$$

Substituting (2.87) into (2.86), we obtain that the decoherence rate of spins A due to the exchange interaction with spins B in this case is

$$W_{decoh} = Re\{W1\} = Re\{ p_{Aex}K_D C_B\} = K_D C_B. \quad (2.88)$$

This is a very remarkable result. It turns out that if the condition $|(\omega_A - \omega_B)|\tau_c > 1$ of nonequivalent spin exchange is fulfilled, the decoherence rate constant for the spins due to the exchange interaction is equal to the diffusion rate constant of bimolecular collisions K_D. Note that in case of equivalent spin exchange with $J\tau_c > 1$, the spin decoherence rate constant is $(1/2)K_D$ (2.63 and 2.85). We remind that all these results are obtained within the model of sudden collisions and disregarding the repeated collisions of particles of pairs.

The module of p_{Bex} (2.86) determines the efficiency of the quantum coherence transfer, in this case, to spin A from the collision partner B. In terminology of the third Newton law, p_{Bex} characterizes the reaction of B to the action of A.

The imaginary part of W1 (2.86) determines the resonance frequency shift of spins A as a result of the exchange interaction during the collisions of A and B. Using (2.87), we have, in the case of nonequivalent spin exchange, $(\omega_A - \omega_B)\tau_c > 1$, for the strong exchange interaction, $|J| \gg |\omega_A - \omega_B|$, the following resonance frequency shift

$$\Delta\omega_A = -(K_D C_B)/((\omega_A - \omega_B)\tau_c). \quad (2.89)$$

In the case of quai-equivalent spin exchange, the resonance frequency shift is expected. Indeed, in the case $(\omega_A - \omega_B)\tau_c < 1$, $J^2 >> (\omega_A - \omega_B)^2$ and when the exchange interaction is strong $J^2\tau_c^2 \gg 1$, the imaginary part of W1 gives the frequency shift for spin A

$$\Delta\omega_A = -(1/4)(\omega_A - \omega_B)\tau_c K_D C_B;$$
$$\Delta\omega_B = (1/4)(\omega_A - \omega_B)\tau_c K_D C_A. \tag{2.90}$$

Similarly, we can find the frequency shift of spin B (see 2.90). This shift was predicted in [27]. It can be seen that the simultaneous effect of the exchange interaction and the difference in the initial resonance frequencies of the colliding spins of the pair can give a shift in the spin frequencies. It can be noted that the resonance frequency shifts of spins A and B have opposite signs: the frequencies of A and B spins are shifted toward the center of the EPR spectrum. This is an expected result. It can be obtained from the following considerations.

At the time of the collision, $J^2 \gg (\omega_A\text{-}\omega_B)^2$ due to the sufficiently large exchange interaction, resonance frequency, e.g., of spins A becomes equal to the mean frequency $(1/2)(\omega_A + \omega_B)$. Thus, there is frequency exchange: in the interval between collisions, spin A has the resonant frequency ω_A, and in the collision state the frequency is $(1/2)(\omega_A + \omega_B)$. The difference of these two frequencies is $(1/2)(\omega_A\text{-}\omega_B)$. In the considered situation of $(\omega_A\text{-}\omega_B)\tau_c \ll 1$, the collapse of two lines into one line in their center of gravity should occur. As a result, the resonance occurs at the average frequency

$$\Omega_A = \omega_A(2/K_D C_B)/(2/K_D C_B + \tau_c) + (1/2)(\omega_A + \omega_B)\tau_c/(2/K_D C_B + \tau_c).$$

Given that the condition $K_D C_B \tau_c \ll 1$is satisfied, we obtain that the resonance frequency is $\Omega_A \approx \omega_A - (1/4)(\omega_A - \omega_B)\tau_c K_D C_B$ [14], see also (2.90).

The discussed resonance frequency shift was studied experimentally in a number of papers [43–47]. In these works, to determine the frequency shift due to the spin dynamics in the course of collision of particles from the analysis of EPR spectra of nitroxide radicals, interesting methodological techniques were used. For example, collisions between identical nitroxide radicals were studied, but ^{14}N or ^{15}N-containing radicals were used in different experiments. In this situation, the exchange integral, the collision radius and practically the diffusion coefficient, and hence the duration of the collision, should coincide. These radicals differ only in the hyperfine interaction constants (the quantities $(\omega_A\text{-}\omega_B)$, see (2.90) and statistical weights of the components of the hyperfine structure of the EPR spectrum, but these values are known. This made it possible to check the correctness of the shift of lines predicted theoretically [46]. Testing of the theory of additional spin frequency shift caused by the spin evolution at the meeting of particles was also carried out using nitroxide radicals with two equivalent ^{14}N nuclei [45].

In fact, the situation with frequency shifts may be unexpected. A detailed analysis of Eq. (2.86) shows that in the region of $J\tau_c \sim 1$ the resonant frequency shift changes its sign when changing this parameter (the sign of $Im(p_{Aex})$ in (2.86) changes). Below we will return to the discussion of the resonance frequency shift due to the interference of the exchange interaction and the initial difference in the spin frequencies in connection with the discussion of the role of re-encounters in the formation of this additional shift in the spin resonance frequencies.

An important parameter of the problem is the quantity $q \equiv (\omega_A\text{-}\omega_B)\tau_c$. When this parameter is zero, the discussed resonance frequency shift of spins A and B is also zero. Equations (2.90) show the dependence of the resonance frequency shifts on the parameter q in the case of a sufficiently large value of the exchange integral. Usually under the conditions of experiments on the study of spin exchange, the conditions $q < 1$ and $J^2 > (\omega_A\text{-}\omega_B)^2$ are fulfilled. Therefore, the theory of spin exchange is commonly developed for this case of equivalent spin exchange. In this situation, in good approximation according to (2.86) we have $p_{Aex} = p_{Bex} = \langle \sin^2(Jt/2) \rangle = (1/2) J^2\tau_c^2/(1 + J^2\tau_c^2)$.

The situation changes fundamentally if the condition $q > 1$ is implemented. Under the condition $J^2 > (\omega_A\text{-}\omega_B)^2$ we have $R \approx J$ and from (2.86) we obtain

$$\begin{aligned} Re(p_{Aex}) &\approx 1 - (1/2)\left(2 + J^2\tau_c^2 + q^2/4\right)/\left(\left(1 + J^2\tau_c^2\right)\left(1 + q^2/4\right)\right); \\ Re(p_{Bex}) &\approx (1/2)J^2\tau_c^2/\left(\left(1 + J^2\tau_c^2\right)\left(1 + q^2/4\right)\right). \end{aligned} \quad (2.91)$$

In contrast to the situation of equivalent spin exchange (2.85) at sufficiently large values of the parameter q with the growth of the exchange integral $Re(p_{Aex})$ and Re (p_{Bex}) (2.91) tend not to the same value 1/2. The limiting values of exchange efficiencies (2.91) are given by

$$\begin{aligned} Re(p_{Aex})1 - 2/\left(4 + q^2\right) > 1/2; \\ Re(p_{Bex})2/\left(4 + q^2\right) < 1/2. \end{aligned} \quad (2.92)$$

It can be seen that at $q \gg 1$

$$\begin{aligned} Re(p_{Aex}) \to 1, \\ Re(p_{Bex}) \to 0. \end{aligned} \quad (2.93)$$

To illustrate the possible effect of the difference in the resonant frequencies of individual particles on the spin exchange parameters, Fig. 2.5 shows the results of calculations of the spins A decoherence efficiency ($Re\{p_{Aex}\}$), the efficiency of the spin coherence transfer ($Abs\{p_{Bex}\}$) and the parameter which characterizes an additional resonance frequency shift of spins A (see $Im\{p_{Aex}\}$). Calculations are made according to the formulas given in (2.86).

Curves on the same horizontal correspond to the specific value of the parameter q. Values of q are indicated above the curves in Fig. 2.5. Thick lines in the first column refer to p_{Aex}, and all thin lines refer to p_{Bex} (see 2.86). The first column of curves represents $Re(p_{Aex})$ and $Re(p_{Bex})$. Note that in the first column the top three figures refer to the case $q < 1$, and therefore $Re(p_{Aex}) \approx$ Re (p_{Bex}), so that the thick and thin curves merge. The second column represents the curves $Im((p_{Aex}))$. The third column of curves represents the absolute value of the complex number p_{Bex}.

Curves in Fig. 2.5 clearly show that if the condition $q < 1$ is fulfilled the efficiency of decoherence and the efficiency of spin coherence transfer are satisfactorily

$t_c=10^{-11}$s, $\omega_A-\omega_B=1.76 \cdot 10^8$ rad/s, q=1.76 10^{-3}

$t_c=10^{-10}$s, $\omega_A-\omega_B=1.76 \cdot 10^8$ rad/s, q=1.76 10^{-2}

$t_c=10^{-10}$ s, $\omega_A-\omega_B=5.22 \cdot 10^8$ rad/s, q=5.22 10^{-2}

$t_c=10^{-10}$s, $\omega_A-\omega_B=1,76 \cdot 10^{10}$ rad/s, q=1.76

$t_c=10^{-9}$s, $\omega_A-\omega_B=1,76 \cdot 10^{10}$ rad/s, q=17.6

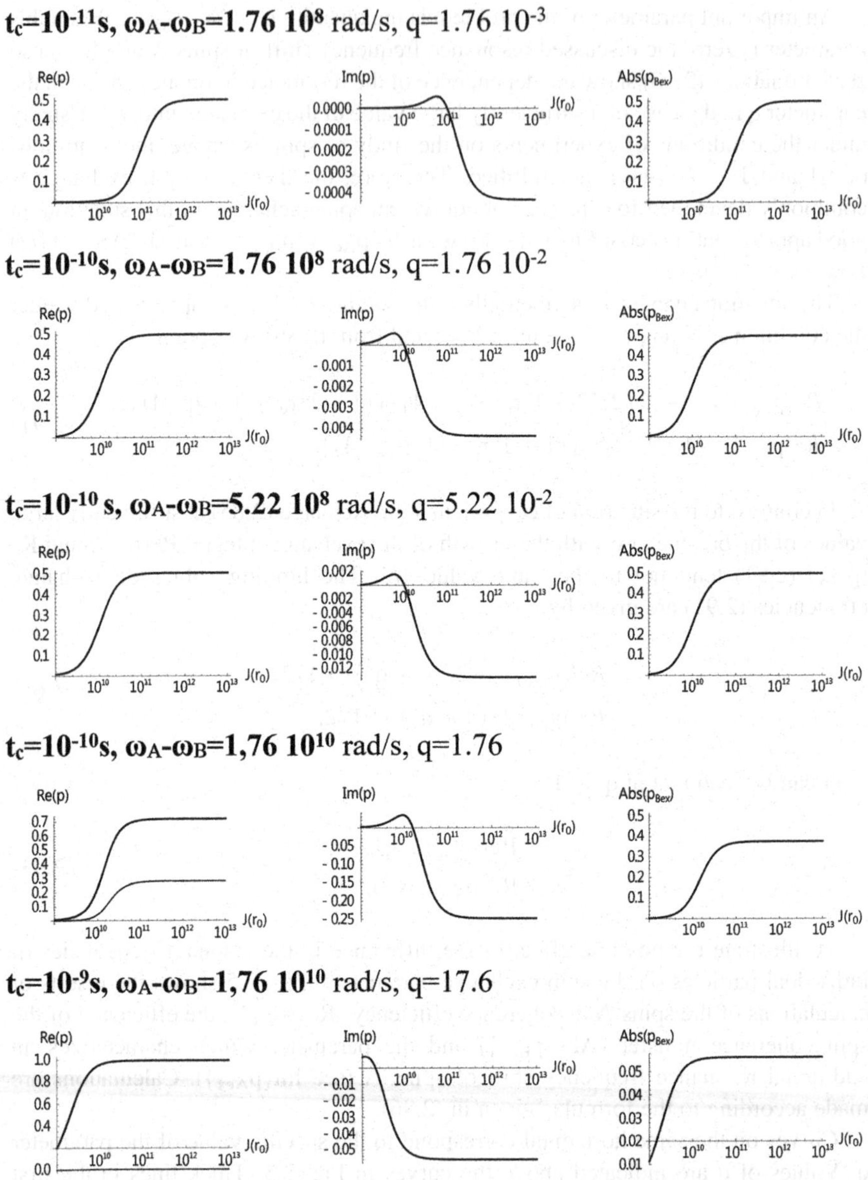

Fig. 2.5 Dependence of real and imaginary parts of the efficiences of spin decoherence, spin coherence transfer and frequency shift of the spins on the parameters $J\tau_c$ and $q = (\omega_A-\omega_B)\tau_c$. In all panels the exchange integral at the closest approach of particles is shown in the x-axis in the logarithmic scale

described as equivalent spin exchange, $p_{ex} = (1/2)\ J^2\tau_c^2/(1 + J^2\tau_c^2)$, and the frequency shift of the spins is a linear function of $q = (\omega_A\text{-}\omega_B)\tau_c$. When moving from the upper panels of curves to the lower panels, the parameter q takes the values: $q = 0.0017$, 0.017, 0.052, 1.76, 17.6, accordingly. Figure 2.5 shows that the maximum frequency shift of the spin in the region $q < 1$ increases linearly with q when moving from the top line to the second and third line from the top. In another limiting situation, $q > 1$, (see curves in the two lower panels of Fig. 2.5) the frequency shift passes through a maximum with increasing q (in the limit $q \gg 1$ the frequency shift decreases as 1/q), the efficiency of depolarization and spin transfer coherence can vary greatly, and with the growth $J\tau_c$ in the limit, we have $Re(p_{Aex})$-->1, $Re(p_{Bex})$-->0 (see also 2.93).

Thus, in the model of sudden collisions of the exchange interaction only in the case of equivalent exchange, the spin decoherence rate constant, K_{sd}, and the spin coherence transfer from the collision partner rate, K_{sct}, are equal and given by Eq. (2.85). In the case of strong exchange when $J\tau_c > 1$, these rate constants tend to half of the diffusion collisions rate constant: $K_{sd} = K_{sct}$->1/2)K_D. In the situation of nonequivalent spin exchange, when at the time of collision along with the exchange interaction it is necessary to take into account the difference of the resonance frequencies of isolated particles, these two constants cease to be the same. The decoherence rate constant K_{sd} becomes *greater* than $K_{sd} = (1/2)\ J^2\tau_c^2/(1 + J^2\tau_c^2)$ K_D for the case of equivalent spin exchange. The coherence transfer rate constant K_{sct} becomes *smaller*,

$$K_{sct} \le (1/2)\ J^2\tau_c^2/(1 + J^2\tau_c^2)K_D.$$

When $q > 1$, in the limiting situation of strong spin exchange

$$K_{sd} \to K_D, K_{sct} \to 0. \tag{2.94}$$

In this limiting situation (2.94), the exchange interaction in bimolecular collisions will *not cause* spin-exchange-induced characteristic transformations of the EPR spectrum with increasing spin concentration or increasing molecular mobility: shifts of the spectrum lines, merging of all lines into one homogeneously broadened line in the center of gravity of the spectrum, then exchange narrowing of the spectrum. Instead, each k-th line of the spectrum will simply broaden by the value of K_{sd} C (1-φk). The fact that here the concentration of spins with the same frequencies is subtracted from the total concentration of spins is due to the fact that the scalar exchange interaction does not change the state of the spin sub-ensemble with the same resonance frequencies.

Nitroxide stable free radicals are often used as spin probes, for which, as a rule, $q < 1$ is fulfilled. Therefore, in the study of spin exchange between nitroxide radicals, the situation of equivalent spin exchange is practically implemented: mutual flip-flops of spins operate, $K_{sd} \approx K_{sct}$. But even for nitroxide radicals, the theory predicts the resonance frequency shift of spins due to the combined action of the exchange interaction and spin-dependent interactions in each paramagnetic particle. The discussion of this frequency shift will be continued in the next section.

It is seen from the curves in the two lower panels in Fig. 2.5 that at $q > 1$ the exchange interaction in molecular collisions does not cause equivalent spin exchange. Other examples of nonequivalent spin exchange will be given below. This can be implemented for triplet excitons [48, 49] because of the relatively high spin-spin interaction energy (the so-called splitting energy in the zero magnetic field) in individual triplet excited molecules. Nonequivalent spin exchange also may be implemented when stable free radicals collide with fast relaxing paramagnetic complexes [38, 50, 51].

2.8.2 *On the Interpretation of the Change in the Sign of the Spin Frequency Shift (S = 1/2) Due to the Spin Dynamics Caused by the Influence of Both the Exchange Interaction and the Difference in the Frequencies of Isolated Spins*

The calculations given in the previous section show that in the region of weak spin exchange $J\tau_c < 1$, the spin frequencies diverge, and in the case of strong spin exchange $J\tau_c > 1$, the spin frequencies approach each other. This change of the sign of the frequency shift is clearly visible in Fig. 2.5 (see curves in the second column). The convergence of frequencies in the region of strong spin exchange can be explained quite well if we take into account that in this situation there is a jump of spins not between two states with frequencies ω_A, ω_B, but between three frequencies ω_A, ω_B and $(1/2)(\omega_A + \omega_B)$ (see 2.90).

To illustrate the difference between the manifestations of the exchange interaction in cases of weak and strong spin exchange, we consider another example of the elementary act of changing the spin state of particles in the collision taking into account the exchange interaction and the difference between their resonance frequencies ω_A and ω_B.

We consider the following model situation. Suppose that at the moment of sudden switching on of the exchange interaction the ensemble of pairs of spins is described by the density matrix

$$\rho(0) = (1/4)\begin{Bmatrix} 1 & 0 & 1 & 0 \\ 0 & 1 & 0 & 1 \\ 1 & 0 & 1 & 0 \\ 0 & 1 & 0 & 1 \end{Bmatrix}. \qquad (2.95)$$

In this state, the average values of the spin projections are

$$< S_{Ax} >=1/2,\ < S_{Ay} >=< S_{Az} >= 0;$$
$$< S_{Bx} >= < S_{By} >=< S_{Bz} >= 0.$$

In fact, this means that the spins A are in the eigenstate of the S_{Ax} operator with the eigenvalue 1/2, $(|1/2> +|-1/2>)/2^{1/2}$. Thus, spins A occupy the coherent state. Spins B occupy the state $|1/2>$ or $|-1/2>$ with the same probability of 1/2, and they have no quantum coherence.

In the course of collision the spin Hamiltonian of pair of spins is Eq. (2.41). By the time t of the end of the collision, the density matrix is

$$\rho(t) = \exp(-i\mathbf{H}t/\hbar)\rho(0)\exp(i\mathbf{H}t/\hbar). \tag{2.96}$$

Using (2.95 and 2.96)) it is possible to find the transverse spin projection

$$< \mathbf{S}_{Ax} - i\mathbf{S}_{Ay} >= \mathrm{Tr}\{(\mathbf{S}_{Ax} - i\mathbf{S}_{Ay})\rho(t)\}.$$

Provided $J > > (w_A - w_B)$, we have

$$\begin{aligned} &< \mathbf{S}_{Ax} - i\mathbf{S}_{Ay} > \\ &\qquad \approx (1/2)exp(-i\omega_0 t) \\ &\qquad \times [cos^2(Jt/2)(1 - i((\omega_A - \omega_B)/(2J)))sin(Jt)]. \end{aligned} \tag{2.97}$$

Here $\omega_0 = (1/2)(\omega_A + \omega_B)$.

The collision duration is assumed to have the Poisson distribution. For the Poisson distribution we have

$$\begin{aligned} &< cos^2(Jt/2) >=(1/2)(1 + 1/(1 + J^2 t_c^{\,2})) \rightarrow 1/2, \text{when } J\tau_c > 1, \\ &< sin(Jt) >=J\tau_c/(1 + J^2\tau_c^{\,2}) \rightarrow 0, \text{when } J\tau_c > 1. \end{aligned} \tag{2.98}$$

Equations (2.97 and 2.98) show that the collision changes the state of the spins qualitatively differently as a function of the $J\tau_c$ value. Indeed, in the case of strong exchange $J\tau_c > 1$

$$< \mathbf{S}_{Ax} - i\mathbf{S}_{Ay} >\approx (1/4)exp(-i\omega_0 t). \tag{2.99}$$

This result shows that in the case of strong spin exchange in good approximation, the spins are precessed with a frequency equal to the average frequency of the colliding pair. In this case, the spin exchange effect can be considered within the model of jumps of spins between states with frequencies ω_A, ω_B, $\omega_0 = (\omega_A + \omega_B)/2$, which was previously used for qualitative interpretation of the resonant frequency shift of spins due to spin exchange in the strong exchange limit (see 2.90).

In this case, the resonance frequencies of the spins are shifted to each other and the difference in the spin resonance frequencies decreases.

A completely different result is obtained in the case of weak spin exchange. At $J\tau_c < 1$, Eq. (2.97) can be approximated as

$$< S_{Ax} - iS_{Ay} > \approx (1/2)exp(-i\varphi t)\left[cos^2(Jt/2)\right],$$

where we introduced the precession phase [27].

$$\varphi = \omega_0 t + ((\omega_A - \omega_B)/J)tg(Jt/2) \approx \omega_A t + (\omega_A - \omega_B)J^2t^3/24.$$

It follows from this equation that the exchange interaction slows down the spin precession with a lower frequency and simultaneously accelerates the spin precession with a higher frequency: during the collision, the spin frequencies are pushed apart and the resonance frequency difference increases.

The analysis of the dynamics of spins during the collision shows that in contrast to the situation of strong spin exchange, the weak spin exchange situation cannot be reduced to the model of the jumps of spins between states with frequencies ω_A, ω_B, $\omega_0 = (\omega_A + \omega_B)/2$ since jumping of spins between the states with different frequencies leads to merging of the spectrum to a single exchange-narrowed line.

In the strong spin exchange limit, the resonance frequency shift was studied in a number of publications (see, e.g., [43–47]). To the best of my knowledge, it was not detected in the weak spin exchange limit.

2.8.3 *The Role of Re-Encounters on Spin Exchange (S = 1/2) at Sudden Switching on of the Exchange Interaction at the Collision Radius*

The efficiency of the spin state change as a result of a single sudden collision of two paramagnetic particles on the collision radius with the sudden switching on of the exchange interaction was given in previous sections.

As noted above (Sect. 2.1), two molecules, which encountered at their closest approach distance r_0, can re-encounter at the same distance after some diffusion walk. To illustrate the role of re-encounters in the bimolecular process, we present results for a simple phenomenological model. We denote by p_r the probability that two molecules re-encouner at a distance r_0. This probability is determined by the kinematics of molecular diffusion, namely, the average length of the displacement of molecules in the elementary act of diffusion and the size of molecules [5, 6]. The total probability of all re-collisions is $q = p_r + p_r^2 + p_r^3 + \ldots = p_r/(1 - p_r)$. The total number of collisions of two molecules at a distance r_0 is

$$n = 1 + q = 1/(1 - p_r). \quad (2.100)$$

The experiment measures the rate of a certain process occurring at bimolecular meetings. As already noted, the measurement of bimolecular spin exchange (see 2.1) seems to be very promising for studying the bimolecular collisions.

On the one hand, the spin exchange rate can be found using the electron paramagnetic resonance methods, since this process is manifested in a characteristic way in the shape of EPR spectra [1, 2]. On the other hand, it is quite a convenient process for the theoretical description. Let us first consider the situation of equivalent spin exchange, when only the exchange interaction can be taken into account during the meeting of two paramagnetic particles.

For certainty, we consider the elementary process of the mutual flip-flop of the colliding spins under the action of the exchange interaction, i.e., we consider the transitions between the states $|1\rangle = \alpha_A\beta_B$ and $|2\rangle = \beta_A\alpha_B$ (see 2.1). Within the model of sudden collisions it is assumed that the exchange interaction is switched on only in a very narrow layer between the spheres with radii r_0 and $r_0 + \delta$. We denote by p_{ex} the probability of the mutual flip-flop of the spins in a single collision. Suppose that in the interval $\{r_0, r_0 + \delta\}$ the exchange integral for a pair AB is J and the average time of the single collision is τ_c. We assume that the exchange integral does not depend on the mutual orientation of the colliding paramagnetic particles. For this model of sudden switching on of the exchange interaction of paramagnetic particles in a collision, the averaged probability of the mutual flip-flop of spins (2.1) is (see 2.62)

$$p_{ex} = (1/2)J^2\tau_c{}^2/\left(1 + J^2\tau_c{}^2)\right) \tag{2.101}$$

After the first collision of the pair A and B with the probability of p_{ex} the spins of A and B exchange their states according to Eq. (2.1) and with the probability $1 - p_{ex}$, the spin states remain unchanged. Spin exchange can also occur in the course of re-encounters of A and B in the pair. Re-encounters in the pair lead to the increase in the time τ_e, which the molecules A and B, which met, spend in the $\{r_0, r_0 + \delta\}$ region with the exchange interaction J. On average, this time is $\tau_e = n\tau_c$. If in the intervals between re-encounters there is no spin evolution, including spin decoherence, then the average probability of spin exchange for one encounter is given by the expression (2.101) with the replacement of the average time of one collision τ_c by the average total duration of all collisions for one meeting $\tau_e = n\tau_c$, so that

$$p_{ex} = (1/2)J^2\tau_e{}^2/\left(1 + J^2\tau_e{}^2\right) = (1/2)J^2n^2\tau_c{}^2/\left(1 + J^2n^2\tau_c{}^2\right). \tag{2.102}$$

However, in the real situation in the intervals between re-encounters, quantum coherence of spin states may not be preserved. Therefore, in general, it is impossible to take into account the effect of all re-encounters by a simple replacement the average time of one collision by the average total duration of all collisions $\tau_e = n\tau_c$ in Eq. (2.101).

To illustrate the contribution of all re-encounters and the role of spin decoherence in the intervals between re-encounters of the pair to spin exchange, we calculate the probability of spin exchange for one meeting for the following model. Let us make the following important assumption: in the time between two collisions the quantum coherence of the states |1> and |2> (see 2.1) is totally destroyed. This means that in the course of every collision, i.e., every re-encounter at the collision radius of spins of the A…B pair starts from the initial state, which has zero quantum coherence and the spin state is completely described by populations of spin states. Of course, at the moment of collision the exchange interaction of the spins forms the quantum coherence of spin states, so that at the end of each collision spins are in a quantum coherent state, the superposition of states |1> and |2>.

To calculate the total contribution of all re-collision of a pair of spins A and B at the collision radius, we introduce the matrix P of the probabilities of transitions between states |1> and |2> (see 2.1 and 2.44):

$$P = \begin{pmatrix} 1 - p_{ex} & p_{ex} \\ p_{ex} & 1 - p_{ex} \end{pmatrix} \tag{2.103}$$

This matrix makes it possible to calculate the change in the populations of the states |1> and |2> during one collision as a result of the coherent motion of two spins under the action of the exchange interaction.

Eigenvalues and eigenstate of **P** equal to 1 and $1–2p_{ex}$, $|\psi_1\rangle = (|1\rangle + |2\rangle)/\sqrt{2}$ and $|\psi_2\rangle = (|1\rangle - |2\rangle)/\sqrt{2}$, respectively. In the eigenstates basis, the matrix P is:

$$V_r = \begin{pmatrix} 1 & 0 \\ 0 & 1 - 2p_{ex} \end{pmatrix} \tag{2.104}$$

In the same basis, the spin states ↑↓ and ↑↓ (see (2.1)) are described by vectors $u_1 = \{1,1\} / \sqrt{2}$ and $u_2 = \{1, -1\}/\sqrt{2}$. The probability that spin exchange occurs after the first collision of two paramagnetic particles and then these particles separate in the volume of the solution having no more collisions is: $p_1 = (1 - p_r)\langle u_2 | V_r | u_1\rangle$. The probability of spin exchange as a result of the first and one more re-encounter is: $p_2 = (1 - p_r)\langle u_2 | V_r p_r V_r | u_1\rangle = (1 - p_r)\langle u_2 | p_r V_r^2 | u_1\rangle$. If during the meeting k-1 re-encounters occur, then the probability for spin exchange is $p_k = (1 - p_r)\langle u_2 | p_r^{k-1} V_r^k | u_1\rangle$.

Summing up the contribution of all possible implementations of re-encounters, we find the probability of the mutual flip-flop of spins of a pair of paramagnetic particles A and B in one meeting:

$$p_t = \sum p_k = \frac{np_{ex}}{1 + 2(n-1)p_{ex}} = \frac{1}{2}\frac{nJ_0^2\tau_0^2}{1 + nJ_0^2\tau_0^2} \tag{2.105}$$

The comparison of (2.101) and (2.105) shows that in the approximation of sudden collisions for the model under consideration, the spin exchange efficiency of a pair of molecules in one meeting is determined by the effective time: $\tau_{eff} = (\sqrt{n})\,\tau_c$,

$$P_{ex} = \frac{1}{2}\frac{J_0^2\tau_{eff}^2}{1 + J_0^2\tau_{eff}^2}. \quad (2.106)$$

It should be noted that in the considered phenomenological model, the effective time determining the spin exchange efficiency is less than the total residence time of two spins in the exchange interaction region on average for one meeting,

$$\tau_{\text{eff}} = \left(\sqrt{n}\right)\tau_c < n\tau_c.$$

In general, when spins loose their quantum coherence between the states 1 > and 2>l in the interval between re-encounters, the effective spin exchange time is within the interval

$$\left(\sqrt{n}\right)\tau_c \le \tau_{\text{eff}} \le n\tau_c. \quad (2.107)$$

Thus, for the model considered, in the situation of equivalent spin exchange, the role of re-encounters is reduced only to the increase in the effective average residence time of the pair in the exchange interaction region. This effective time is always greater than the single collision duration, but it is less than the total residence time of the pair of particles in the exchange interaction region due to the possible decoherence of spins between re-encounters of spins in the same pair due to its own spin dynamics and its own spin relaxation of each of the paramagnetic particles.

The role of re-encounters and spin evolution in the intervals between re-encounters was theoretically studied in [51]. It is assumed that the exchange integral differs from zero only within a thin spherical "reaction"layer of the thickness $\delta \ll r_0$, r_0 is the closest approach distance between the particles in a bimolecular collision (2.4). The diffusion of molecules is described by a continuous diffusion model. For this model, the average residence time of the ensemble of pairs of particles in the reaction layer is known and it is $\tau_c = \delta r_0/D$, D is the mutual diffusion coefficient of colliding molecules. We emphasize that this is the total residence time in the "reaction" layer, it includes the residence time of a pair of particles in the reaction layer during the first and all subsequent re-encounters of colliding molecule in the pair. There is one more characteristic time of the meeting of a pair of particles, this is the total duration of all re-encounters. This characteristic time is $\tau_D = r_0^2/D$.

The role of re-encounters was discussed in [51, 53]. For nondiagonal elements of the spin density matrix, e.g., for spin A, the kinetic equation is obtained (cf. 2.79)

$$(\partial/\partial t)\rho^A{}_{1/2,-1/2} = -i\omega_A\rho^A{}_{1/2,-1/2} - K_D C_B\left(p\,\rho^A{}_{1/2,-1/2} - p'\rho^B{}_{1/2,-1/2}\right).$$

In general, the spin exchange efficiency parameters p and p⊇ are complex values. The imaginary part of p characterizes the frequency shift

$$\Delta\omega_A = K_D C_B Im\{p\},$$
$$p = (J^2\tau_c{}^2/4)z(1+z)/\left[z^2 + (J^2\tau_c{}^2/4)(1+z)^2\right], \tag{2.108}$$
$$z = 1 + (-i\Delta\omega\,\tau_D)^{1/2} - i\Delta\omega\,\tau_c, \Delta\omega = \omega_A - \omega_B.$$

The frequency shift caused by the exchange interaction depends on the difference of spin frequencies. The frequency shift (2.108) is determined by the motion of the spins during the collision of particles in the reaction layer (see term $-i\Delta\omega\,\tau_c$ in the expression for z), and the "free" movement of spins in the intervals between the re-encounters, when the exchange interaction was supposed to be switched off (see the term $(-\,i\,\Delta\omega\,\tau_D)^{1/2}$ in the expression for z).

When spin exchange is weak, $J\tau_c < 1$, then

$$\Delta\omega_A \approx K_D C_B\left(J^2\tau_c{}^2/4\right)\left[\Delta\omega\tau_c + sign(\Delta\omega)\,\left(|\Delta\omega|\tau_D/2\right)^{1/2}\right]/|z|^2, \tag{2.109}$$

where

$$|z|^2 = \left[1 + \left(|\Delta\omega|\tau_D/2\right)^{1/2}\right]^2 + \left[\Delta\omega\tau_c + sign(\Delta\omega)\,\left(|\Delta\omega|\tau_D/2\right)^{1/2}\right]^2.$$

When spin exchange is strong, $J\tau_c > 1$, then

$$\Delta\omega_A \approx -K_D C_B(1/4)\left[\Delta\omega\tau_c + sign(\Delta\omega)\,\left(|\Delta\omega|\tau_D/2\right)^{1/2}\right]/|1+z|^2, \tag{2.110}$$

where

$$|1+z|^2 = \left[2 + \left(|\Delta\omega|\tau_D/2\right)^{1/2}\right]^2 + \left[\Delta\omega\tau_c + sign(\Delta\omega)\,\left(|\Delta\omega|\tau_D/2\right)^{1/2}\right]^2.$$

The frequency shift for spins B is obtained from these equations by replacing A with B and B with A.

These expressions for the frequency shift are greatly simplified when the conditions $|\Delta\omega|\tau_c < 1$, $|\Delta\omega|\tau_D < 1$ are fulfilled.

These conditions are fulfilled, e.g., for nitroxide radicals.

In this situation we get

$$\begin{aligned}\Delta\omega_A &\approx K_D C_B\left(J^2\tau_c{}^2/4\right)\left[\Delta\omega\tau_c + sign(\Delta\omega)\,\left(|\Delta\omega|\tau_D/2\right)^{1/2}\right], \text{when } J\tau_c < 1,\\ \Delta\omega_A &\approx -\,K_D C_B(1/4)\left[\Delta\omega\tau_c + sign(\Delta\omega)\,\left(|\Delta\omega|\tau_D/2\right)^{1/2}\right], \text{when } J\tau_c > 1.\end{aligned} \tag{2.111}$$

From these equations we obtain that the sign of the spin frequency shift changes in the region of $J\tau_c \sim 1$. As a result, in the region of weak spin exchange, the spin frequencies are repulsed, and in the region of strong spin exchange, the spin frequencies approach each other. These results are in agreement with the results obtained by taking into account a single collision of particles when re-encounters are ignored [1, 2, 27] (see the previous section). Re-encounters are very significant: the spin dynamics in between re-encounters, makes, as a rule, the main contribution to the discussed frequency shift, since $\tau_D > \tau_c$ (see 2.110 and 2.111).

It is interesting to note that the manifestation of the "free" movement of spins in the intervals between the re-encounters, when the exchange interaction is turned off, is determined by the parameter $(|\Delta\omega| \tau_D/2)^{1/2}$, and not just by $\Delta\omega\ \tau_D$, as one would expect by analogy with the parameter $\Delta\omega\tau_c$ (see 2.110and 2.111)). These differences in the manifestation of the difference in the frequency shift are associated with the difference in the Poisson distribution (2.61) of the lifetime of a pair of particles in the collision state, i.e., in the reaction layer with the exchange interaction switched on, and the distribution of the duration of time between re-encounters of a pair of particles (2.2). When calculating the spin exchange efficiency, averaging over the ensemble of pairs of the function of the form sin(Ωt) is carried out.

For the Poisson time distribution we have

$$< \sin(\Omega t) >= (1/\tau_c) \int_0^\infty sin(\Omega t)\, exp(-t/\tau c)dt = \Omega\tau_c/(1 + \Omega^2 {\tau_c}^2)$$

$$\approx \Omega\tau_c, \text{when}\, \Omega\tau_c < 1.$$

For the distribution (2.2)

$$< \sin(\Omega t) >= \int_0^\infty sin(\Omega t) f(t)dt = \text{m0}\,(2\pi)^{1/2}(\Omega\tau_D)^{1/2\sim}(\Omega\tau_D)^{1/2}.$$

The distribution f(t) (2.2) over the interval between re-encounters decreases as $1/t^{3/2}$ at large t. This decline is much slower than for the Poisson distribution of the lifetimes of the pair in the colliding state. Because of this feature of the distribution of f(t), the spin dynamics in the interval between re-encounters of particles can have a great effect on the result of the meeting of two particles.

In real experiments, paramagnetic particles can have more than one frequency. For example, EPR spectra of organic radicals have a hyperfine structure. The above results can simply be generalized to such systems. Assume that the EPR spectrum of isolated particles has a set of components. The whole ensemble of particles in the solution can be divided into sub-ensembles, which give individual components in the EPR spectrum. The particles of the selected k-th component experience spin exchange when colliding with particles belonging to other sub-ensembles. Summing up the contributions of bimolecular spin exchange with all particles and using Eqs. (2.110 and 2.111), for the spins of the k-th sub-ensemble we obtain the following results for the frequency shift:

$$\Delta\omega_k \approx K_D C(J^2\tau_c{}^2/4) \sum_{n(n\neq k)} \varphi_n X_{nk},$$
$$X_{nk} = (\omega_k - \omega_n)\tau_c + \text{sign}(\omega_k - \omega_n)\,(|\omega_k - \omega_n|\tau_D/2)^{1/2}], \text{ when } J\tau_c < 1,$$
$$\Delta\omega_k \approx -(1/4)K_D C \sum_{n(n\neq k)} \varphi n X n k, \text{ when } J\tau_c > 1. \tag{2.112}$$

In the intermediate situation $J\tau_c \approx 1$, the frequency shift $\Delta\omega k$ should pass through zero (see also Fig. 2.5, second column of curves).

Thus, the exchange interaction at the meeting of two paramagnetic particles in the solution leads not only to the spin excitation transfer, spin decoherence transfer and the coherence transfer of electron spins. Along with these effects, there is a spin frequency shift, which is linearly dependent on the concentration of paramagnetic particles. The frequency shift under discussion is due to the fact that at the meeting of two particles the spin motion is determined not only by the exchange interaction, but also by all spin-dependent interactions, e.g., Zeeman interaction of each spin with an external magnetic field, hyperfine interaction of unpaired electrons with magnetic nuclei, etc.

Within the phenomenological theory, the resonance frequency shift under consideration can be described by introducing a complex value for the effective radius r_{eff} into Eqs. (2.10 and 2.78) for the spin exchange rate constant. The imaginary part of r_{eff} gives the resonance frequency shift. It is very interesting that the sign of this frequency shift is different for situations of weak and strong spin exchange.

The study of this frequency shift has good potential to reveal fine details of the particle collision, in particular, to reveal the role of repeated collisions of particles in the pair. The contribution of the "free" spin dynamics in the intervals between repeated collisions leads to the fact that the frequency shift should be proportional to the square root of the frequency difference, and the contribution of the spin dynamics in the pair of colliding particles leads to the fact that the frequency shift should be proportional to the frequency difference. This makes it possible, in principle, to obtain unique information about re-encounters of particles in the pair, studying the dependence of the resonance frequency shift on the difference of frequencies of isolated particles. If the difference in frequencies is due to the difference in g-factors of paramagnetic particles, the frequency shift can be controlled by changing the induction of the external magnetic field. If the frequency difference is due to the hyperfine interaction, the frequency difference can be changed by the isotope substitution.

In EPR spectroscopy, the spin frequency shift is determined by the position of absorption lines in the spectrum. The frequency shift under consideration was observed in the experiments with solutions of nitroxide radicals and thoroughly investigated in [43–47]. It should be noted that there are also other mechanisms of shift of position of the lines observed in a spectrum (see below). This circumstance creates problems for the extraction of the contribution of the spin dynamics during the elementary act of the spin exchange in the total shift of the resonance frequency observed in the EPR experiment.

2.8.4 Spin Exchange Between Triplet Excitons

The electron excited states with two unpaired electrons with a total spin $S = 1$ among the paramagnetic particles (quasiparticles) are of great interest. For example, triplet excited molecules play an important role in photochemistry [52]. Relatively long-lived localized electron excited states-triplet excitons appear in molecular crystals (see, e.g., [35, 48, 49, 54–59]). Localized triplet excitons are formed in semiconductors (electron-hole pairs of small radius, which are called Frenkel excitons). Under the diffusion motion, triplet excitons can “collide” with each other. The study of spin exchange caused by the exchange interaction between triplet excitons in collisions by EPR spectroscopy methods has attracted much attention of researchers (see, e.g., [35, 48, 49, 54–59]).

The spin Hamiltonian of an isolated triplet exciton in the presence of an external constant magnetic field B_0 has the form

$$\mathbf{H} = \hbar(\omega_0 \mathbf{S}_z + D\mathbf{S}_z{}^2 + E(\mathbf{S}_x{}^2 - \mathbf{S}_y{}^2)). \tag{2.113}$$

Here ω_0 is the Zeeman spin frequency in the field $\mathbf{B}_0$, the parameters D and E are the principal values of the so-called tensor of the splitting in the zero magnetic field, and they reflect the spin-spin interaction in the triplet. In (2.113), it is assumed that the external constant magnetic field is directed along the z-axis of the zero field level-splitting tensor. In sufficiently strong magnetic fields, the non-secular term containing E (2.113) can be neglected. Note that this term vanishes in the case of the axial symmetry of the splitting tensor in the zero field.

Below, to illustrate the spin exchange between triplet excitons, we consider the simplest situation. We consider a molecular crystal. So, the principal axes of the zero field level-splitting tensor coincide for all excitons. For the spin Hamiltonian $\mathbf{H}_0 = \hbar(\omega_0 \mathbf{S}_z + D\mathbf{S}_z{}^2)$, there are three non-equidistant energy levels that correspond to S_z projections of -1, 0, 1, respectively:

$$E_{-1} = \hbar(\omega_0 + D), E_0 = 0, E_1 = \hbar(-\omega_0 + D).$$

In the EPR experiment, when a transverse microwave field is applied, two dipole transitions with frequencies ω_0-D, ω_0 + D are allowed, so that two lines are observed in the EPR spectrum, the difference in the resonant frequencies of these lines is 2D.

Spin exchange between triplet excitons within the sudden collision model was theoretically analyzed in [1, 35, 39, 48].

The spin Hamiltonian of two non-interacting triplet excitons is chosen as

$$\mathbf{H}_0 = \hbar(\omega_0(\mathbf{S}_{1z} + \mathbf{S}_{2z}) + \mathrm{D}(\mathbf{S}_{1z}{}^2 + \mathbf{S}_{2z}{}^2)).$$

When triplet excitons “collide”, the exchange interaction is switched on.

Therefore, the spin Hamiltonian of the pair of excitons during the collision has the form

$$\mathbf{H} = \hbar(\omega_0(\mathbf{S}_{1z} + \mathbf{S}_{2z}) + \mathrm{D}(\mathbf{S}_{1z}{}^2 + \mathbf{S}_{2z}{}^2) + \mathbf{J}\mathbf{S}_1\mathbf{S}_2). \tag{2.114}$$

The spin exchange characteristics were calculated in the same way as described in the previous section for spin exchange between free radicals.

The collision matrix in the considered model situation takes the form

$$\begin{aligned} \mathbf{S}(\tau) &= \exp(-i\hbar^{-1}\mathbf{H}\tau)\exp(i\hbar^{-1}\mathbf{H}_0\tau) \\ &= \exp(-i(\mathrm{D}(\mathbf{S}_{1z}{}^2 + \mathbf{S}_{2z}{}^2) + \mathbf{J}\mathbf{S}_1\mathbf{S}_2)\tau)\exp(i\mathrm{D}(\mathbf{S}_{1z}{}^2 + \mathbf{S}_{2z}{}^2)\tau). \end{aligned} \tag{2.115}$$

It can be seen that the Zeeman interaction in this case does not affect the spin exchange efficiency in the collision. This is due to the fact that the spin Hamiltonian of the Zeeman interaction of two identical excitons commutes with the spin Hamiltonian of the exchange interaction and the spin Hamiltonian of the zero-field splitting. According to the above arguments, the contribution of bimolecular collisions to the change in the spin density matrix of the triplet excitons is given by the following equation within the model considered

$$(\partial\rho(1)/\partial t)_{ex} = -\mathrm{K_D C}(\rho(1) - \mathrm{Tr}_2 < \mathbf{S}(\tau)\rho(1)\rho(2)\mathbf{S}^{-1}(\tau) > . \tag{2.116}$$

Here <...> means averaging over the duration of the collision, C is the concentration of triplet excitons.

The elements of the S-matrix (2.115) in the representation of the direct product of the eigenfunctions of $\mathbf{S}_{1z}$ and $\mathbf{S}_{2z}$, $|m_1,m_2> = |m_1>|m_2>$, are [1, 39].

$$\begin{aligned} S_{11,11} &= S_{-1-1,-1-1} = exp(-iJ\tau_c), \\ S_{01,01} &= S_{0-1,0-1} = cos(J\tau_c), \\ S_{01,10} &= S_{0-1,-10} = -i\,sin(J\tau_c), \\ S_{00,00} &= c_1{}^2 exp(-i\lambda_1\tau_c) + c_2{}^2 exp(-i\lambda_2\tau_c), \\ S_{00,-11} &= -\left(c_1c_2/2^{1/2}\right)(exp(-i\lambda_1\tau_c) - exp(-i\lambda_2\tau_c)), \\ S_{1-1,00} &= exp(i2D\tau_c)\,S_{00,-11}, \\ S_{1-1,1-1} &= (1/2)\{exp(iJ\tau_c) + c_2{}^2 exp(i(2D - \lambda_1)\tau_c) + c_1{}^2(i(2D - \lambda_2)\tau_c)\}, \\ S_{1-1,-11} &= S_{1-1,1-1} - exp\,(iJ\tau_c), \end{aligned} \tag{2.117}$$

where

$$c_1 = cos(\varphi), c_2 = sin(\varphi), cos(2\varphi) = (D - J/2)/R, R^2 = (D - J/2)^2 + 2J^2,$$
$$\lambda_1 = D - J/2 - R, \lambda 2 = D - J/2 + R.$$

From Eqs. (2.116 and 2.117) we obtain that the contribution of the exchange interaction to the change of populations of exciton energy levels is described by the following linearized equations [35, 39, 48, 58].

$$\begin{aligned} \partial\sigma_{00}/\partial t &= -(2/3)K_D C p_{ex}(2\sigma_{00} - \sigma_{11} - \sigma_{-1-1}), \\ \partial\sigma_{11}/\partial t &= \partial\sigma_{-1-1}/\partial t = -(1/2)\partial\sigma_{00}/\partial t, \\ p_{ex} &= < |S_{00,-11}|^2 > = 2J^2\tau_c^2/[1 + 4((D - J/2)^2 + 2J^2)\tau_c^2]. \end{aligned} \quad (2.118)$$

The contribution of collisions to the change of non-diagonal elements of the density matrix is described by the equations

$$\begin{aligned} \partial\sigma_{01}/\partial t &= -K_D C p_1 \sigma_{01} + K_D C p_2 \sigma_{-10}, \\ \partial\sigma_{-10}/\partial t &= -K_D C p_1 * \sigma_{-10} + K_D C p_2 * \sigma_{01}, \end{aligned} \quad (2.119)$$

where

$$p_1 = 1 - (1/3) \sum_{k=-1,0,1} < S_{0k,0k} S^{+}{}_{1k,k1} + S_{0k,k0} S^{+}{}_{k1,1k} > ,$$

$$p_2 = (1/3) \sum_{k=-1,0,1} < S_{-1k,-1k} S^{+}{}_{0k,0k} + S_{-1k,k-1} S^{+}{}_{k0,0k} > .$$

The p_1 and p_2 values in limiting cases were calculated using these equations (see Table 2.3).

Similarly to the particles with spin ½, the “kinetic” coefficients in the equations for quantum spin coherence are complex numbers. The real part of p_1 describes spin decoherence caused by bimolecular collisions. The imaginary part of p_1 describes the resonance frequency shift of triplet excitons caused be their spin dynamics in course of an elementary act of spin exchange.

Absolute value of p_2 describes the spin coherence transfer efficiency. Similarly to the case of particles with spin $S = 1/2$, equivalent spin exchange is implemented for triplet excitons in the limiting case when the conditions $J^2 > D^2$, $D^2\tau_c^2 \ll 1$ are satisfied. For triplet excitons, equivalent spin exchange is described by the rate constant

$$K_{ex} = p_{ex} K_D = 2J^2\tau_c^2/(1 + 9J^2\tau_c^2)\, K_D. \quad (2.120)$$

From Eq. (2.120), the limiting value of the equivalent spin exchange rate is

$$K_{ex} = (2/9)K_D.$$

Table 2.3 Parameters of the spin exchange efficiency [39]

	$J > D$, $D\tau_c < 1$	$J > D$, $D\tau_c > 1$	$J < D$
p_{ex}	$2J^2\tau_c^2/(1 + 9J^2\tau_c^2)$	$2/9$	$2J^2\tau_c^2/(1 + 4D^2\tau_c^2)$
Rep_1	$2J^2\tau_c^2/(1 + 9J^2\tau_c^2)$	$1/2$	$(1/3)J^2\tau_c^2/(1 + J^2\tau_c^2) + (2/3)J^2\tau_c^2/(1 + 4J^2\tau_c^2)$
Rep_2	$2J^2\tau_c^2/(1 + 9J^2\tau_c^2)$	$5/(16D^2\tau_c^2)$	$-(J/(6D))[1/(1 + J^2\tau_c^2) - 1/(1 + 4J^2\tau_c^2)]$
Imp_1	$(2D\tau_c/9)(1 + 1/(1 + 9J^2\tau_c^2))$	$2/(3D\tau_c)$	$-(J\tau_c/3)[1/(1 + J^2\tau_c^2) - 1/(1 + 4J^2\tau_c^2)]$
Imp_2	$-2D\tau_cJ^2\tau_c^2/(1 + 9J^2\tau_c^2)$	$-1/(4D\tau_c)$	$-(J^2\tau_c/(6D))[1/(1 + J^2\tau_c^2) + 2/(1 + 4J^2\tau_c^2)]$

In the same situation, when $J^2 > D^2$, $D^2\tau_c^2 \ll 1$, the theory predicts the shift of EPR lines to the center of the spectrum by

$$\Delta\omega = (2/9)\left(D\tau_c/\left(1 + 1/\left(1 + 9J^2\tau_c^2\right)\right)\right)K_DC, \tag{2.121}$$

where C is the concentration of excitons.

The analysis of Table 2.3 shows that the efficiency of the processes induced by the exchange interaction of triplet excitons in bimolecular collisions depends on the parameters J, D and τ_c qualitatively similar to the dependence of the spin exchange efficiency between particles with spin $S = 1/2$ on parameters J, $\omega_A - \omega_B$ and τ_c (see Fig. 2.5). Similar to the situation with particles with spin $S = 1/2$, in the approximation of the sudden switching on of the exchange interaction at the collision radius, the spin exchange rate constant of triplet excitons reaches the limiting value of $2K_D/9$ (see third column in Table 2.3). In the case $J > D$, $D\tau_c < 1$, the theory predicts the frequency shift proportional to $D\tau_c$. As a result, the splitting between the components of the EPR spectrum of triplet excitons becomes less than $2|D|$ and is

$$\Delta\omega_0 = \omega_{0\to 1} - \omega_{-1\to 0} = 2\,|\,D - K_DCIm\{p_1\}\,| - (K_DC)^2|p_2|^2/\,|\,D - K_DCIm\{p_1\}\,|. \tag{2.122}$$

Features of nonequivalent spin exchange between triplet excitons in the case of $J > D$, $D\tau_c > 1$ (see the second column of Table.2.3) are also qualitatively similar to nonequivalent spin exchange between particles with spin $S = 1/2$. Note that the maximum spin decoherence efficiency is greater than the maximum spin excitation transfer efficiency: $\max\{Re(p_1)\} = 1/2 > \max\{p_{ex} = 2/9\}$. At the same time, in the case of nonequivalent spin exchange, the coherence transfer efficiency is small, Abs $(p_2) = 1/(16D^2\tau_c^2) \ll 1$ (cf. the second column of Table 2.3 and Fig. 2.5).

Apparently, the frequency shift under discussion was first observed in [54]. In this work, the authors showed that the manifestations of the exchange interaction of localized triplet excitons in collisions cannot be described by the introduction of a single parameter: the spin exchange rate contant. Following widely accepted protocol for the case of equivalent spin exchange, it was found [54] that, e.g., the spin exchange rate constant found from the line shift appeared to be larger than the spin exchange rate constant found from the exchange broadening. To interpret the experimental data, the authors [54] suggested that the resonance frequence shift observed in the EPR experiment may be related to the temperature dependence of the splitting parameter D in the zero magnetic field. In ref. [39] (see also [1, 2]), another explanation for the additional resonant frequence shift observed in [54] was proposed. The experimental data can be explained as the manifestation of the frequency shift due to the interference of the exchange interaction and the zero-field splitting energy in the course of an elementary act of spin exchange (2.122). Thus, when interpreting experimental data [54] one has to consider the possibility of the realization of nonequivalent spin exchange.

In the case when the exchange integral is less than the splitting energy in the zero magnetic field, $J^2 < D^2$, but $J^2\tau_c^{\ 2} \gg 1$, we have $D^2\tau_c^{\ 2} \gg 1$, the spin decoherence rate constant and its maximum value are greater than those in the case of equivalent spin exchange ((2.120), see Table 2.3, second column). At the same time, the spin coherence transfer rate constant is less than that in the case of equivalent spin exchange (2.120), and its maximum value with the incease in $D^2\tau_c^{\ 2}$ tends to zero as $(1/16)K_D/(D^2\tau_c^{\ 2})$, (see Table 2.3, second column). Such changes in the rate constants of processes associated with the exchange interaction of excitons are qualitatively similar to changes of the corresponding values in the case of free radicals (cf. Table 2.3 and Fig. 2.5).

2.8.5 *Equivalent Spin Exchange Between Paramagnetic Particles with Spin 1/2 and Particles with Arbitrary Spin S*

In the previous sections, examples of spin exchange between spin 1/2 particles, typical of which are free radicals with one unpaired electron, and spin 1 particles, representative of which are localized triplet excitons, were considered. Another important class of paramagnetic particles are paramagnetic ion complexes. It suffices to note that such complexes act as active centers in heterogeneous catalysis, including enzymatic catalysis. Such complexes are involved in redox reactions with electron transport, etc. Paramagnetic complexes are also interesting objects from the point of view of studying bimolecular spin exchange. In these systems, it is possible to control the exchange interaction of particles in collisions by changing the ligands of the complexes.

In this section, the focus is the consideration of the dependence of spin exchange on the spin value of one of the partners in the collisions. It was shown by the example of triplet excitons above that the spin-spin interaction in isolated particles with spin $S \geq 1$ can affect spin exchange. Moreover, in paramagnetic complexes, the electron paramagnetic relaxation times may be comparable with or even less than the characteristic time of the meeting of two particles.

Here we consider the following model situation. We suppose that the paramagnetic relaxation times are significantly longer than the duration of the meeting of two particles and therefore the effect of paramagnetic relaxation on the spin exchange can be neglected, and that the spin-spin interaction in isolated complexes and/or solution viscosity are low enough so that $\langle D^2\tau_c^2\rangle \ll 1$ and this spin-spin interaction is not manifested during the collision. For this model situation, the spin exchange rate constants in solutions of two types of paramagnetic particles were calculated in [36, 38] within the approximation of the sudden switching on of the exchange interaction, the spin of one particles $S_1 = 1/2$ and the spin of others was considered arbitrary (S).

We consider a binary solution containing paramagnetic particles A and B with spins $S_A = 1/2$, $S_B = S$. The Spin Hamiltonian of a pair of particles A and B in a magnetic field B_0 has the form

$$\begin{aligned} \mathbf{H}_0 &= \mathbf{H}_{0A} + \mathbf{H}_{0B} = \hbar(\omega_A \mathbf{S}_{Az} + \omega_B \mathbf{S}_{Bz}) \\ &= \hbar\omega_0(\mathbf{S}_{Az} + \mathbf{S}_{Bz}) + \hbar\delta(\mathbf{S}_{Az} - \mathbf{S}_{Bz}), \end{aligned} \tag{2.123}$$

where $\omega_A = g_A\beta B_0/\hbar$, $\omega_B = g_B\beta B_0/\hbar$ are Zeeman frequencies of spins A and B, $g_{A,B}$ are g-spectroscopic splitting factors for A and B particles, β is the electron Bohr magneton, $\delta = (\omega_A - \omega_B)/2$, $\omega_0 = (\omega_A + \omega_B)/2$. At the moment of the collision, the exchange interaction is assumed to suddenly turn on, and taking into account the exchange interaction the spin Hamiltonian of the pair of particles takes the form

$$\mathbf{H} = \mathbf{H}_0 + \mathbf{V}_{ex} = \mathbf{H}_0 + \hbar[J_0\mathbf{S}_{Az}\mathbf{S}_{Bz} + J(\mathbf{S}_{Ax}\mathbf{S}_{Bx} + \mathbf{S}_{Ay}\mathbf{S}_{By})]. \tag{2.124}$$

Here, the exchange interaction is written in the anisotropic form in order to clearly show the role of the secular, $J_0\mathbf{S}_{Az}\mathbf{S}_{Bz}$, and non-secular, $J(\mathbf{S}_{Ax}\mathbf{S}_{Bx} + \mathbf{S}_{Ay}\mathbf{S}_{By})$, parts of the exchange interaction in the exchange processes. In principle, the parameters J_0 and J can be really different. In all calculations below the case $J_0 = J$ is considered.

Within the model of sudden collisions, the kinetic equations for spin density matrices have the form (2.71), and the collision matrix is given by Eq. (2.70). Using the expression for the collision matrix (2.70) and the spin Hamiltonian of the pair (2.124), we obtain

$$\begin{aligned} \mathbf{S}(t_c) &= \exp(-i\mathbf{H}t/\hbar)\exp(i\mathbf{H}_0t_c/\hbar) \\ &= \exp(-i\mathbf{V}_{ex}t_c/\hbar - i\delta t_c(\mathbf{S}_{Az} - \mathbf{S}_{Bz}))\exp(i\delta t_c(\mathbf{S}_{Az} - \mathbf{S}_{Bz})). \end{aligned} \tag{2.125}$$

Here t_c is the duration of a single collision of two particles. In Eq. (2.125), it is taken into account that the exchange interaction does not change the total projection of the spins of the pair, so the term $\hbar\omega_0(\mathbf{S}_{Az} + \mathbf{S}_{Bz})$ in the spin Hamiltonian commutes with all terms in the spin Hamiltonian. As a result, the collision matrix does not depend on the average spin frequency, but only on the difference in the spin frequencies of the pair. Using (2.125), we obtain the following expressions for nonzero elements of the $\mathbf{S}(t_c)$ matrix in the representation of eigenfunctions $|m_A, m_B> = |m_A>|m_B>$, while $m_A = \pm 1/2$, $m_B = S, S-1, \ldots -S$:

$$< 1/2, S|\mathbf{S}(t_c)|1/2, S >= \exp(-i(\delta(1/2 - S) + J_0 S/2)t_c);$$
$$< -1/2, -S|\mathbf{S}(t_c)| - 1/2, -S >= \exp(-i(\delta(-1/2 + S) + J_0 S/2)t_c);$$
$$< m_A, M - m_A|\mathbf{S}(t_c)|m'_A, M - m'_A >= \exp(i(\delta M + J_0/4)t_c)\{\cos(Z_M t_c)\delta_{mA, m'A} - i[\text{sign}(m_A)((J_0 M/2 + \delta)/Z_M)\delta_{mA, m'A} + (J/(2Z_M))((S + 1/2)^2 - M^2)^{1/2}\delta_{mA, -m'A}]\sin(Z_M t_c). \tag{2.126}$$

Here the following notations are introduced

$$M = m_A + m_B, Z_M{}^2 = (J^2/4)\left((S + 1/2)^2 - M^2\right) + (J_0 M/2 + \delta)^2,$$

while $M \neq 1/2 + S, -1/2 - S$.

In the model situation under consideration, the collision does not change the total z-projection of the spins of the pair. Therefore, for sub-ensembles of collisions with a given total projection of the spins of the pair, $m_A + m_B = M$, the collision matrix can be represented in the form 2×2 matrix in the indices m_A, m'_A:

$$< m_A, M - m_A|\mathbf{S}(t_c)|m'_A, M - m'_A > = \exp(i(J_0/4)t_c)\exp(i\delta M t_c)\left\{\begin{matrix} S^M_{1/2,1/2} & S^M_{1/2,-1/2} \\ S^M_{-1/2,1/2} & S^M_{-1/2,-1/2} \end{matrix}\right\}, \tag{2.127}$$

where

$$S^M_{1/2,1/2} = \left(S^M_{-1/2,-1/2}\right)* = [\cos(Z_M t_c) - i((J_0 M/2 + \delta)/Z_M)\sin(Z_M t_c)],$$

$$S^M_{1/2,-1/2} = S^M_{-1/2,1/2} = -i(J/(2Z_M))\left((S + 1/2)^2 - M^2\right)^{1/2}\delta_{mA, -m'A}]\sin(Z_M t_c).$$

Using the collision matrix (2.127), we obtain parameters of kinetic equations for spin density matrices for A and B particles [36, 38, 40].

To identify clearly the effect of the spin quantum number of paramagnetic particles on spin exchange, we consider the simplest situation of equivalent spin exchange, when $\delta t_c \ll 1$, and in the calculation of the collision matrix, we can assume that during elementary act of spin exchange the motion of spins is determined only by the exchange interaction. We also assume that the exchange interaction is described by the scalar product of the spin moment operators of the colliding pair of particles, and therefore $J_0 = J$. In this case, we have

$$\mathrm{S}(\mathrm{t_c}) = \exp(-i\mathrm{JS_AS_Bt_c}/\hbar), \mathrm{S}^{-1}(\mathrm{t_c}) = \exp(i\mathrm{JS_AS_Bt_c}/\hbar) = \mathrm{S}*(\mathrm{t_c}). \tag{2.128}$$

Substituting (2.128) into kinetic Eq. (2.71), we obtain a system of nonlinear equations for single-particle spin density matrices. As already noted, under typical conditions of studying spin exchange by EPR spectroscopy, spin systems are in states close to equilibrium, so spin matrices can be represented as the sum of the equilibrium matrix ρ_0 and a small additive σ, in which the matrix elements are much smaller than the population of spin states in equilibrium, $\rho = \rho_0 + \sigma$. Substituting this decomposition of spin matrices in (2.71) and leaving only the terms linear in σ, it is possible to obtain linear kinetic equations for single-particle spin density matrices taking into account bimolecular collisions.

In the considered model system, the B spins have arbitrary value. This means that, in principle, in the nonequilibrium situation, there may be matrix elements $\sigma_{mm'}$ that relate the two states with any possible spin projection values $m' \neq m$. Nonzero nondiagonal elements of the density matrix characterize quantum coherence in the system. If the spins absorb or emit one quantum, the magnetic quantum number m changes by ± 1. Therefore, matrix elements $\sigma_{m,m\,\pm\,1}$ characterize coherence, which is called single-quantum one. Matrix elements $\sigma_{m,m\,\pm\,2}$ characterize two-quantum coherence, etc. Thus, for particles with spin $S > 1/2$, e.g., in multi-pulse EPR experiments, it is necessary to take into account not only single-quantum coherence. But in the common EPR experiments, when recording the stationary absorption spectrum in the linear response regime, the multi-quantum coherences can be neglected. Therefore, we assume that in (2.71) $\sigma_{m,m\,\pm\,k} = 0$ if $k > 1$. Taking into account the above considerations from (2.71), we obtain kinetic equations that describe equivalent spin exchange between particles A with spin 1/2 and particles B with arbitrary spin S.

As was already pointed out, in EPR experiments under the conditions of linear response of the system, spin matrices can be represented as $\rho = \rho_0 + \sigma$, where ρ_0 is the equilibrium density matrix. We explicitly present the σ-linearized equations describing the contribution of spin exchange to the changes of the spin density matricies

$$\begin{aligned}
&\partial\sigma^{\mathrm{A}}{}_{-1/2,\,1/2}/\partial t = i\omega_{\mathrm{A}}\sigma^{\mathrm{A}}{}_{-1/2,\,1/2} - \\
&\mathrm{K_DC_Bp_1}\sigma^{\mathrm{A}}{}_{-1/2,\,1/2} + \mathrm{K_DC_B}\sum_{m_A,\,m_B} p_{12}(m_A, m_B)\sigma^B_{m_B-m_A-1/2,\,m_B-m_A+1/2}; \\
&\partial\sigma^B_{m_B,\,m_B+1}/\partial t = i\omega_{\mathrm{B}}\,\sigma^B_{m_B,\,m_B+1} - \mathrm{K_DC_Ap_2}\,\sigma^B_{m_B,\,m_B+1} + \mathrm{K_DC_A} \\
&\sum_{m\neq m_{1A}} p_{22}(m_A, m_{1A}, m_B)\sigma^B_{m_A-m_{1A}+m_B,\,m_A-m_{1A}+m_B+1} + \mathrm{K_DC_Ap_{21}(m_B)}\,\sigma^{\mathrm{A}}{}_{-1/2,\,1/2}.
\end{aligned} \tag{2.129}$$

Here the following notations are introduced:

$$p_1 = 1 - < \sum_{m_B} S^{m_B-1/2}_{-1/2,-1/2}\left(S^{-1}\right)^{m_B+1/2}_{1/2,1/2}\left(\rho^B_0\right)_{m_B,m_B} >,$$

$$p_{12}(m_A, m_B) = < \sum_{m_A=-1/2}^{1/2} \left(1/\left(1 + J_0^2 m_A^2 \tau_c^2\right)\right)\left(\rho^A_0\right)_{m_A,m_A}$$

$$+ i \sum_{m_A=-1/2}^{1/2} \left(J_0 m_A \tau_c/\left(1 + J_0^2 m_A^2 \tau_c^2\right)\right)\left(\rho^A_0\right)_{m_A,m_A} \cdot >,$$

$$p_2(m_B) = 1 - < \sum_{m_A} S^{m_A+m_B}_{m_A,m_A}\left(S^{-1}\right)^{m_A+m_B+1}_{m_A,m_A}\left(\rho^A_0\right)_{m_A,m_A} >,$$

$$p_{22}(m_A, m1_A, m_B) = < S^{m_A+m_B}_{m_A,m1_A}\left(S^{-1}\right)^{m_A+m_B+1}_{m1_A,m_A}\left(\rho^A_0\right)_{m1_A,m1_A} >,$$

$$p_{21}(m_B) = < \sum_{m_A} S^{m_A+m_B}_{m_A,-1/2}\left(S^{-1}\right)^{m_A+m_B+1}_{1/2,m_A}\left(\rho^B_0\right)_{m_A+m_B+1/2,\,m_A+m_B+1/2} >.$$

The absorption of microwave power in EPR experiments is determined by the transverse magnetization, which can be characterized by the quantities

$$\begin{aligned} M_A{}^+ &= M_{Ax} + iM_{Ay} = \delta\beta\left[C_A \mathrm{Tr}_A\left(\sigma^A\left(S_{Ax} + iS_{Ay}\right) = g\beta CA\sigma^A{}_{-1/2,1/2};\right.\right. \\ M_B{}^+ &= M_{Bx} + iM_{By} = g\beta C_B \mathrm{Tr}_B\left(\sigma^B\left(S_{Bx} + iS_{By}\right) = \right. \\ & g\beta C_B \mathrm{Tr}_B\left(\sum_{m_B} \sigma^B_{m_B,m_B+1}\left((S + m_B + 1)(S - m_B)\right)^{1/2}\right). \end{aligned} \quad (2.130)$$

Using the linearized Eq. (2.129), we can obtain equations for the transverse magnetizations of spins, M_{A+} and M_{B+}. In two limiting cases $J > \delta$, $\delta\tau_c \gg 1$ or $J < \delta$, the equations for $M_A^+$and M_B^+ take a fairly simple form

$$\begin{aligned} &\partial M_A^+/\partial t = i\omega_A M_A^+ - K_D C_B p_A M_A^+ + K_D C_A p_{AB} M_B^+ = i(\omega_A - K_D C_B Im\{p_A\}) M_A^+ \\ &K_D C_B Re\{p_A\} M_A^+ + K_D C_A p_{AB} M_B^+; \\ &\partial M_B^+/\partial t = i\omega_B M_B^+ - K_D C_A p_B M_B^+ + K_D C_B p_{BA} M_A^+ = i(\omega_B - K_D C_A Im\{p_B\}) M_B^+ - \\ &K_D C_A Re\{p_B\} M_B^+ + K_D C_B p_{BA} M_A^+. \end{aligned} \quad (2.131)$$

In Eqs. (2.131), the parameters p_A, p_{AB}, p_B, p_{BA} characterize the efficiency of the processes of the change of the coherence of spins in the collision. The terms containing the parameters p_A and p_B describe the frequency shift and decoherence of spins A and B in the collision, respectively, and the terms with p_{AB} and p_{BA} appear due to the reaction effect in the collision.

As already mentioned in the discussion of spin exchange between particles with spin $S = 1/2$ (see, e.g., Fig. 2.3), the reaction effect in a collision is negligible under the conditions of the weak exchange interaction when $J < \delta$, or strong interference of the exchange interaction with the intrinsic spin-dependent interactions of isolated paramagnetic particles at $J > \delta$, and $\delta\tau_c > 1$. For example, the p_A and p_B values at $J < \delta$ are

$$p_A = -\sum_{m_B=-S}^{S} \left(1/\left(1+J_0^2 m_B^2 \tau_c^2\right)\right)\left(\rho_0^B\right)_{m_B,m_B} + i\sum_{m_B=-S}^{S} \left(J_0 m_B \tau_c/\left(1+J_0^2 m_B^2 \tau_c^2\right)\right)\left(\rho_0^B\right)_{m_B,m_B},$$

$$p_B = 1 - \sum_{m_A=-1/2}^{1/2} \left(1/\left(1+J_0^2 m_A^2 \tau_c^2\right)\right)\left(\rho_0^A\right)_{m_A,m_A} + i\sum_{m_A=-1/2}^{1/2} \left(J_0 m_A \tau_c/\left(1+J_0^2 m_A^2 \tau_c^2\right)\right)\left(\rho_0^A\right)_{m_A,m_A}.$$

$$p_{AB} \sim 0(J/\delta),\ p_{BA} \sim 0(J/\delta). \tag{2.132}$$

Thus, when $J < \delta$, the effect of the exchange interaction is fundamentally different from the situation of equivalent spin exchange: the reaction in the collision is negligible, i.e., in (2.131) the terms containing p_{AB} and p_{BA} are negligible; the spin decoherence rate constant tends to the diffusion collision constant K_D with increasing $J_0\tau_c$, (according to (2.129) $p_A \rightarrow 1$, $p_B \rightarrow 1$ with $J_0\tau_c > 1$). In the situation under consideration, the non-secular part of the exchange interaction makes a negligible contribution to the efficiency of mutual flip-flops of spins in the collision. The decoherence of spins is due to the additional random phase that each spin acquires as a result of the precession in the exchange field of the other spin it encountered. In this case, the paramagnetic shift proportional to the population difference of spin levels is also expected (paramagnetic shift). In EPR experiments on spin exchange, this shift is negligible.

As already noted, the reaction in the collision is negligible also at $J > \delta$, $\delta\tau_c > 1$. In this case

$$\begin{aligned} p_A &\approx 1, p_B \approx 1, \\ p_{AB} &\approx 0,\ \ p_{BA} \approx 0, \end{aligned} \tag{2.133}$$

so that the result of the collision is reduced only to the spin decoherence, and the decoherence rate constant is equal to the diffusion collision rate constant of particles.

Thus, in the cases $J < \delta$ or $J > \delta$, $\delta\tau_c > 1$, in the model of sudden switching on of the exchange interaction in the collision of particles, the spin decoherence rate constant can reach the maximum possible value (in the framework of sudden collisions model) equal to the diffusion collision rate constant K_D. The condition of achieving this maximum value of the spin decoherence rate constant can be formulated as ($J = J_0$)

$$min\{J, \delta\}\tau_c > 1. \tag{2.134}$$

However, this condition is quite strict. In EPR experiments, the frequency difference $\delta \sim 10^7 - 10^{10}$ rad/s. Therefore, the conditions $J < \delta$, $J\tau_c > 1$ can be implemented simultaneously only in highly viscous media or in systems, in which two paramagnetic particles form a relatively long-lived complexes upon collision.

When the condition (2.134) is fulfilled, spin decoherence (phase relaxation) occurs due to a random change of the spin phase during the collision, i.e., both partners lose their phase irreversibly.

A fundamentally different mechanism of phase relaxation is implemented in another limiting case of equivalent spin exchange $J > \delta$, $\delta\tau_c \ll 1$. In this case, the difference in spin frequencies can be neglected at the moment of collision. The total spin and the projection of the total spin on any quantization axis are preserved at each collision, so that collisions, unlike the above considered cases of nonequivalent spin exchange (2.132, 2.133 and 2.134), do not produce spin decoherence (irreversible phase relaxation). The coherence of spins is transferred reversibly without loss between partners with different Zeeman frequencies due to the reaction in the collision. However, such an equivalent coherence transfer between paramagnetic particles ultimately leads to irreversible decoherence of the spins. This occurs because, due to spin exchange during meetings, spectral diffusion operates and phase of precession for spins becomes stochastic process.

In the case of equivalent spin exchange, the reaction in the course of bimolecular meetings in solution has very interesting consequences. For example, the effect of the exchange narrowing of the EPR spectra is due solely to the reaction in the collision.

In the case of equivalent exchange, we have

$$\begin{aligned} p_A = p_{BA} &= (2/3)[J^2\tau_c^{\,2}S(S+1) + i2 < S_{Bz} >_0 J\tau_c]/[1 + J^2\tau_c^{\,2}(S+1/2)^2]; \\ p_B = p_{AB} &= (1/2)[J^2\tau_c^{\,2} + i(3/2) < S_{Az} >_0 J\tau_c]/[1 + J^2\tau_c^{\,2}(S+1/2)^2]. \end{aligned} \tag{2.135}$$

Here $<S_{A(B)z}>_0 \sim 10^{-3}$ denotes the equilibrium polarization of the spin. Therefore, in the situation of equivalent spin exchange, we can neglect the frequency shift, which is given by the imaginary parts of p_A and p_B. Then we have

$$\begin{aligned} p_A = p_{BA} &= (2/3)J^2\tau_c^{\,2}S(S+1)/\left[1 + J^2\tau_c^{\,2}(S+1/2)^2\right]; \\ p_B = p_{AB} &= (1/2)J^2\tau_c^{\,2}/\left[1 + J^2\tau_c^{\,2}(S+1/2)^2\right]. \end{aligned} \tag{2.136}$$

Substituting (2.136) into the kinetic equation for the magnetization of spins (2.131), we obtain the equations

$$\begin{aligned} (\partial \mathbf{M}_A/\partial t)_{ex} &= -K_{eA}C_B\mathbf{M}_A + K_{eB}C_A\mathbf{M}_B; \\ (\partial \mathbf{M}_B/\partial t)_{ex} &= -K_{eB}C_A\mathbf{M}_B + K_{eA}C_B\mathbf{M}_A. \end{aligned} \tag{2.137}$$

Here

$$\begin{aligned} K_{eA} &= (2/3)\left[J^2\tau_c{}^2S(S+1)/\left(1+J^2\tau_c{}^2(S+1/2)^2\right)\right]K_D, \\ K_{eB} &= (1/2)\left[J^2\tau_c{}^2/\left(1+J^2\tau_c{}^2(S+1/2)^2\right)\right]K_D. \end{aligned} \tag{2.138}$$

Note that in the case of $S = 1/2$ Eqs. (2.138) are reduced to the well-known result $K_{eB} = K_{eA} = (1/2)[J^2\tau_c{}^2/(1 + J^2\tau_c{}^2)]K_D$. In the case of the collision of particles with the same spins $S = 1/2$, equivalent spin exchange is described by equations that are similar to the chemical kinetics equations for a reversible reaction when the rate constants of the forward and reverse reactions coincide. In the case of the collision of particles with spin ½ with particles with spin $S > 1/2$, equivalent spin exchange is described by kinetic equations similar to the equations of chemical kinetics for reversible reactions when the rate constants of direct and reverse reactions are not the same. Note that the ratio of the rate constants given by (2.138) is

$$K_{eA}/K_{eB} = (4/3)S(S+1) \geq 1. \tag{2.139}$$

The dependence of the exchange efficiency on the spin S value was established in the EPR experiments by Zamaraev et al. [34, 36]. The concentration dependence of the EPR line width was studied in dilute aqueous solutions containing free radicals of 2,2′,6,6′-tetramethyl-4′-oxopiperidin-1-oxil (TOPO, $S_A = 1/2$) and aqua complexes of Mn^{2+} ($S_B = 5/2$). It was found that $K_{eA} = (2.1 \pm 0.2)10^9 s^{-1}$, $K_{eB} = (2.3 \pm 0.4)\,10^8 s^{-1}$. According to Eq. (2.139), the theoretically expected ratio of these constants is 11.7. The ratio of these constants found in the experiment is 9 ± 3, i.e., within the accuracy of the measurements it is in good agreement with the theory.

In the case of nonequivalent exchange, i.e., under the conditions $J < \delta$ or $J > \delta$, $\delta\tau_c > 1$, the reaction effect in the collision is negligible and therefore in (2.131) $p_{AB} \approx 0$, $p_{BA} \approx 0$. Table 2.4 summarizes the results obtained for the spin decoherence efficiency in the collision for radical, p_A, (spin½) and for paramagnetic complexes, p_B, in extreme situations.

When using paramagnetic particles as spin probes, it is necessary to select particles, for which efficient spin exchange is expected and therefore the increased sensitivity of EPR spectra to spin exchange is expected. Free nitroxide radicals are widely used as spin probes. In this case, the EPR spectra have a hyperfine structure caused by the hyperfine interaction of unpaired electrons with magnetic nuclei of nitrogen, hydrogen, deuterium. With these considerations in mind, we generalize the above results for equivalent spin exchange between radicals taking into account the isotropic hyperfine interaction with magnetic nuclei. The ensemble of radicals is divided into sub-ensembles, each of which includes radicals that have a given orientation of nuclear spins and give a certain hyperfine component in the EPR spectrum. We denote by M_k the magnetization vector of the spins of the k-th sub-ensembles of the radicals, and resonance frequency of the spins of the sub-ensembles is denoted as ω_k, and the total hyperfine splitting of the EPR spectrum is denoted as Δ. In the case of equivalent spin exchange, when $J > \Delta$, $\Delta\tau_c < 1$,

Table 2.4 Spin decoherence efficiency in the collision of particles A with spin $S_A = 1/2$ with particles with arbitrary spin $S_B = S$ at $J_0 = J$ (2.129 and 2.131)

J vs δ	$\delta\tau_c$	$J\tau_c$	*Re*(p_1)	*Re*(p_2)	Mechanim of decoherence
$J < \delta$	$\delta\tau_c < 1$	$J\tau_c < 1$	$(1/3)S(S+1)J^2\tau_c^2$	$(1/4)J^2\tau_c^2$	The loss of phase of the spin precession at collision
	$\delta\tau_c > 1$	$J\tau_c < 1$	$(1/3)S(S+1)J^2\tau_c^2$	$(1/4)J^2\tau_c^2$	
$J < \delta$	$\delta\tau_c > 1$	$J\tau_c > 1$	1, if S is half-integer 2S/(2S + 1), if S is integer	1	The loss of phase of the spin precession at collision
$J > \delta$	$\delta\tau_c > 1$	$J\tau_c > 1$	1	1	The loss of phase of the spin precession at collision
$J > \delta$	$\delta\tau_c < 1$	$J\tau_c < 1$	$(2/3)S(S+1)J^2\tau_c^2$	$(1/2)J^2\tau_c^2$	Spectral diffusion induced by equivalent spin exchange
$J > \delta$	$\delta\tau_c < 1$	$J\tau_c > 1$	$(2/3)S(S+1)/(S+1/2)^2$	$(1/2)(1/(S+1/2)^2$	Spectral diffusion induced by equivalent spin exchange

equations for transverse and longitudinal projections of the magnetization taking into account the hyperfine structure have the form

$$\partial M_{k-}/\partial t = -\,i(\omega_k + \Delta\omega_k)M_{k-} - K_D C p_{ex}\left(M_{k-} - \varphi_k \sum_n M_{n-}\right),$$
$$\partial M_{kz}/\partial t = -\,K_D C p_{ex}\left(M_{kz} - \varphi_k \sum_n M_{nz}\right). \tag{2.140}$$

where $\Delta\omega_k$ is the frequency shift of the spins of the k-th sub-ensembles due to the exchange interaction with the spins of all other sub-ensembles (2.112), M_{kz} is the longitudinal projection of the magnetization of the spins of the k-th sub-ensemble, and φ_k is the statistical weight of radicals of the k-th sub-ensemble, $p_{ex} = (1/2)\,J^2\tau_c^2/(1 + J^2\tau_c^2)$.

In conclusion, it is worth to note a feature of the kinetics of relaxation of the non-diagonal elements of the spin density matrix for particles with spin $S_B > 1/2$.

To illustrate, we consider the change in the spin density matrix of particles for the case of $S_A = 1/2$, $S_B = 5/2$ in the case of equivalent spin exchange. Note that the problem of the effect of the spin of the paramagnetic complex on spin exchange with the radical was theoretically studied for the first time in [34, 36] for this particular case in connection with the interpretation of experimental EPR data for binary solutions containing free radicals, $S_A = 1/2$, and bivalent manganese complexes, $S_B = 5/2$. In this case, the kinetic equations for the elements of spin density matrices (2.71) take the form

$$\begin{aligned}
&\partial\sigma^{A}{}_{-1/2,1/2}/\partial t = i\omega_A\sigma^{A}{}_{-1/2,1/2} - K_D C_B p_A \sigma^{A}{}_{-1/2,1/2} + \\
&\quad K_D C_B p_B \left(5^{1/2}\sigma^{B}{}_{3/2,5/2} + 2^{3/2}\sigma^{B}{}_{1/2,3/2} + 3\,\sigma^{B}{}_{-1/2,1/2} + 2^{3/2}\sigma^{B}{}_{-3/2,-1/2} + 5^{1/2}\sigma^{B}{}_{-5/2,-3/2}\right); \\
&\partial\sigma^{B}{}_{3/2,5/2}/\partial t = i\omega_B\sigma^{B}{}_{3/2,5/2} - \\
&\quad K_D C_A p_B \left(5\sigma^{B}{}_{3/2,5/2} - 10^{1/2}\sigma^{B}{}_{1/2,3/2} - \left(5^{1/2}/3\right)\sigma^{A}{}_{-1/2,1/2}\right)\Big); \\
&\partial\sigma^{B}{}_{1/2,3/2}/\partial t = i\omega_B\sigma^{B}{}_{1/2,3/2} - \\
&\quad K_D C_A p_B \left(-10^{1/2}\sigma^{B}{}_{3/2,5/2} + 8\sigma^{B}{}_{1/2,3/2} - 3\,2^{1/2}\sigma^{B}{}_{-1/2,1/2} - \left(2^{3/2}/3\right)\sigma^{A}{}_{-1/2,1/2}\right)\Big); \\
&\partial\sigma^{B}{}_{-1/2,1/2}/\partial t = i\omega_B\sigma^{B}{}_{-1/2,1/2} - \\
&\quad K_D C_A p_B \left(-3\,2^{1/2}\sigma^{B}{}_{1/2,3/2} + 9\sigma^{B}{}_{-1/2,1/2} - 3\,2^{1/2}\,\sigma^{B}{}_{-3/2,-1/2} - \sigma^{A}{}_{-1/2,1/2}\right)\Big); \\
&\partial\sigma^{B}{}_{-3/2,-1/2}/\partial t = i\omega_B\sigma^{B}{}_{-3/2,-1/2} - \\
&\quad K_D C_A p_B \left(-3\,2^{1/2}\sigma^{B}{}_{-1/2,1/2} + 8\sigma^{B}{}_{-3/2,-1/2} - 10^{1/2}\,\sigma^{B}{}_{-5/2,-3/2} - \left(2^{3/2}/3\right)\sigma^{A}{}_{-1/2,1/2}\right)\Big); \\
&\partial\sigma^{B}{}_{-5/2,-3/2}/\partial t = i\omega_B\sigma^{B}{}_{-5/2,-3/2} - \\
&\quad K_D C_A p_B \left(-10^{1/2}\sigma^{B}{}_{-3/2,-1/2} + 5\sigma^{B}{}_{-5/2,-3/2} - \left(5^{1/2}/3\right)\sigma^{A}{}_{-1/2,1/2}\right)\Big).
\end{aligned} \tag{2.141}$$

These kinetic equations clearly show that the exchange interaction in collisions induces the coherence transfer not only between different spins A and B, but also the one-quantum coherence transfer between different pairs of spin states of S_B, when $S_B > 1/2$. All single quantum coherences of S_B, $\sigma^{B}{}_{m,m+1}$, correspond to the same frequency ω_B. So there is the degeneracy of coherences, which correspond to transitions between different pairs of B spin states. Due to this, in the subspace of spin B coherences, one has to find, first, "eigenstates" of the spin B coherence, which are not "mixed" by collisions, they are such combinations of $\sigma^{B}{}_{m,m+1}$, each of which changes in collisions only due to the exchange of coherence with spins A. To find these characteristic, generalized modes of B spin quantum coherences, one has to consider (2.141), first, keeping only degenerate coherences, i.e., neglecting all terms with A spin coherence in (2.141). For the system of equations for $\sigma^{B}{}_{m,m+1}$ obtained from (2.141) in this approximation, the characteristic complex numbers are

$$\begin{aligned}
\lambda_1 &= i\omega_B - K_D C_A p_B, \lambda_2 = i\omega_B - 3K_D C_A p_B, \lambda_3 = i\omega_B - 6K_D C_A p_B, \\
\lambda_4 &= i\omega_B - 10K_D C_A p_B, \lambda_5 = i\omega_B - 15K_D C_A p_B.
\end{aligned} \tag{2.142}$$

This characteristic numbers correspond to certain linear combinations of single-quantum coherences $\sigma^{B}{}_{m,m+1}$. For example, λ_1 corresponds to the generalized coherence mode

$$\begin{aligned}
\xi_1 = {} & 5^{1/2}\sigma^{B}{}_{3/2,5/2} + 2^{3/2}\sigma^{B}{}_{1/2,3/2} + 3\sigma^{B}{}_{-1/2,1/2} + 2^{3/2}\sigma^{B}{}_{-3/2,-1/2} \\
& + 5^{1/2}\sigma^{B}{}_{-5/2.-3/2}.
\end{aligned} \tag{2.143}$$

For ξ_1 we have the equation

$$\xi_1 = (i\omega_B - K_D C_A p_B)\xi_1. \qquad (2.144)$$

Similar equations can be found for each characteristic number. In fact, we do not need them when interpreting the EPR data. The reason is that the microwave field in EPR excites only one generalized coherence mode, namely, ξ_1 (2.143).

An interesting situation could arise if the spin energy levels in the case under discussion were not equidistant. In principle, this can be implemented due to the level-splitting in a zero magnetic field or the hyperfine interaction with magnetic nuclei. Then it would be possible in EPR experiments to selectively excite individual transitions and observe lines of different widths which correspond to the individual transitions $m \leftrightarrow m \pm 1$ (coherences $\sigma^B{}_{m,m\pm1}$). In this case, the exchange broadening of EPR spectrum lines under the selective observation of individual transitions would be 5, 8, 9 times greater (see 2.141) than the exchange broadening of the complex spectrum in the case of the equidistant EPR spectrum of the paramagnetic complex with $S_B = 5/2$. It would be interesting to observe this effect in experiment.

2.8.6 The Influence of the Paramagnetic Relaxation of Spins on the Spin Exchange Efficiency

The above results are applicable only when the paramagnetic relaxation times of spins, T_1 and T_2, are much larger than the collision time τ_c. This condition is satisfied, e.g., for organic radicals, complexes of divalent manganese, trivalent chromium, divalent vanadium, trivalent iron, and divalent copper.

Transition metal ions can have short paramagnetic relaxation times, shorter than the collision time τ_c. For example, Mn(III), Co(II), and Ni(II) complexes have the T_1, T_2 times on the order of picoseconds, while the collision lasts about 0.1 ns. In such a situation, the effect of the exchange interaction with the paramagnetic complex S_2 on the spin state of the free radical decreases, the rapid "flips" of S_2 spins in the course of the paramagnetic relaxation tend to average to zero the reaction due to the exchange interaction JS_1S_2 on the movement of the spin S_1 of radicals.

The influence of fast spin-lattice relaxation on the spin exchange efficiency of radicals in solutions containing free radicals (spin $S_1 = 1/2$) and paramagnetic complexes of transition metal ions, in the approximation of sudden switching on of the exchange interaction has been studied in [36, 38, 48, 49] (see also [1, 2]). The decoherence efficiency of spins of radicals induced by the exchange interaction in the course of their collisions with fast-relaxing paramagnetic complexes, which have $T_1 = T_2$, is (2.145)

$$p_{ex}(S_1) = (2/3)\, S_2(S_2+1)\, J^2 T_1 \tau_c / \left(1 + (2/3)\, S_2(S_2+1)\, J^2 T_1 \tau_c\right). \qquad (2.145)$$

The comparison of (2.145) with the result (2.136) obtained for the case of collisions of radicals with slowly relaxing paramagnetic particles shows that in the weak exchange region, when $J^2\tau_c^2 < 1$, the fast paramagnetic relaxation of the complex reduces the efficiency of the spin-exchange-induced decoherence of the radical spin by a factor $T_1/\tau_c \ll 1$. This conclusion is consistent with experimental observations by Molin et al. ([50, 51], see also Table 2.5). However, in the strong exchange situation at $J^2T_1\tau_c > 1$ or for sufficiently large particles collision time, the fast paramagnetic relaxation of the complex increases the the radical spin exchange decoherence efficiency in $3(1/2 + S_2)^2/(2S_2(1 + S_2)) \geq 2$ times (see Eqs.(2.138 and 2.145) in the strong exchange limit).

When free radicals collide with paramagnetic complexes with short times of paramagnetic relaxation, the spin coherence transfer from the complex to the radical is not expected, i.e., in this case there is no reaction from the paramagnetic complex to the radical in the collision. Collisions lead only to decoherence (phase relaxation) of the radical spins, which is described by the equation

$$\partial M_{A\text{-}}/\partial t = -i\omega_A M_{A\text{-}} - p_{sd}K_D C_B M_{A\text{-}}. \quad (2.146)$$

Here $M_{A\text{-}}$ is the transverse magnetization of the spin magnetization, C_B is concentration of rapidly relaxing paramagnetic complexes. Table 2.5 shows the phase relaxation efficiency of radical spins in the collision with complexes.

It can be seen from Table 2.5 that the decoherence rate constant of the spins of radicals as a result of the exchange interaction in the collision with rapidly relaxing complexes of paramagnetic ions, $K_{sd} = p_{sd}K_D$, can reach the value of the diffusion collision rate constant K_D only at sufficiently large exchange integral values and sufficiently long collisions. Note that no equivalent spin exchange occurs under collisions with paramagnetic particles with short paramagnetic relaxation times. As a result, there is no spectral diffusion, as in the case of equivalent spin exchange.

The theoretical conclusions on the effect of the paramagnetic relaxation rate of the complex on the effect of their exchange interaction with radicals are consistent with the corresponding experimental data. For example, in [50] the concentration broadening of EPR spectra of solutions of TMOPO nitroxide radicals were measured and the rate constants of dephasing spins of the radicals due to their collisions with cobalt complexes were determined. The results obtained in [50] are shown in Table 2.6.

It can be seen from Table 2.6 that only in the case of a complex that has chlorine atoms as ligands, the spin decoherence efficiency is close to 1 ($p_{sd} = 2.3/3.2 = 0.72$), and the spin decoherence rate constant is close to the rate constant of diffusion collisions (cf. the last two columns in Table 2.6). For complexes without chlorine $K_{sd} \ll K_D$, and this is to be expected, as cobalt complexes have the short paramagnetic relaxation time. In the case of a complex that has chlorine atoms on the periphery as ligands, the effect of the short paramagnetic relaxation time is practically not manifested. This is due to the fact that the unpaired electron density on chlorine is more than an order of magnitude larger than the unpaired electron density

Table 2.5 The efficiency of collision of radicals with complexes that have a short time of spin-lattice relaxation time compared with the collision duration, $T_1 \ll \tau_c$

JT_1	$J^2T_1\tau_c$	p_{sd}	Collision effect
$JT_1 \ll 1$	$J^2T_1\tau_c \ll 1$	$(2/3)\ J^2T_1\tau_c\ S_B(S_B + 1)$	Partial dephasing of spins of radicals
$JT_1 \ll 1$	$J^2T_1\tau_c \gg 1$ (large τ_c)	1	Full dephasing of spins of radicals
$JT_1 \gg 1$	$J^2T_1\tau_c \gg 1$ (large J)	1	Full dephasing of spins of radicals

Table 2.6 The rate constant of dephasing spins of the radicals due to their collisions with cobalt complexes

Complex	Radical	Solvent	$K_{sd}\ 10^{-9}$ ($M^{-1}\ s^{-1}$)	$K_D\ 10^{-9}$ ($M^{-1}\ s^{-1}$)
Co $(C_3H_7OH)_2Cl_2$	TMOPO	Propanol	2.3	3.2
$[Co(H_2O)_6]^{2+}$	TMOPO	Water	0.9	7
$Co(acac)_2Py_2$	TMOPO	Clorophorm + piridin	0.4	12.0

on the periphery of the other two complexes. Therefore, there is a strong difference in the properties of the complex containing chlorine atoms.

2.8.7 *Quenching of Positronium with Paramagnetic Particles*

An interesting example of the process depending on spin exchange is the exchange quenching of positronium atom, Ps, with paramagnetic additives [60]. This process can be called spin catalysis of quenching of positronium. The quenching mechanism consists in the fact that the exchange interaction in the collision of the Ps atom with a paramagnetic particle induces the transition of the long-lived ortho positronium (lifetime $\tau_o = 143$ ns) into the short-lived para state (lifetime $\tau_p = 125$ ps). In fact, due to the exchange interaction, there is a mutual flip-flop of the spin of the electron of positronium and the spin of a paramagnetic particle (radical or complex). The efficiency of such a process was discussed in the previous section. The rate constant for exchange quenching of positronium with paramagnetic additives was calculated in [61] and the role of the relaxation rate of paramagnetic complexes in the efficiency of the exchange quenching of positronium was discussed. The results coincide qualitatively with the above results on the effect of the paramagnetic relaxation rate of particles on the spin exchange efficiency.

Below is presented the rate constant of quenching of positronium with paramagnetic particles with spin S in the case when the Ps converters have the long spin relaxation times, i.e., the inequalities are T_1, $T_2 > \tau_c$. Suppose also that $\tau_p > \tau_c$. In this case, obtained [61].

$$K = (1/6)\{W\tau_c/(1 + W\tau_c)\}K_D. \tag{2.147}$$

The rate of the "quasi-monomolecular" process of electron spin flip or flop in the positronium atom caused by the exchange interaction of a paramagnetic particle with the positronium electron in a collision is introduced [61].

$$W = J^2\tau_c S(S + 1)/\left(1 + (\Delta + J/2)^2{\tau_c}^2\right). \tag{2.148}$$

The parameter Δ means singlet-triplet splitting in the positronium atom, which is caused by the contact interaction of the electron with the positron. For an arbitrary relation between the time of one collision and the lifetime of a pair, the positronium quenching rate constant is given by [61].

$$\begin{aligned} K &= \left(W_p\tau_c/\left(1 + W_p\tau_c(1 + 5\tau/\tau_c)\right)\right)K_D, \\ W_p &= (1/6)J^2\tau/S(S + 1)/\left(1 + (\Delta + J/2)^2\tau^2\right), 1/\tau = 1/\tau_c + 1/\tau_p. \end{aligned} \tag{2.149}$$

The above results are applicable to the analysis of Ps quenching by paramagnetic additives such as organic free radicals, Cu(II), Fe(III), Mn(II), Cr(III) ion complexes, which have relatively long paramagnetic relaxation times. But there are paramagnetic additives with short relaxation times T_1, $T_2 \sim 10^{-11} - 10^{-13}$ s. Examples are such complexes as Co(II), Fe(II),Ni(II). The fast paramagnetic relaxation of the spin of complex randomly modulates and averages to zero the exchange interaction that induces the spin flip/flop of the positronium electron. Therefore, such complexes should be less effective for quenching of positronium. Note that this is analogous to a decrease in the efficiency of spin exchange for radicals colliding with particles with short paramagnetic relaxation times. The experiments confirm [62] the similar behavior of the Ps quenching rate constant and the spin exchange rate constant found from the analysis of the shape of EPR spectra of solutions containing free radicals and paramagnetic complexes.

2.8.8 *Effect of Rotational and Translational Diffusion of Particles on Spin Exchange in the Case of Anisotropic Spin Density Distribution in Paramagnetic Particles*

In all the above calculations of the effective radius and the spin exchange rate constant, it is assumed that the exchange integral does not depend on the mutual orientation of the colliding paramagnetic particles. This assumption is justified if the spin density is uniformly distributed over the "surface" of colliding paramagnetic particles, for example, for a ferricyanide ion [63]. But in stable nitroxide radicals, which are often used as spin probes, the spin density is concentrated on the oxygen atom and the main contribution to the efficiency of spin exchange is given by collisions in which it is the NO group of the radical that maximally approaches the paramagnetic particles. The anisotropic spin density distribution on the surface of a paramagnetic particle can be described approximately using the "steric" parameter f, which specifies the fraction of the particle surface on which the spin density is concentrated [37, 51, 53]. For example, in the case of nitroxide radical steric factor f can be estimated as the ratio of half of the van der Waals surface of oxygen to the total surface of the radical. We obtain $f \approx 0.08$ [63].

Rotational and translational diffusion of particles randomly changes the mutual orientation of colliding particles and thus changes their exchange interaction. Within the model of sudden switching on of the exchange interaction and without taking into account re-encounters of particles, the effect of rotational diffusion of particles on the efficiency of spin exchange was first theoretically investigated in [37]. Rotational diffusion is characterized by the correlation time τ_1 for the particle orientation ([23], (VII.106))

$$\tau_1 = 4\pi\eta a_{vW}{}^3/(3kT) \qquad (2.150)$$

η-viscosity of the solution, a_{vW}-van der Waals radius of the particle.

In the situation of equivalent spin exchange within the model of sudden collisions, the effect of the anisotropic spin density distribution is easy to take into account if during the collision the rotational diffusion does not have time to change the mutual orientation of the colliding particles, $\tau_c < \tau_1$. In this case, in the above expressions for the rate constant (or efficiency, or effective radius) of the equivalent spin exchange, a steric factor should be added. For example, for two identical radicals, the exchange efficiency should not be described as given by (2.62) but by (2.151)

$$p_{ex} = (1/2)\, f^2 J^2 \tau_c{}^2/\left(1 + J^2 \tau_c{}^2\right), \qquad (2.151)$$

where f^2 is the probability that the paramagnetic particles will collide with those atoms on which the spin density is concentrated. In [37] was studied the situation when in a colliding pair of particles with spins $S = 1/2$ only one of the particles has an anisotropic spin density distribution. The effect of rotational diffusion during one collision on the efficiency of spin exchange is considered. For example, in the limit case of relatively fast random rotations with $\tau_1 < \tau_c$ and $d(J)\tau_1\tau_c < 1$, where d(J) is the dispersion of J, the following value of the spin exchange efficiency is obtained

$$p_{ex} = (1/2)(<J>^2\tau_c{}^2 + d(J)\,\tau_1\tau_c(1 + d(J)\tau_1\tau_c))/(1 + d(J)\tau_1\tau_c)^2 + <J>^2\tau_c{}^2). \tag{2.152}$$

In this model, the average value of the exchange integral $<J> = fJ$, and the dispersion of the exchange integral in the collision $d(J) = f(1-f)J^2$. In the limit of strong exchange interaction when $f^2J^2\tau_c{}^2 \gg 1$, $p_{ex} = 1/2$: and $p_{ex} \geq 1/2$, i.e. coincides with the efficiency of spin exchange between particles with isotropic spin density distribution.

In [51, 53] the role of re-encounters and the role of translational diffusion in the intervals between re-encounters in the efficiency of spin exchange was studied, when one of the particles involved in the collision has an anisotropic spin density distribution. This anisotropy is described by the steric parameter f_{12}. It is shown in [51, 53] that the averaging of the effect of anisotropy of spin density distribution on the efficiency of spin exchange as a result of translational diffusion of particles can be estimated using the formulas obtained for the isotropic spin density distribution, if the effective collision time τ_c is appropriately determined. For the isotropic spin density distribution, the total time that two particles spend in the exchange interaction region during one meeting is given by (2.65): $\tau_c = r_0 a/D_{AB}$. In the case of anisotropic distribution of the spin density for the description of averaging of the anisotropic exchange interaction by translational diffusion of molecules in intervals between re-encounters it was suggested to use effective "time" of collision:

$$\tau_{c,\,ef.} = f_{12}{}^{1/2}\, r_0 a/D_{AB}. \tag{2.153}$$

When the same particles collide with the anisotropic spin density distribution with the steric factor f, $f_{12} = f^2$, therefore (2.153) takes the form

$$\tau_{c,\,ef} = f\, r_0 a/D_{AB}. \tag{2.154}$$

It should be noted that this suggestion (2.1530 [51, 53]) is in a good correspondence with speculations concerning this problem presented above (see 2.100, 2.101, 2.102, 2.103, 2.104, 2.105, 2.106 and 2.107).

Taking into account translational and rotational diffusion of particles in the work (see [51], Eq. (18)) a more complex expression is proposed to find the effective total time that two particles spend in the area of exchange interaction during one meeting

$$\tau_{c,\,ef.} = f_{12}{}^{1/2}\,(ba/D_{AB})/\chi,$$
$$\chi = \left(1 + f_{12}{}^{1/2}v\right)/(1 + v) \leq 1,\ v = \left(b^2/D\tau_1\right)^{1/2} + \tau_c/\tau_1. \tag{2.155}$$

For the spin exchange rate contant is proposed [51]

$$\begin{aligned}K_{ex} = 4\pi bD\,(f_{12})^{1/2}\chi\,(1/2)\,J^2\tau_{c,\,эфф}{}^2/\left(1+J^2\tau_{c,\,эфф}{}^2\right) = \\ 4\pi bD\,(f_{eff})\,(1/2)\,J^2\tau_{c,\,эфф}{}^2/\left(1+J^2\tau_{c,\,эфф}{}^2\right),\end{aligned} \tag{2.156}$$

where $f_{eff} = (f_{12})^{1/2}\,\chi \le (f_{12})^{1/2}$ (compare with Eq. (2.107)).

The parameter χ depends on the rotational relaxation time. If τ_1 is less than the duration of the collision, $\chi \to f_{12}{}^{1/2}$, with slow rotational diffusion of the radical $\chi \to 1$.

The role of the steric factor in spin exchange was studied in detail in [51] for the collision of a nitroxide radical with a $Fe(CN)_6{}^{3-}$complex. For the nitroxide radical, the characteristic rotational diffusion time was estimated as $\tau_1 \approx 1.5\ 10^{-11}$ sec and the times of paramagnetic relaxation for this complex are $T_1 = T_2 = 6\ 10^{-12}$ s [63]. In this case, the spin density is distributed uniformly over the surface of the complex. In a nitroxide radical, the spin density is concentrated on NO and the steric factor can be estimated as $f = 0.08$, as noted above. In this case $f_{12} = f$. The steric factor of the nitroxide radical was determined in [51] from the analysis of the broadening of the EPR spectrum of the radical caused by spin exchange with the $Fe(CN)_6{}^{3-}$ complex. The value of the parameter $\chi \approx 1$was found, so that the steric factor was found to be $f \approx 0.07$ and exchange integral $J \approx 10^{12}$ rad/s. The found values of the steric factor and the exchange integral are in satisfactory agreement with their expected values.

Manifestation of steric factor in spin exchange between nitroxide stable radicals, 3β-doxyl-5α-cholesterol (CSL), was experimentally investigated in [64]. Measurements were carried out in different n-alkanes and at different temperatures. The spin exchange rate constant was determined from the analysis of EPR spectra. It turned out that the temperature dependence of the spin exchange rate constant is described by the temperature dependence of the mutual diffusion coefficient of radicals, i.e. $K_{ex} \sim D \sim T/\eta$, where η is the viscosity of the solution. Such temperature dependence of the spin exchange rate constant within the model of sudden collisions can serve as a criterion that a strong spin exchange is implemented. Indeed, within this model, in the extreme case of strong exchange, the exchange efficiency reaches the limit value of 1/2 and for particles with an anisotropic spin density distribution this limit rate of the spin exchange equals

$$K_{ex} = (1/2) f_{ef} K_D. \tag{2.157}$$

Here f_{ef} is a steric factor for bimolecular spin exchange. It is associated with anisotropy of spin density distribution on the radical surface. For the radical studied in [64], the geometric estimate of the fraction of the radical surface on which the spin density is concentrated, i.e. the steric factor f of the radical with respect to the spin exchange, gives $f \approx 0.59$. It is assumed that when the radicals collide, the exchange integral takes the value J with the probability f, and the exchange integral with the probability 1-f can be considered negligible. According to the theoretical calculations above, in this situation it is expected that $f_{12} = f^2$, and $f_{ef} = f\chi$. Hence, the expected value is $0.35 < f_{ef} < 0.59$. From the analysis of EPR spectra in [64] it is found that $f_{ef} = 0.49$. This value is consistent with the theoretically expected value of

f_{eff}. Note that in [64] the value of $f_{ef} = 0.49$ is confirmed by an independent method. In parallel, spin exchange between fully deuterated nitroxyl radicals 2,2,6,6-tetramethyl-4-oxyl (pDT) was also studied. In this radical, the spin density is distributed evenly over the surface, so $f_{ef} = 1$. Based on a detailed comparison of spin exchange rate constants for CSL and pDT radicals, the authors [64] found that for CSL the steric factor $f_{ef} = 0.49$.

Thus, in the approximation of the sudden collision, the effective spin exchange radius was calculated for a number of model situations. The obtained theoretical results allowed us to interpret the influence of the spin value, the rate of paramagnetic relaxation of spins, the effect of anisotropy of spin density distribution, the viscosity of the solvent on the efficiency of spin exchange. However, the approach of sudden inclusion of the exchange interaction does not take into account the extended nature of the exchange interaction. Spin exchange can also occur at distant approaches of molecules, when the distance r between the colliding particles is much greater than the radius of their collision r_0, $r > r_0$. Moreover, the "colliding" particles of the pair can leave the effective spin exchange region (the effective exchange interaction region) and return to this region again.

To elaborate a consistent theory of bimolecular spin exchange, kinetic equations for single-particle spin density matrices were obtained taking into account bimolecular collisions and exchange interaction [40]. The derivation of the kinetic equations taking into account the extended nature of the exchange interaction and taking into account all possible re-encounters of two particles during their meeting in solution is presented below briefly. The results of the theory of spin exchange obtained by these kinetic equations are also presented.

2.9 Kinetic Equations for Single-Particle Spin Density Matrices in Dilute Solutions of Paramagnetic Particles in the General Case

General kinetic equations for a single-particle spin density matrix taking into account bimolecular collisions were formulated in a consistent way in [40, 65]. Later they were developed in [66–68], (see also [1, 2]).

We consider a dilute solution of two types of paramagnetic particles (spin probes) A and B. The spin Hamiltonian of such a system, taking into account only the Heisenberg exchange interaction between molecules A and B, can be written as

$$\mathbf{H} = \mathbf{H}_A + \mathbf{H}_B + \sum \mathbf{V}(|\mathbf{r}_{Ak} - \mathbf{r}_{Bn}|) = \mathbf{H}_A + \mathbf{H}_B + \sum \mathbf{V}_{AB}(r_{kn}(t)). \quad (2.158)$$

For simplicity, we omit the interaction of identical A-A and B-B molecules. The contribution of these interactions can be calculated analogously as it is done for the interaction of A and B molecules. Single-particle spin Hamiltonians $\mathbf{H}_A$ and $\mathbf{H}_B$ contain the interaction between the electron spins with the external magnetic field and the hyperfine interaction with magnetic nuclei. For paramagnetic particles with spin $S > 1/2$, the spin-spin interaction in an isolated particle (the so-called zero-field splitting) should be taken into account.

In the simplest case, $\mathbf{H}_A$ and $\mathbf{H}_B$ are the Zeeman energies of the interaction of the magnetic spin moment of paramagnetic particles with a constant external magnetic field

$$\begin{aligned} H_A &= \hbar\omega_{0A} S_{AZ}, \\ \mathbf{H}_B &= \hbar\omega_{0B} S_{BZ}, \end{aligned} \qquad (2.159)$$

where $\omega_{0A,B}$ is the Larmor frequencies of A and B spins. Note that in the case of (2.159) the exchange interaction of the A-A and B-B type does not change the state of spins, so we can ignore bimolecular collisions of particles of the same type, A + A and B + B.

We assume that the exchange interaction between two paramagnetic particles A_k and B_n depends only on the distance r_{kn} between them and has the form:

$$\mathbf{V}_{AB}(r_{kn}(t)) = \hbar\, J_0 \exp(-\kappa(r_{kn}(t) - r_0))\, (\mathbf{S}_A \mathbf{S}_B). \qquad (2.160)$$

Here J_0 is the value of the exchange integral at the distance r_0 of the closest approach of spin probes, κ sets the slope of the decline of the exchange integral, $\mathbf{S}_A$ and $\mathbf{S}_B$ are spin operators.

The radius vectors $\mathbf{r}_{kn}$ connecting spins A(k) and B(n) change over time as a result of the thermal motion of molecules. In the theory of spin exchange in solutions, $\mathbf{r}_{kn}$ are considered as external parameters that change randomly over time due to the diffusion of molecules.

A complete description of the spin system of N spin probes is given by the multiparticle spin density matrix $\rho_N(t)$. To describe the experiments, one-particle or two-particle density matrices are usually sufficient, which can be obtained by the convolution of $\rho_N(t)$ on the spin states of N-1 or N-2 particles, respectively [69]. It is well known from statistical mechanics that partial density matrices for a multiparticle system satisfy a chain of coupled Bogolyubov-Born-Green-Kirkwood-Yvon- equations (see, e.g., [71]). The chain of BBGKY equations expresses the evolution of the s-partial density matrix using the (s + 1) partial density matrix.

For the spin Hamiltonian (2.158) a single-particle spin density matrix, e.g., for the particle A satisfies the equation of motion

$$\begin{aligned} \partial\rho_A(t)/\partial t = &-(i/\hbar)[\mathbf{H}_A, \rho_A(t)] \\ &- (1/V)\mathrm{Tr}_B \sum \int (i/\hbar)[\mathbf{V}_{AB}(\mathbf{r}_n(t)), \rho_2(n,t)]d^3\mathbf{r}_n. \end{aligned} \qquad (2.161)$$

Here $\rho_2(n,t)$ is a two-particle spin density matrix of the pair, which consists of an arbitrarily selected particle A and a particle B with n-th number ($n = 1, 2, \ldots N_B$, N_B is the total number of particles B). Summation in (2.161) is performed for all B particles and integration is over the entire volume V. Tr_B means the convolution on spin variables of the particles B. In the thermodynamic limit, when V, N_B -->∞, provided N_B/V--> C_B, every particle B gives on average the same contribution to the change in the state of the selected particle A. As already mentioned in connection with the model of sudden collisions, the one-particle density matrix varies in the time scale of diffusion of particles between collisions with different particles, while the last term in (2.161) varies in a substantially smaller time scale of collision. Let us introduce one-particle density matrixes σ_A, σ_B averaged in the time interval of the collision. Averaging (2.161) on the collision time scale, we obtain

$$\partial < \sigma_A(t)/\partial t >= -(i/\hbar)[\mathbf{H}_A, < \sigma_A(t) >] - C_B \, Tr_B \int (i/\hbar) < [\mathbf{V}_{AB}(r(t)), \sigma_2(r,t)] > d^3\mathbf{r}. \tag{2.162}$$

In this equation, $\sigma_2(r,t)$ is the density matrix of an "impersonal" pair of particles A and B, <. . .> means averaging over all possible realizations of the random process r (t). For spins B the equation for a single-particle spin matrix σ_B is obtained from (2.162) replacing A with B and vice versa.

Similarly, we can obtain an equation for the pair density matrix $\sigma_2(r,t)$, which "engages" the three-particle, e.g., A + 2 B, density matrix [71].

$$\partial < \sigma_2(1,t)/\partial t >= (-i/\hbar) < [\mathbf{H}_A + \mathbf{H}_B + \mathbf{V}_{AB}(r_1(t)), \sigma_2(1,t)] > + < \mathbf{F}(\sigma_3(1,2,t)) > . \tag{2.163}$$

The last term in the right-hand side contains a three-particle density matrix $\sigma_3(1,2,t)$, three particles A, B(1), B(2) are considered (one particle A and two B particles with numbers 1 and 2), r_1, r_2 are distances between particles in pairs A . . . B(1) and A . . . B(2), respectively. We do not discuss the functional F here in detail, since it describes three-partcles collisions. The probability of finding one particle B in the region of the exchange interaction with particle A can be estimated as the product of the volume of the exchange interaction region on the concentration of particles B, $p = (4/3)\pi r_{EX}^3 C_B$. We consider dilute solutions for which the condition

$$p = (4/3)\pi r_{EX}^3 \, C_B \ll 1, \tag{2.164}$$

should be satisfied. Under this condition, the contribution of three-particle collisions, which are realized with the probability p^2, can be neglected. So the chain of BBGKY equations is cut and for the pair density matrix takes the equation

$$\partial < \sigma_2(t)/\partial t >= (-i/\hbar) < [\mathbf{H}_A + \mathbf{H}_B + \mathbf{V}_{AB}(r(t)), \sigma_2(t)] > . \tag{2.165}$$

This equation depends on the random process r(t). It is necessary to solve this equation for each possible realization of the random process r(t) and then to average the results. However, the assumption that r(t) is an external parameter, i.e., it does not depend on the state of spins, allows this problem to be reformulated so that instead of the stochastic Eq. (2.164), it is necessary to solve equations with time-independent coefficients [33, 40, 67–70]. This reformulation of the problem is essentially based on the ergodic hypothesis of the equivalence of averaging over a large period of time and over an ensemble. Therefore, the ensemble of pairs of particles A and B is divided into sub-ensembles of pairs with a given distance r between the partners of the pair A and B. For each sub-ensemble of pairs, we introduce a partial two-particle spin density matrix $\sigma_2(t|r)$. This two-particle spin density matrix changes for two reasons. On the one hand, the state of a pair of spins changes due to the spin dynamics under the action of the spin Hamiltonian $\mathbf{H}_{AB}(r) \equiv \mathbf{H}_A + \mathbf{H}_B + \mathbf{V}_{AB}(r)$ of the given pairs with a given distance r between the partners of the pair A...B. The spin dynamics induced change in the pair spin density matrix is given by Eq. (2.166) where a distance r between the particles of the pair is no longer a stochastic process, but a given external parameter

$$(\partial\sigma_2(t|r)/\partial t)_{\text{spin dynam..}} = (-i/\hbar)[\mathbf{H}_A + \mathbf{H}_B + \mathbf{V}_{AB}(r), \sigma_2(t|r)]. \qquad (2.166)$$

On the other hand, due to the diffusion of molecules, each pair can "flow" from one sub-ensemble of pairs to another. The changes of $\sigma_2(t|r)$ caused by the diffusion of molecules can be found from the continuity equation

$$(\partial\sigma_2(t|r)/\partial t)_{\text{diffusion}} + div_{\mathbf{r}}(\boldsymbol{j}(\sigma_2(t|r)) = 0. \qquad (2.167)$$

Here $\boldsymbol{j}(\sigma_2(t|r))$ is a flux of pairs in the space. For the continuous diffusion model in the case of neutral molecules:

$$\boldsymbol{j}(\sigma_2(t|r)) = -D_{AB}\nabla_{\mathbf{r}}\sigma_2(t|r). \qquad (2.168)$$

In solutions of charged particles A and B in electrolytes, the flow is induced not only by the diffusion of particles, but also by the drift of particles under the action of the electrostatic interaction [9, 34].

The total change of the partial spin density matrix is given by the sum of two contributions

$$\partial\sigma_2(t|r)/\partial t = (-i/\hbar)[H_{AB}(r), \sigma_2(t|r)] - \text{div}_{\mathbf{r}}(\boldsymbol{j}(\sigma_2(t|r)). \qquad (2.169)$$

Note that for the Markov random process, Eqs. (2.165 and 2.166) can be obtained by directly summing the solution of the stochastic Eq. (2.160) for all possible implementations of the random process (see, e.g. [40, 66, 70]).

The corresponding equation for solutions of charged paramagnetic particles is considered in [34].

To obtain the kinetic equation for a single-particle spin density matrix, the two-particle (pair) density matrix should be expressed via single-particle density matrices. Kinetic equations describe the behavior of the system at the times longer than a duration of collision of the particles, $t > \tau_c$. Following [69], at these times the solution of (2.165) we search in the form

$$\sigma_2(r,t) \approx \mathbf{G}(r)\,\sigma_A(t)\,\sigma_B(t). \tag{2.170}$$

In this expression, G(r) is the desired linear operator.

Substitute (2.170) into (2.169). The left-hand side of (2.169) takes the form

$$Y = \mathbf{G}(r)\,(\partial\sigma_A(t)/\partial t)\,\sigma_B(t) + \mathbf{G}(r)\,\sigma_A(t)\,(\partial\sigma_B(t)/\partial t). \tag{2.171}$$

If we now substitute here the expressions for $\partial\sigma_{A,\,B}/\partial t$ (see 2.162), there appear quadratic terms containing $\sigma_A(t)\sigma_B(t)$ and cubic terms containing $\sigma_A(t)\sigma_B(t)\sigma_B(t)$ and $\sigma_A(t)\sigma_A(t)\sigma_B(t)$. Cubic terms correspond to trimolecular collisions, which we neglect for sufficiently dilute solutions. Therefore, in the expression for Y (2.171) we assume

$$\begin{aligned}\partial\sigma_A(t)/\partial t &\approx -(i/\hbar)[\mathbf{H}_A, \sigma_A(t)] \equiv -i\mathbf{Q}_{0A}\sigma_A(t),\\ \partial\sigma_B(t)/\partial t &= -(i/\hbar)[\mathbf{H}_B, \sigma_B(t)] \equiv -i\mathbf{Q}_{0B}\sigma_B(t).\end{aligned} \tag{2.172}$$

As a result, we obtain the following equation for the operator **G**(r) [40].

$$\mathbf{W}(r)\mathbf{G}(r) + [\mathbf{Q}_0, \mathbf{G}(r)] + D_{AB}\Delta\mathbf{G}(r) = 0. \tag{2.173}$$

Here superoperators **W**(r) and $\mathbf{Q}_0$ are introduced:

$$\begin{aligned}\mathbf{W}(r)_{ik,\,lm} &= -\,iJ(r)\left[(\mathbf{S}_A\mathbf{S}_B)_{mk}\delta_{li} - (\mathbf{S}_A\mathbf{S}_B)_{il}\delta_{km}\right],\\ \mathbf{Q}_0 &= \mathbf{Q}_{0A} + \mathbf{Q}_{0B},\\ \mathbf{Q}_{0ik,\,lm} &= -\,i\left[(\mathbf{H}_A + \mathbf{H}_B)_{mk}\delta_{li} - (\mathbf{H}_A + \mathbf{H}_B)_{il}\delta_{km}\right].\end{aligned} \tag{2.174}$$

In Eq. (2.173), the first term describes the contribution of the spin-spin interaction, the second term is associated with the motion of spins in an external magnetic field, and the third term takes into account the diffusion-induced transitions of pairs between the sub-ensembles with different distances between the spins in the pair. Note that (2.173) is written for the model of the continuous diffusion of particles. For the model of sudden exchange interaction considered above, Eq. (2.173) is reduced to (2.74).

We need to formulate the boundary conditions for **G**(r). When separating partners of the pair, the correlation between their states should weaken [69]. Therefore, we impose a condition that **G**(r) with the increase in r tends to a unit matrix

$$\mathbf{G}(r) \;{-}{-}{>}\; \mathbf{E} \quad \text{when } r \;{-}{-}{>}\; \infty. \tag{2.175}$$

At the radius of the closest approach the particles are completely reflected, hence we have the second boundary condition: the flow of $\sigma_2(r,t)$ through the sphere of radius $r = r_0$ should be zero. For neutral particles we have

$$\nabla_r \mathbf{G}(r = r_0) = 0. \tag{2.176}$$

We introduce a collision superoperator **P** that describes the change in the pair density matrix in the interaction region and, according to Eqs. (2.161 and 2.170) it is defined by the expression:

$$\mathbf{P} = \int \mathbf{W}(r)\mathbf{G}(r)d^3\mathbf{r}. \tag{2.177}$$

Note that in some specific cases when in (2.173) one can neglect the term $i[\mathbf{Q}_0,\mathbf{G}(r)]$ or when this term is zero, the collision superoperator, (2.177) can be represented in another form [40, 65].

$$\mathbf{P} = -D_{AB}\int \Delta\mathbf{G}(r)d^3\mathbf{r}. \tag{2.178}$$

Thus, for a single-particle spin density matrix we have a kinetic equation

$$\partial\sigma_A(t)/\partial t = -(i/\hbar)[\mathbf{H}_{0A}, \sigma_A(t)] + C_B \mathrm{Tr}_B(\mathbf{P}\sigma_A(t)\sigma_B(t)). \tag{2.179}$$

A similar equation is obtained for $\sigma_B(t)$

$$\partial\sigma_B(t)/\partial t = -(i/\hbar)[\mathbf{H}_{0B}, \sigma_B(t)] + C_A \mathrm{Tr}_A(\mathbf{P}\sigma_A(t)\,\sigma_B(t)). \tag{2.180}$$

Note that the model considered in Sect. (2.7.3) for the sudden switching on of some constant exchange interaction when a pair of particles enters a certain "cell" at the collision distance, r_0, and its switching off at the exit from the "cell" implies a jump passage of the interaction region. In this case, pairs of spins can be divided into two sub-ensembles: in one sub-ensemble the spins are in the "cell"at the collision distance, in another sub-ensemble the distance between the spins is greater than the collision distance, so that the pairs are outside the "cell". For this model, the kinematic flow of the pair density matrix, the kinetic equations are [35].

$$\begin{aligned}\partial\sigma_2(t|r_0)/\partial t = {} & (-i/\hbar)[\mathbf{H}_A + \mathbf{H}_B + \mathbf{V}_{AB}(r_0), \sigma_2(t|r_0)] \\ & - (1/\tau_c)(\sigma_2(t|r_0) - \sigma_A(t)\,\sigma_B(t)).\end{aligned} \tag{2.181}$$

In this case instead of (2.173) we obtain

$$\mathbf{W}(r_0)\mathbf{G}(r_0) + [\mathbf{Q}_0, \mathbf{G}(r_0)] - (1/\tau_c)(\mathbf{G}(r_0) - \mathbf{E}) = 0, \tag{2.182}$$

where **E** is a unit operator. This equation is the same as (2.74).

Equations (2.179 and 2.180) are kinetic equations for single-particle spin density matrices of paramagnetic particles in dilute solutions taking into account the bimolecular spin exchange. At the same time, in the course of deriving kinetic equations for single-particle density matrices, the equations for calculating the effective spin exchange radius and the spin exchange rate constant are derived (2.173 and 2.177). These equations make it possible to calculate the effective spin exchange radii taking into account the extended nature of the exchange interaction. The presented approach makes it possible to calculate the effective radii of the bimolecular spin exchange with arbitrary kinematics of the passage of the exchange interaction region in the solution by the colliding particles. Similar arguments were used to obtain kinetic equations for spin-dependent recombination of radicals [6, 71, 72], for spin-dependent triplet-triplet annihilation of excitons [68], for spin exchange between charged paramagnetic particles [34] and for spin exchange between particles with arbitrary spins [67].

The resulting kinetic equations are essentially simplified if the systems are considered in a state close to equilibrium. For small deviations from the thermodynamic equilibrium spin density matrices of particles can be represented as the sum of the equilibrium density matrix and a small additive. For the considered model of a solution of two types of paramagnetic particles we have

$$\begin{aligned}\sigma_A(t) &= \sigma_{Aeq} + \Delta\sigma_A(t),\\ \sigma_B(t) &= \sigma_{Beq} + \Delta\sigma_B(t).\end{aligned} \tag{2.183}$$

Suppose that the components of $\Delta\sigma_{A,B}$ are small additives, the absolute value of each matrix element of these additives is much less than 1. Then the kinetic equations can be linearized. To do this, substitute (2.183) into (2.179 and 2.180) and discarding quadratic terms of the small additives. In the thermodynamic equilibrium, the right-hand side of the kinetic equations, e.g., (2.179 and 2.180) is zero. As a result, to describe the temporal behavior of the system near the equilibrium, we obtain linear kinetic equations

$$\begin{aligned}\partial\Delta\sigma_A(t)/\partial t &= -(i/\hbar)[\mathbf{H}_{0A}, \Delta\sigma_A(t)] + C_B\,\mathrm{Tr}_B(\mathbf{P}\Delta\sigma_A(t)\,\sigma_{Beq} + \boldsymbol{P\sigma_{Aeq}D\sigma_B(t)}),\\ \partial\Delta\sigma_B(t)/\partial t &= -(i/\hbar)[\mathbf{H}_{0B}, \Delta\sigma_B(t)] + C_A\,\mathrm{Tr}_A(\boldsymbol{P\,\Delta\sigma_A(t)\,\sigma_{Beq}} + P\sigma_{Aeq}\Delta\sigma_B).\end{aligned} \tag{2.184}$$

Special attention should be paid to the terms in (2.184), which are in bold italics. It is these terms that describe the exchange of quantum states as a result of bimolecular collisions. They describe the reaction of a partner in a « collision ». Populations of spin states are described by diagonal elements of the density matrix. The change of the diagonal elements of the matrices of the particle density describes the transfer of energy between the spins in bimolecular collisions. Nondiagonal elements of the spin density matrix characterize the quantum coherence of spins. Thus, in bimolecular collisions, the exchange interaction induces decoherence and spin coherence transfer and spin energy transfer between paramagnetic particles.

The averaged properties for the whole ensemble of molecules are measured in the experiment. Usually, the direction of the constant external magnetic field $\mathbf{B}_0$ is selected for the z-axis of the coordinate system. In the vast majority of cases, in EPR experiments one observes the magnetization component of the system perpendicular to the direction of the constant magnetic field, i.e., M_x, M_y components of the magnetization vector. The procedure of the derivation of the equations for the magnetization of the system is illustrated here for a solution of paramagnetic particles A and B with one unpaired electron.

Suppose that all spins have the same g-factor. Consider a model situation in which the spin Hamiltonians of particles A and B are given by (2.159). In this case, the eigenfunctions of the operators $\mathbf{H}_A$, $\mathbf{H}_B$ are the eigenfunctions of $\mathbf{S}_{AZ}$, $\mathbf{S}_{BZ}$. In the basis of eigenfunctions of the spin operator S_z, the spin density matrix has the form

$$\rho = \left\{\left\{\rho_{1/2,1/2}, \rho_{1/2,-1/2}\right\}, \left\{\rho_{-1/2,1/2}, \rho_{-1/2,-1/2}\right\}\right\}.$$

$$\rho = \left\{ \begin{matrix} \rho_{1/2,1/2} & \rho_{1/2,-1/2} \\ \rho_{-1/2,1/2} & \rho_{-1/2,-1/2} \end{matrix} \right\}$$

Diagonal elements of the density matrix characterize populations of spin states, and nondiagonal elements characterize quantum spin coherence. For the considered situation of particles with spin $S = 1/2$ we have

$$\begin{aligned} < \mathbf{S}_X - i\mathbf{S}_Y > &= \mathrm{Tr}((\mathbf{S}_X - i\mathbf{S}_Y)\,\rho) = \rho_{1/2,-1/2}. \\ < \mathbf{S}_Z > &= \mathrm{Tr}(\mathbf{S}_Z\rho) = (1/2)\left(\rho_{1/2,1/2} - \rho_{-1/2,-1/2}\right). \end{aligned} \tag{2.185}$$

Macroscopic magnetization of particles A and B is expressed through single-particle spin matrices. For example, we have

$$\begin{aligned} M_{A-} &= g\beta C_A \mathrm{Tr}_A\left(\left(\mathbf{S}_{Ax} - i\mathbf{S}_{Ay}\right)\rho_A(t)\right), \\ M_{B-} &= g\beta C_B \mathrm{Tr}_B\left(\left(\mathbf{S}_{Bx} - i\mathbf{S}_{By}\right)\rho_B(t)\right). \end{aligned} \tag{2.186}$$

Multiply the first equation in (2.184) by C_A and the second equation by C_B. Then, in the first equation, we take a trace on the spin states of A, and in the second equation, trace on the spin states of B. As a result, the kinetic equations for the magnetizations of spins are obtained. They have the simplest form in the situation when the product $|(\omega_{0A} - \omega_{0B})|$ by the meeting duration of a pair of particles is much less than unity and in the course of bimolecular meetings of the particles A and B the equivalent spin exchange is realized. In this case we obtain

$$\begin{aligned}\partial \mathrm{M}_{\mathrm{A}-}/\partial t &= -\,i\omega_{0\mathrm{A}}\mathrm{M}_{\mathrm{A}-} - \mathrm{K}_{\mathrm{ex}}\mathrm{C}_{\mathrm{B}}\mathrm{M}_{\mathrm{A}-} + \mathrm{K}_{\mathrm{ex}}\mathrm{C}_{\mathrm{A}}\mathrm{M}_{\mathrm{B}-},\\ \partial \mathrm{M}_{\mathrm{B}-}/\partial t &= -\,i\omega_{0\mathrm{B}}\mathrm{M}_{\mathrm{B}-} + \mathrm{K}_{\mathrm{ex}}\mathrm{C}_{\mathrm{B}}\mathrm{M}_{\mathrm{A}-} - \mathrm{K}_{\mathrm{ex}}\mathrm{C}_{\mathrm{A}}\mathrm{M}_{\mathrm{B}-},\end{aligned} \tag{2.187}$$

These equations clearly show that the equivalent exchange of transverse magnetization components occurs between the spin subsystems.

A similar equation is obtained for the longitudinal component of the magnetization vectors

$$\begin{aligned}\partial \mathrm{M}_{\mathrm{Az}}/\partial t &= -\,\mathrm{K}_{\mathrm{ex}}\mathrm{C}_{\mathrm{B}}\mathrm{M}_{\mathrm{Az}} + \mathrm{K}_{\mathrm{ex}}\mathrm{C}_{\mathrm{A}}\mathrm{M}_{\mathrm{Bz}},\\ \partial \mathrm{M}_{\mathrm{Bz}}/\partial t &= +\,\mathrm{K}_{\mathrm{ex}}\mathrm{C}_{\mathrm{B}}\mathrm{M}_{\mathrm{Az}} - \mathrm{KexC}_{\mathrm{A}}\mathrm{M}_{\mathrm{Bz}}.\end{aligned} \tag{2.188}$$

Equations (2.187 and 2.188) reproduce well-known Eq. (2.53) suggested by Kivelson [22].

Note that according to (2.187 and 2.188) under relevant conditions the Heisenberg exchange interaction induces a mutual flip-flop of two colliding spins, as a result of which the equivalent spin exchange occurs. These equations show that the exchange interaction in the course of bimolecular encounters of the molecules causes depolarization, i.e., reduces the polarization (see the terms with the minus sign in the right-hand side of Eqs. (2.187 and 2.188), but also implement the transfer of polarization from partners in the collision (see the terms with the plus sign in the right-hand sides of (2.187 and 2.188). So in the course of the spin exchange a reaction of a partner operates.

We have already seen above in the discussion of the theory of spin exchange in the approximation of sudden collisions, in the general case the exchange interaction in bimolecular collisions is not reduced to equivalent spin exchange, which is described by equations of the type (2.187 and 2.188).

Note that the phenomenological kinetic Eqs. (2.187 and 2.188) were written first by analogy with the equations of chemical kinetics for bimolecular reactions [22]. However, in this case the spin exchange rate constants are phenomenological parameters. The above approach based on the kinetic equations for spin density matrices creates a mathematical apparatus for calculating these parameters based on the spin Hamiltonian of the studied system and the kinematics of the diffusion motion of paramagnetic particles. Moreover, the consistent theory of spin kinetics taking into account bimolecular collisions in the general case is not reduced to equations such as the equations of chemical kinetics. In general, kinetic constants in equations of the (2.187 and 2.188) type become complex quantities. In chemical kinetics, we deal only with populations of states. In the analysis of spin exchange, along with the population of spin states, quantum spin coherence is also considered.

Note that in real systems in kinetic equations it is necessary to take into account the contribution of the spin-spin dipole-dipole interaction, the interaction of electrons with magnetic nuclei, spin-lattice interaction, etc. The contribution of the dipole-dipole interaction is discussed below in Chap. 3.

2.10 Calculations of the Effective Spin Exchange Radius Taking into Account the Extended Exchange Interaction

2.10.1 Paramagnetic Particles with Spin S = 1/2

The effective spin exchange radius between neutral particles with spins S = 1/2 was calculated in [65] under the assumption that the exchange integral decreases exponentially with increasing distance between the particles in the pair, and that the continuous diffusion model is applicable.

The spin Hamiltonian of two isolated particles is chosen in the form

$$\mathbf{H}_0 = \hbar\omega_A S_{Az} + \hbar\omega_B S_{Bz}$$

and their exchange interaction is given by the spin Hamiltonian

$$\mathbf{V} = \hbar J(r_0)\exp(-\kappa(r - r_0))\mathbf{S}_A\mathbf{S}_B.$$

The eigenvectors of $\mathbf{H}_0$, |k>, are an external product of the eigenvectors of $\mathbf{S}_{Az}$ and $\mathbf{S}_{Bz}$, $|m_A>$ and $|m_B>$, i.e., $|k> = |m_A> x\, |m_B> = |m_A, m_B>$. Denote the states of two electron spins which give the full and orthogonal basis of possible states as

$$|1>=|1/2, 1/2>;\ |2>=|1/2, -1/2>;\ |3>=|-1/2, 1/2>;\ |4>= |-1/2, -1/2>.$$

The eigenstates of **V** are the singlet $|S>=(|2>-|3>)/\sqrt{2}$ and the states of the triplet $|T_+>=|1>$, $|T_0>=(|2>+|3>)/\sqrt{2}$ and $|T_->=|4>$. In the S-T basis $\mathbf{H}_0$ has only one nonzero nondiagonal element

$$(\mathbf{H}_0)_{STO} = \hbar(\omega_A - \omega_B)/2 = \hbar\delta/2.$$

The singlet-triplet intersystem crossing due to the difference of the resonance frequencies of spins δ fails to reveal itself within the time of the diffuse passage of the interaction region by particles, if the conditions

$$\delta r_0{}^2/D < 1; \delta/\left(D\kappa^2\right) < 1,$$

are fulfilled, where κ is a characteristic of the exchange integral decay. Under these conditions, the effect of δ in the spin exchange process can be neglected. This means that during the collision the populations of the singlet and triplet states are preserved.

In the approximation $\delta = 0$ from (2.178 and 2.184) we find that the change of non-diagonal density matrix elements of A and B due to spin exchange is decribed by [65].

$$(\partial/\partial t)\rho^{A}{}_{1/2.-1/2} = -4\pi DC_{B}\int_{b}^{\infty} drr^{2}\left(\Delta T_{13,kl} + \Delta T_{24,kl}\right)\left(\rho^{A}\rho^{B}\right)_{kl},$$
$$(\partial/\partial t)\rho^{B}{}_{1/2.-1/2} = -4\pi DC_{A}\int_{b}^{\infty} drr^{2}\left(\Delta T_{12,kl} + \Delta T_{34,kl}\right)\left(\rho^{A}\rho^{B}\right)_{kl}. \qquad (2.189)$$

The matrix elements $T_{mn,kl}$ determining the exchange efficiency satisfy a system of coupled equations

$$D\Delta T_{13,13} = (i/2)J(r)\left(T_{13,13} - T_{12,13}\right),$$
$$D\Delta T_{12,13} = (i/2)J(r)\left(T_{12,13} - T_{13,13}\right). \qquad (2.190)$$

It is convenient to solve (2.190) in the S-T basis: in the approximation $\delta = 0$ we obtain uncoupled equations

$$D\Delta T_{pq,rs} = (i/\hbar)(V_{pp}(r) - V_{qq}(r))T_{pq,rs},$$

where $|p>, |q>, |r>, |s> = |S>, |T_{+1}>, |T_{0}>, |T_{-1}>$.

This equation under the boundary conditions (2.175) and (2.176) yields nonzero solutions only for the diagonal elements $T_{pq,pq}$. Linearizing the right-hand sides of (2.189) with the help of $\rho = \rho_0 + \sigma$ and assuming the high temperature limit case, $\rho_{0mm} \approx 1/2$ we arrive at

$$(\partial/\partial t)\sigma^{A}{}_{1/2.-1/2} = -K_{ex}C_{B}\left(\sigma^{A}{}_{1/2.-1/2} - \sigma^{B}{}_{1/2.-1/2}\right),$$
$$(\partial/\partial t)\sigma^{B}{}_{1/2.-1/2} = -K_{ex}C_{A}\left(\sigma^{B}{}_{1/2.-1/2} - \sigma^{A}{}_{1/2.-1/2}\right)$$

with the exchange rate constant

$$K_{ex} = 4\pi r_{ex}D.$$

Here the effective exchange radius is introduced

$$r_{ex} = (1/2)Re\int_{r_0}^{\infty} drr^{2}\Delta T_{T_{+1}S,T_{+1}S}. \qquad (2.191)$$

The value of $T_{T_{+}S,T_{+}S}$ is found from the equation

$$D\Delta T_{T_{+}S,T_{+}S} - iJ(r)\, T_{T_{+}S,T_{+}S} = 0.$$

Hence we obtain

$$T_{T_+S,T_+S} = (c_1/r)I_0(z) + (c_2/r)N_0(z), \tag{2.192}$$

where

$$z = z_0 exp(-\kappa(r - r_0)/2), z_0 = 2\left[-iJ(r_0)/(D\kappa^2)\right]^{1/2}.$$

In Eq. (2.192), I_0 and N_0 are the Bessel functions of the first and second order, respectively.

Substituting the solution (2.192) into (2.191) we obtain an expression for the effective spin exchange radius

$$r_{ex} = r_0/2 + Re[(C + ln(z_0/2))/\kappa - c_1/2], \tag{2.193}$$

where

$$z_0 = 2\left(-iJ(r_0)/(D\kappa^2)\right)^{1/2},$$
$$c_1 = (\pi/\kappa)[N_1(z_0)z_0\kappa\, r_0 - 2N_0(z_0)]/[I_1(z_0)z_0\kappa\, r_0 - 2\, I_0(z_0)],$$

C is the Euler constant, D is the coefficient of mutual diffusion of two colliding particles, J_0 is the exchange integral on the collision radius, and I_k and N_k are Bessel functions of the first and second kind, respectively.

As could be expected, in the limit of relatively small values of the exchange integral J_0 (the case of weak spin exchange) and very sharp decline of the exchange integral with increasing distance between particles, when the condition $|z_0|^2 = 4J(r_0)/(D\,\kappa^2) < < 1$ is satisfied, the effective radius of the weak spin exchange can be represented in the same form as in the approximation of the sudden switching-on of the exchange interaction

$$r_{ex} \approx (1/2)\, r_0 J(r_0)^2 \tau^2, \tag{2.194}$$

where

$$\tau = r_0/(D\kappa). \tag{2.195}$$

Equation (2.194) in form coincides with (2.63) for weak exchange ($J\tau_c < 1$) in the approximation of sudden switching on of exchange interaction. But the rigorous theory on the basis of the above kinetic equations gives an explicit expression for the average effective time of the "pair state" with the exchange interaction switched on during one meeting of the particles. The comparison of (2.195) with (2.65) shows that the time τ (2.195) is the ensemble average time that the selected pair of particles (representing balls) spends in a spherical layer with an inner radius r_0 and a layer thickness of $1/\kappa$ (a layer in which the exchange integral decreases in *e* times). This total time includes the first collision and all re-encounters of a pair of particles in a layer between two spheres with radii r_0 and $r_0 + 1/\kappa$. Note that in this case of weak exchange in the diffusion model it is possible to introduce the efficiency of spin exchange at one meeting of particles in the same way as in the model of sudden collisions

$$p_{ex} = (1/2)J_0{}^2\tau^2, \tau = r_0/(D\kappa). \quad (2.196)$$

In the limit of strong exchange interaction, when 4 $J(r_0)/(D\ \kappa^2) \geq 1$, the spin exchange occurs efficiently not only at the closest approach of particles, but also at distances $r \gg r_0$. In this case from (2.193) we obtain

$$r_{ex} = r_0/2 + (1/(2\kappa))ln(|J(r_0)/(D\kappa^2)|). \quad (2.197)$$

It can be seen that due to the contribution to the spin exchange of those pairs in which the partners did not come closer to the collision radius, the effective spin exchange radius can be greater than r_0.

Note that with an increase in the exchange integral or a decrease in the mutual diffusion coefficient, r_{ex} (2.193) does not reach the limit value, but continues to grow, the logarithmic dependence of the spin exchange radius on the value of the exchange integral at the collision radius and on the mutual diffusion coefficient is manifested, as well as the dependence on the steepness of the decline of the exchange integral with an increase in the distance between the particles in the colliding pair.

In the situation of the slow exchange integral decline, when $\kappa r_0 \ll 1$, the spin exchange represents the integral interaction effect when particles pass through a relatively large area with a radius $r* \sim 1/\kappa$. As a result, as might be expected, the effective spin exchange radius practically ceases to be dependent on the distance of the closest approach of the particles.

For illustration, Fig. 2.6 shows the dependences of the r_{ex} (2.191) and K_{ex} (2.190) on the coefficient of mutual diffusion of particles. These figures confirm the above statements about the possible role of the extended nature of the exchange interaction in the value of the effective radius, and hence the spin exchange rate constant.

The results presented in Fig. 2.6, clearly illustrate that due to the extended nature of the exchange integral for the typical values of the particle parameters, the effective spin exchange radius may be greater than the maximum possible value ($r_0/2 = 3\ 10^{-8}$ cm in the above calculations), which predicts for these systems the theory based on the model of sudden collisions (2.63 and 2.64).

Figure 2.7 shows the dependence of the effective radius and the rate constant of the equivalent spin exchange on the value of the exchange integral at the collision radius r_0 for two values of the mutual diffusion coefficient of particles.

In the weak exchange region, the effective spin exchange radius with the growth of the exchange integral increases proportionally to $J(r_0)^2$ and tends to "saturation" with the value of $r_{ex} \approx (8/15)r_0$. These observations are very close to the results of the theory of spin exchange within the model of sudden collisions. Indeed, in the model

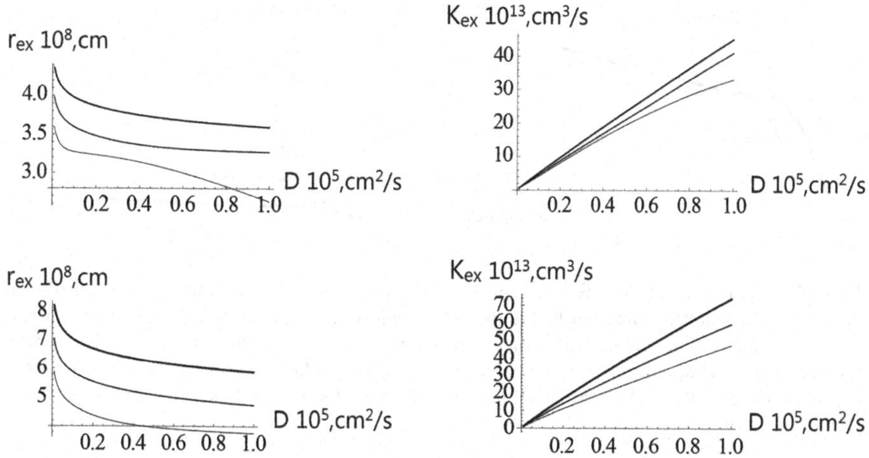

Fig. 2.6 Dependence of the effective radius (left-hand column) and the spin exchange rate constant (right-hand column) on the coefficient of mutual diffusion of particles. The curves calculated for two values of the steepness of the decline of the exchange integral and three values of the exchange integral at the collision radius, $r_0 = 6\ 10^{-8}$ cm. Curves for the slope $\kappa = 3\ 10^8\ \text{cm}^{-1}$ are shown in the top row, for a smaller curvature $\kappa = 10^8\ \text{cm}^{-1}$ are shown in the bottom row. The curves calculated at $J(r_0) = 10^{11}, 10^{12}, 10^{13}$ rad/s are drawn with thin, medium thickness and thick lines, respectively. The coefficient of mutual diffusion is given in units of $10^5\ \text{cm}^2/\text{s}$

of sudden collisions $r_{ex} = (1/2)r_0J(r_0)^2\tau^2/(1 + J(r_0)^2\tau^2)$ and it tends to saturation value $r_{ex} = (1/2)r_0$. The principal difference occurs in the area of strong exchange interaction. Within the model of sudden collisions with growth of $J(r_0)$, the effective spin exchange radius reaches the limit $r_{ex} = (1/2)r_0$ and ceases to depend on the value of the exchange integral. In the model of diffusion passage by spins of the extended exchange interaction region, the effective radius grows linearly in the logarithm of the exchange integral module $J(r_0)$. This means that the diffusion collision rate constant can be under estimated if it is found from the spin exchange rate in the limit of strong exchange using the theory of spin exchange based on the model of sudden collisions.

In the above calculations, the parameters that can be expected for real systems were used. The mutual diffusion coefficient in Fig. 2.6 was varied in the interval $\{10^{-7}, 10^{-5}\ \text{cm}^2/\text{s}\}$. This corresponds to the viscosity range of solvents 1–100 cP, i.e. corresponds to the transition from water to such viscous solvents as, for example, squalane. The exchange integral at the van der Waals radius of paramagnetic particles is expected to be in the range $\{10^{11}, 10^{13}$ rad/s$\}$ [1, 2]. The steepness of the decreasing of exchange integral for atoms is about $3\ 10^8\ \text{cm}^{-1}$ [14]. So in Figs. 2.6 and 2.7 calculations for this value of the steepness parameter of the decreasing exchange integral are given. Calculations are also given for a smaller value of the steepness $10^8\ \text{cm}^{-1}$ of the decrease in the exchange integral with the increase in the distance between the paramagnetic particles. This is possible, in

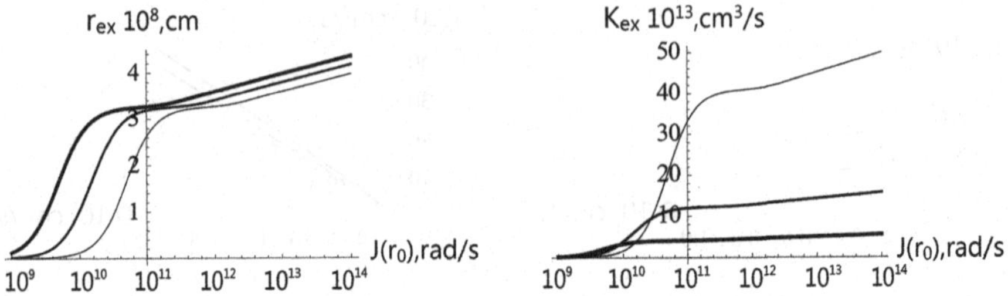

Fig. 2.7 Dependence of the effective radius and rate constant of the equivalent spin exchange on the value of the exchange integral $J(r_0)$ at the closest approach of spins. The calculations are carried out for the values of the mutual diffusion coefficient $D = 10^{-5}$ cm²/s (thin lines), $D = 310^{-6}$ cm²/s (average line thickness) and $D = 10^{-6}$ cm²/s (thick lines). The steepness of the decline of the exchange integral $\kappa = 310^8$ cm^{-1}. In Fig. 2.7, abscissa for the exchange integral J_0 uses the logarithmic scale

principle, due to the effect of delocalization of the spin density of paramagnetic particles on their peripheral atoms.

Figures 2.6 and 2.7 illustrate that the effective spin exchange radius decreases with the growth of the coefficient of mutual diffusion of colliding particles. This is due to the fact that the increase in the diffusion coefficient reduces the total meeting time of the two particles in the solution. Dependence of the spin exchange rate constant on the coefficient of mutual diffusion of particles, as is expected, is different in cases of the weak and strong exchange interaction. In the case of strong exchange (see Figs. 2.6 and 2.7 for $J(r_0)$ in interval 10^{11}–10^{14} rad/s) the spin exchange rate constant increases when the diffusion coefficient increases. In this case the dependence of the spin exchange rate constant K_{ex} on D deviates from the linear one (see Fig. 2.6), as is the case for the bimolecular collision rate constant, $K_D \sim D$ (2.4). But in case of weak exchange (see Fig. 2.7 for $J(r_0) < 10^{10}$ rad/s) the spin exchange rate constant decreases when the diffusion coefficient increases. Thus in this case D dependence of Kex is opposite to D dependence of the diffusion collisions rate constant (2.4). In the case of weak spin exchange, $Kex \sim J^2\tau^2 D \sim 1/D$, since $\tau \sim 1/D$. Here τ is a duration of a meeting of two paramagnetic particles in the exchange interaction region.

2.10.2 *Comparison of Results for Models of Sudden Collisions and Diffusion Passage of the Exchange Interaction Region. Possible Modification of the Model of Sudden Collisions*

In the case of weak equivalent spin exchange, when the exchange integral $J(r_0)$ is not very large and $J(r)$ decreases sharply with increasing distance between spins, the result of the meeting of two particles can be calculated from the perturbation theory.

In this case, the probability of the mutual flip-flop of spins is proportional to the square of the product of the matrix element of the transition caused by the exchange interaction at the time of the meeting. In this situation, there is a complete correspondence between the results of both models: the effective exchange radius can be presented as a product of the distance of the closest approach of particles by the exchange efficiency at one meeting, $r_{ex} = r_0 p_{ex}$, and $p_{ex} \sim J^2 t^2$. The difference between the two models is only in the duration of the elementary act of spin exchange in these models.

In general, these two models are fundamentally different. In contrast to the model of sudden collisions within the diffusion model, the interaction makes a significant contribution to spin exchange not only at the closest approach of particles, but also at large distances between them. This makes the concept of collision efficiency probability meaningless, since the effective radius of the spin can become greater than the distance of the closest approach r_0, and the probability of the spin exchange p_{ex} can become greater than 1 if it is formally introduced as $p_{ex} = r_{ex}/r_0$.

The main difference between the models under discussion is that the model of sudden collisions takes into account the exchange only at the closest approach of paramagnetic particles, while the diffusion model takes into account the exchange at different distances. But spin exchange at different distances between particles can be taken into account within the model of sudden collisions, if it is appropriately modified.

Suppose that the diffusion of molecules occurs by relatively long jumps. In such a situation, the sudden switching on of an extended exchange interaction can occur not only at the smallest distance r_0 between the particles. With this possibility in mind, the following model could be considered. Sudden collisions of particles occur at different distances $r \geq r_0$, and the exchange interaction is switched on, which is given by the exchange integral $J(r) = J_0 \exp(-\kappa(r - r_0))$. The statistical weight of sudden collisions in the layer between spheres with radii r and r + dr can be estimated as the ratio of the volume of this layer to the total volume of ν_{ex}, in which the exchange interaction can be effectively manifested. For simplicity, the duration of the collision will be considered the same regardless of the distance at which the exchange interaction is suddenly switched on. Then the effective spin exchange radius can be found as

$$r_{ex} = (1/\nu_{ex}) \int_{r_0}^{\infty} r 4\pi r^2 dr (1/2) J^2(r)\tau^2 / (1 + J^2(r)\tau^2) \tag{2.198}$$

where τ is the duration of a "collision". In the case of weak exchange, when even with the closest approach of particles at the distance r_0, the condition $J_0\tau < 1$ is satisfied, from (2.198) we have in the case $r_0\kappa > 1$

$$r_{ex} = (1/\nu_{ex}) \int_{r_0}^{\infty} r 4\pi r^2 dr (1/2) J^2(r)\tau^2 \approx (1/2)\,(1/\kappa) J_0{}^2\tau^2. \tag{2.199}$$

In fact, Eq. (2.199) reproduces the result of the model of sudden collisions. The situation changes in the case of a strong exchange. Assume that $J_0\tau > 1$. Let us define the maximum distance r_m, where the condition of strong exchange $J(r_m)\tau \geq 1$ is satisfied. Hence, in the layer between the spheres with radii r_0 and r_m, the collision efficiency is ½. Neglecting the contribution of collisions with $r > r_m$, from (2.198) we obtain an estimate of the effective radius in the strong exchange region

$$r_{ex} \approx (1/2)(r_0 + (1/\kappa)ln(|J_0|\tau). \tag{2.200}$$

This result qualitatively coincides with the result of the theory of spin exchange, when the diffusion passage of an extended area of exchange interaction (cf. (2.200) and (2.197)) is taken into account in a consistent manner.

2.10.3 Paramagnetic Particles with Arbitrary Spins

Effective radii of bimolecular spin exchange for particles with arbitrary spins are calculated using the general kinetic equations (Sect. 2.9) in [67]. The spin Hamiltonian of the exchange interaction is described by (2.26). Diffusion of molecules is considered within the continuous diffusion model. For particles with spins $S > 1/2$, the so-called zero-field splitting energy can play an important role. In [67], it is assumed that in liquids with a sufficiently low viscosity and fast rotational diffusion of molecules, this zero-field splitting can be neglected, which is averaged to zero due to the rotational diffusion of particles. Moreover, it is assumed that in the collision region of particles the exchange interaction is large enough so that all other spin interactions can be neglected. Within the above assumptions, analytical formulas for effective spin exchange radii between particles with arbitrary spins are obtained in [67].

In [67], the following kinetic equations are obtained for the transverse magnetization components of the spins A and B

$$\begin{aligned} \partial M_A - /\partial t &= -i\omega_{0A} M_{A-} - K_A C_B M_{A-} + K_B C_A M_{B-}, \\ \partial M_{B-}/\partial t &= -i\omega_{0B} M_{B-} + K_A C_B M_{A-} - K_B C_A M_{B-}. \end{aligned} \tag{2.201}$$

These equations have the form similar to (2.188). But in contrast to the situation for spin exchange between free radicals with spin $S = 1/2$ (2.188) in the case of arbitrary spins of the colliding particles (2.201), equivalent spin exchange is not defined by a single rate constant, Kex, (see 2.188), but two spin exchange rate constants K_A and K_B operate in (2.201). In the theory of spin exchange within the model of sudden collisions we have already seen that equivalent spin exchange between particles with spin 1/2 and particles with arbitrary spin is also described by two different rate constants (see 2.137). The rate constants K_A and K_B in Eq. (2.201) are

$$
\begin{aligned}
K_A &= 4\pi D_{AB} \sum_{S=mod(SA-SB)+1}^{SA+SB} \frac{N(S)}{SA(SA+1)} rex(S) \equiv 4\pi R_{exA} D_{AB}, \\
K_B &= 4\pi D_{AB} \sum_{S=mod(SA-SB)+1}^{SA+SB} \frac{N(S)}{SB(SB+1)} rex(S) \equiv 4\pi R_{exB} D_{AB},
\end{aligned}
\tag{2.202}
$$

$$
\begin{aligned}
N(S) &= \frac{(S^2-(S^A-S^B)^2)((S^A+S^B+1)^2-S^2)}{S(2S^A+1)(2S^B+1)};\\
r_{ex}(S) &= (1/2)\{r_0 + (1/\kappa)[\ln(|J_0|S/(D_{AB}\kappa^2)) + 2C - C_1']\};\\
C_1' &= \pi \mathrm{Re}[N_1(2x_0)x_0\kappa r_0 - 2N_0(2x_0)]/[I_1(2x_0)x_0\kappa r_0 - 2I_0(2x_0)];\\
x_0 &= \sqrt{(|J_0|S/(D_{AB}\kappa^2))}^{-1/2}\exp(i\pi/4), C \approx 0.57721566.
\end{aligned}
$$

In these equations, I_k and N_k are functions of Bessel and Neumann of the k-th order, respectively, and summation of S is carried out in increments of 1. Equations (2.202) show that the kinetic constants of the spin exchange K_A and K_B relate to each other as the values of the squares of the spins of the partners

$$
K_A/K_B = S_B(S_B+1)/S_A(S_A+1). \tag{2.203}
$$

This coincides with the result obtained for the model of sudden collisions obtained for the collisions of paramagnetic particles with spin S = 1/2 with particles with arbitrary spins (2.139). In the case of a collision of particles with the same spin values ½ or 1, the expressions (2.202) reproduce the result [65, 73].

Note again that the above results (2.202) refer to the situation where the influence of all spin interactions except the exchange interaction can be neglected in the course of bimolecular collisions. For such an equivalent spin exchange case, we give explicit expressions for the spin exchange rate constants for some systems. For spin exchange between radicals ($S_A = 1/2$) and triplets ($S_B = 1$) we have

$$
\begin{aligned}
K_A &= (128/27)\pi D_{AB} r_{ex}(3/2),\\
K_B &= (16/9)\pi D_{AB} r_{ex}(3/2).
\end{aligned}
\tag{2.204}
$$

For spin exchange between triplets we have

$$
K_A = K_B = (4/9)\pi D_{AB}(4\, r_{ex}(1) + 5\, r_{ex}(2)). \tag{2.205}
$$

For spin exchange between particles with spin $S_A = 1/2$ and arbitrary spin S_B we obtain

$$
\begin{aligned}
K_A &= (16/3)\pi D_{AB}\left[S_B(S_B+1)/(S_B+1/2)^2\right] r_{ex}(S_B+1/2),\\
K_B &= 8\pi D_{AB}\frac{1}{(SB+1/2)(2SB+1)} r_{ex}(S_B+1/2).
\end{aligned}
\tag{2.206}
$$

The obtained rate constants for spin exchange can be written in the form

$$
K_A = 4\,\pi\,D_{AB}\,R_{exA}, K_B = 4\,\pi\,D_{AB}\,R_{exB}, \tag{2.207}
$$

where R_{exA}, R_{exB} are effective spin exchange radii.

The effective spin exchange radius depends both on the parameters of the exchange interaction J_0 and κ and on the D_{AB}, coefficient of mutual diffusion of colliding particles, and the radius r_0 of their closest possible approach. Depending on these parameters, the effective spin exchange radius can vary widely. In contrast to the results of the theory based on the model of sudden collisions, in the theory, which takes into account the extended nature of the exchange interaction, the effective spin exchange radius can be much greater than $r_0/2$. At $(|J_0|S/(D_{AB}\kappa^2))^{1/2} >> 1$, the value C_1'--> 0 (2.202), and the effective radius depend on the exchange integral logarithmically

$$
r_{ex}(S) \approx (1/2)\{r_0 + (1/\kappa)[\ln(|J_0|S/(D_{AB}\kappa^2)) + 2C]\}. \tag{2.208}
$$

In this case, the characteristic meeting time of the particles is determined by the time of diffusion passage of the region with a radius of ~$1/\kappa$. Since the exchange integral decreases exponentially in the model under consideration, for large values of J_0 and a sufficiently small steepness of its decrease with the distance, a logarithmic dependence of the effective spin exchange radius on the interaction intensity and diffusion coefficient should be expected. If the exchange integral decreases rapidly, that is, $r_0 > 1/\kappa$, the spin exchange takes place in a thin layer with a thickness of about $1/\kappa$, the characteristic time of the meeting is $\tau_c \approx r_0/(D_{AB}\kappa)$, (see (2.195)). If $(|J_0|S/(D_{AB}\kappa^2))^{1/2} < 1$ efficiency of spin exchange is proportional to $J_0^2 r_0^2/(D_{AB}^2\kappa^2)$.

To illustrate the theoretical results presented above, in Figs. 2.8 and 2.9 dependences of the effective spin exchange radius on the parameters of the exchange integral and the coefficient of mutual diffusion of molecules are given.

These curves allow us to make several observations. As the coefficient of mutual diffusion of paramagnetic particles increases, the effective spin exchange radius always decreases, as expected. Indeed, the greater the mobility of molecules in solutions, the less time a pair of colliding particles spends in the exchange interaction region. As a result, the effect of exchange interaction also decreases. The effective spin exchange radius increases with the growth of the exchange integral on the collision radius, J_0 and with the decrease of the slope of the decline of the exchange integral with the increase in the distance between the colliding particles. Qualitatively, these are the expected results. The theory allows to find quantitatively these dependences.

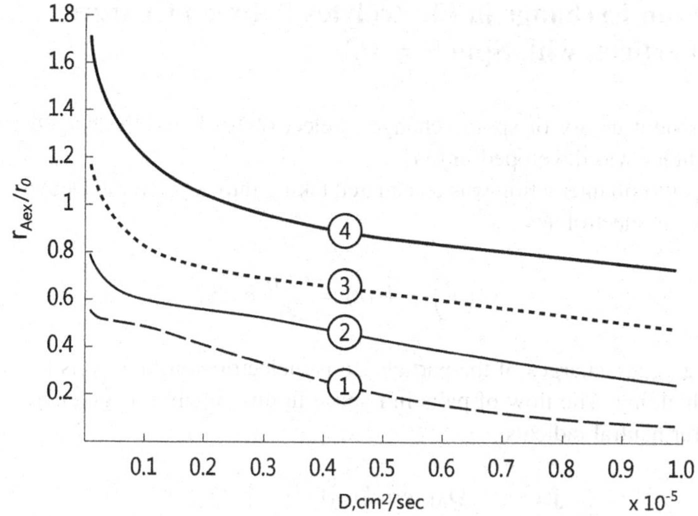

Fig. 2.8 Dependence of r_{Aex}/r_O (see Eqs. 2.202 and 2.206) on the mutual diffusion coefficient D_{AB}. Different curves refer to different values of the slope κ of the decline of the exchange integral, $\kappa = 2*10^8$ 1/cm (curve 1), $\kappa = 1*10^8$ 1/cm (curve 2), $\kappa = 0.6*10^8$ 1/cm (curve 3), $\kappa = 0.4*10^8$ 1/cm see (curve 4). Calculations are carried out for the case of $S_A = 1/2$, $S_B = 1$, $r_O = 0.5$ nm, D_{AB} varies in the interval (10^{-6}, 10^{-5} cm^2/s), $J_O = 10^{10}$ rad/s. (Adapted with permission from [67])

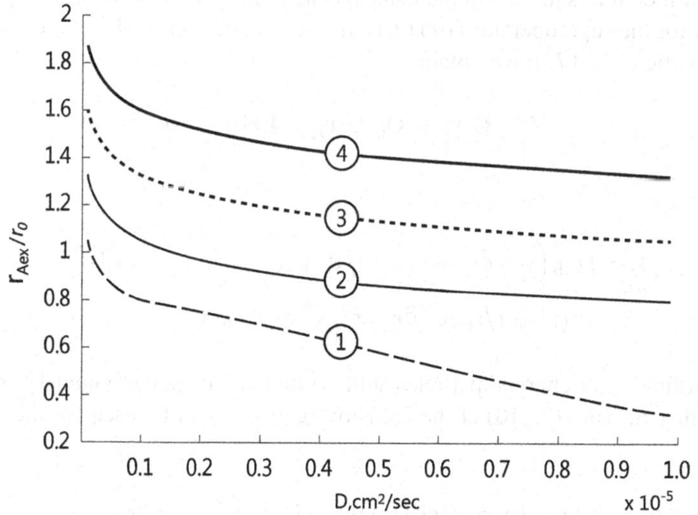

Fig. 2.9 Dependence of r_{Aex}/r_O (see Eqs. 2.202 and 2.206) on the mutual diffusion coefficient D_{AB}. Different curves refer to different values of the exchange integral J_O: $J_O = 10^{10}$ rad/s (curve 1), $J_O = 10^{11}$ rad/s (curve 2), $J_O = 10^{12}$ rad/s (curve 3), $J_O = 10^{13}$ rad/s (curve 4). The calculations are performed for the case of $S_A = 1/2$, $S_B = 1$, $r_O = 0.5$ nm, D_{AB} varies in the interval (10^{-6}, 10^{-5} cm^2/s), $\kappa = 10^8$ 1/cm. (Adapted with permission from [67])

2.11 Spin Exchange in Electrolytes Between Charged Particles with Spin S = 1/2

The consistent theory of spin exchange in electrolytes between charged paramagnetic particles was developed in [34].

The Coulomb interaction was considered taking into account the Debye shielding of charges in electrolytes

$$U\left(\vec{r}\right) = U(r) = \frac{q_A q_B}{\varepsilon r} e^{-r/r_D} \tag{2.209}$$

Here $q_{A,B}$ are charges of the particles, ε is dielectric constant, r_D is the radius of Debye shielding. The flow of pairs in r space in this situation is given as ((cf. with (2.167) for neutral radicals)

$$\mathbf{j}(\mathbf{r}) = -D_{AB}\, grad_r(\rho(t|r)) + \rho(t|r)\boldsymbol{\nu}(r), \tag{2.210}$$

where $\boldsymbol{\nu}(r) = \mu\mathbf{F}(r) = -(D_{AB}/kT)\,\mathrm{grad}\,(U(r))$.

For this situation, kinetic equations for spin density matrices were formulated using the reasonings detailed in Sect. (2.2). The only difference is that due to the Coulomb interaction, the flux of pair density matrices should also include the drift of charged particles in the local electric field. Therefore, in the continuity Eq. (2.166), the expression (2.210) should be used as the flux for charged particles instead of (2.167), which refers to neutral molecules. Due to the change in the flux of pairs, the equation for the superoperator **G**(r) (2.173) also changes. Instead of the equation for neutral particles (2.173), we obtain

$$\mathbf{W}(r)\mathbf{G}(r) + [\mathbf{Q}_0, G(r)] + \mathbf{L}\mathbf{G}(r) = 0, \tag{2.211}$$

where

$$\begin{aligned}\mathbf{L} = D_{AB}\left(\partial^2/\partial r^2 + (2/r)\partial/\partial r\right) - D_{AB}\left(q_A q_B/(\varepsilon k T r^2)\right)\\ \times\left((1 + r/r_D)\partial/\partial r - r/{r_D}^2\right)exp(-r/r_D).\end{aligned}$$

Accordingly, for charged particles at the collision radius, the boundary condition of zero flux of pairs (2.210) at the collision radius should be used for the operator **G**(r)

$$\partial G(r_0)/\partial r = \left(q_A q_B/(\varepsilon k T r^2)\right)(1 + r_0/r_D)exp(-r_0/r_D). \tag{2.212}$$

Using the above equations there were numerically calculated [34] the effective spin exchange radius for charged particles with spin ½ for different values of the exchange interaction parameters (exchange integral at the collision radius J_0 and the slope of the decay of the exchange integral κ, coefficient of the mutual diffusion of paramagnetic particles D_{AB} and electrostatic interaction parameters (dielectric permittivity of the solution ε, Debye screening radius r_D of the Coulomb interaction).

Figure 2.10 illustrates the dependence of the effective radius on the value J_0 of the exchange interaction at the collision radius. The analysis of curves given in Fig. 2.10a–c shows that the effective spin exchange radius increases with the decrease in the coefficient of the mutual diffusion of paramagnetic particles. This is explained by the fact that with the decrease in the mobility of molecules (the decrease in the diffusion coefficient) the time of the stay of two colliding particles in the region of the exchange interaction increases. The effective spin exchange radius for oppositely charged particles is larger than that for neutral particles, and the effective spin exchange radius for particles of like charge is less than that for neutral particles (see Fig. 2.10). These are expected results qualitatively. In the case of neutral particles the effective spin exchange radius r_{eff} comes to the linear dependence on $\ln|J_0|$ with the increase in the exchange integral J_0 on the collision radius.

The numerical calculations show that similar to neutral radicals (see 2.208) also for charged particles the effective spin exchange radius r_{eff} comes to the linear dependence on $\ln|J_0|$ in the strong exchange region (see Fig. 2.10). This observation can be interpreted as follows. If the effective spin exchange radius is larger than a half of the closest approach radius, the situation of the strong spin exchange takes place. Let us assume that the situation of the strong spin exchange is realized, when the action of the exchange interaction, i.e., the product of the exchange integral $J(r_{eff})$ by the time τ_{eff} of the stay of the particles in the interaction region is on the order of 1 (note that the exchange integral is given in the units of rad/s). From the condition, $J_0 \exp(-kr_{eff})\tau_{eff} \approx 1$, one obtains the estimate of the effective radius of the strong spin exchange $r_{eff} \approx (1/\kappa) \ln (J_0\tau_{eff})$. In fact, τ_{eff} in turn depends on r_{eff}, but this dependence is not exponential, it could be, e.g., quadratic $\tau_{eff} \approx r_{eff}^2/D_{AB}$ or linear $\tau_{eff} \approx r_{eff}/\kappa D_{AB}$.

The effective time of the stay of particles in the effective exchange interaction region depends on the slope of the exchange integral decay with the increase in the distance between particles. The faster the decay of the exchange integral, the less is the effective interaction region, and as a result, the less is the effective spin exchange radius. This qualitative expectation is completely confirmed by the numerical calculations (see Fig. 2.11).

Thus, the spin exchange radius for charged particles is depending on parameters κ and D_{AB} qualitatively similar to the spin exchange between neutral particles (compare Figs. 2.10 and 2.11 with Figs. 2.8 and 2.9).

In the case of the spin exchange between charged particles, the Debye screening radius is an important parameter. With the decrease in the Debye screening radius, the role of the Coulomb interaction should decrease and the effective spin exchange radius should tend to the spin exchange radius for neutral particles. Numerical calculations completely confirm these qualitative statements [34]. Figure 2.12a, c illustrate that at $r_D = 1.6\ 10^{-8}$ cm (when the Debye screening radius is much less

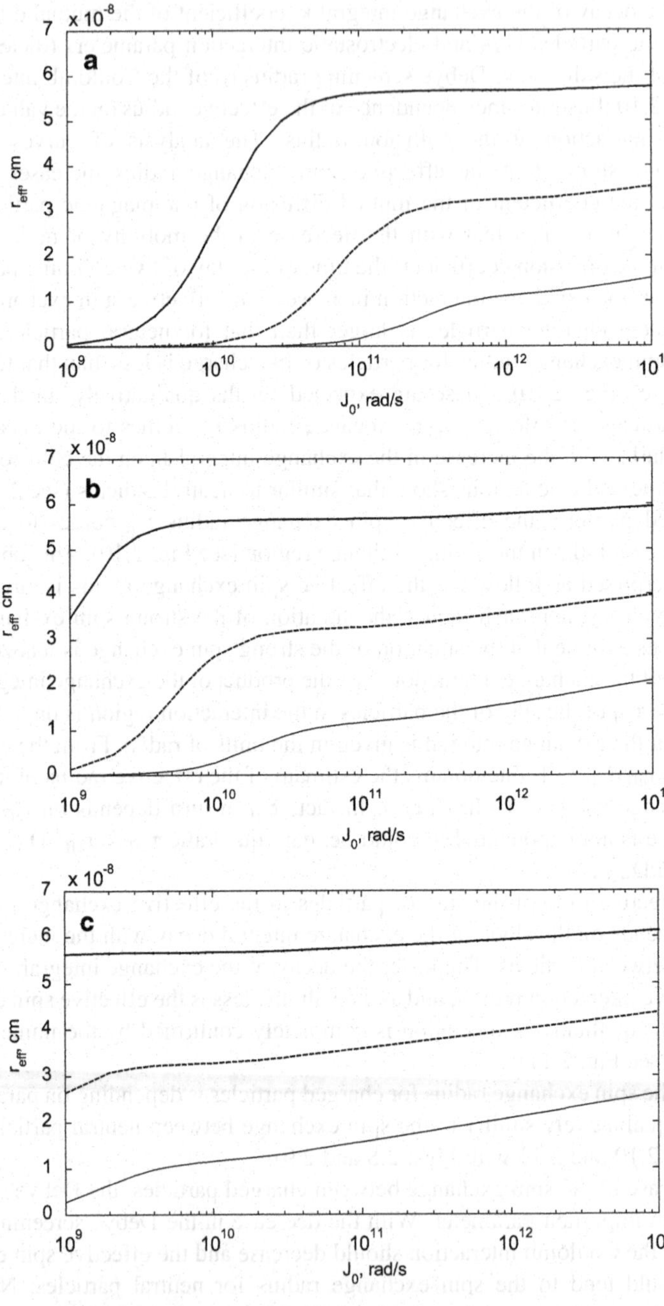

Fig. 2.10 Dependence of the effective spin exchange radius on the exchange integral at the collision radius for different values of the diffusion coefficient of paramagnetic particles for particles of like charge (thin solid curve), for neutral particles (thick dotted curve) and for oppositely charged particles (thick solid curve). Calculations were performed for the following parameter

than the collision radius) the effective spin exchange radius between charged particles almost coincides with the radius of the spin exchange for neutral particles. In another example, when $r_D = 10^{-7}$ cm, the Coulomb interaction quite strongly affects the spin exchange radius in comparison with the case of neutral particles (see Fig. 2.12b, d); for oppositely charged particles the effective spin exchange radius becomes to be larger than the collision radius $r_0 = 6\ 10^{-8}$ cm, which was selected for these calculations.

The numerical calculations presented show that at rather high values of the coefficient of mutual diffusion of particles and rather large slope of the exchange integral decay, the spin exchange is no longer strong even for collisions of oppositely charged particles, which can be kept in the interaction region by the Coulomb attraction for the relatively long time.

Tables 2.7 and 2.8 summarize the calculated values of the effective spin exchange radius for charged paramagnetic particles for a series of selected values of parameters, which affect the value of this radius [34].

Data given in these tables may be useful as references during the interpretation of experimental data. Data presented in these Tables and Figs. 2.10, 2.11 and 2.12 give an idea about the scale of the possible variations of the effective spin exchange radius during the variation of the molecular-kinetic parameters, and also parameters, which characterize the exchange and electrostatic interaction between charged paramagnetic particles. Spin exchange between charged particles was comprehensively studied experimentally, e.g., in [34, 73].

The theory presented was successfully applied for analyzing the results obtained when studying experimentally the spin exchange between charged stable radicals in [34]. Transformations of the EPR spectrum for water solutions of the 3-carboxy-2,2,5,5-tetramethyl-1-pyrrolidinyloxy (3-Carboxy-Proxyl, $C_9H_{16}NO_3$) (3CP) radical have been studied experimentally as a function of the concentration of paramagnetic particles, temperature and ionic strength of the solution [34]. The concentration broadening of the EPR spectrum lines has been found from experimental data. It has been shown that in the studied system the spin exchange gives the main contribution to the concentration broadening of the EPR spectrum lines.

In work [34] the charged particles are produced by changing the acidity of the water solution during the addition of NaOH or HCl. The concentration of ion-radicals was less than 10 mM/L. Experiments were performed in systems with the ratio $C_{NaOH}/C_{3CP} = 2.5$ and $C_{HCl}/C_{3CP} = 5$, the ionic strength of the solution was controlled by adding NaCl [34]. Samples were studied in the temperature range of 283–328 K with a step of 5 K. The viscosity of the medium in the experiment varied

Fig. 2.10 (continued) values: $r_0 = 6\ 10^{-8}$ cm,, ε=20, $r_D = 10\ 10^{-8}$ cm, a) $D_{AB} = 1.38\ 10^{-5}$ cm²/s, b) $D_{AB} = 10^{-6}$ cm²/s, c) $D_{AB} = 10^{-7}$ cm²/s. Abscissa for the exchange integral J_0 uses the logarithmic scale to present the variation of the effective spin exchange radius in the wide interval of J_0 {10^9 rad/s, 10^{13} rad/s}. (Adapted with permission from [34])

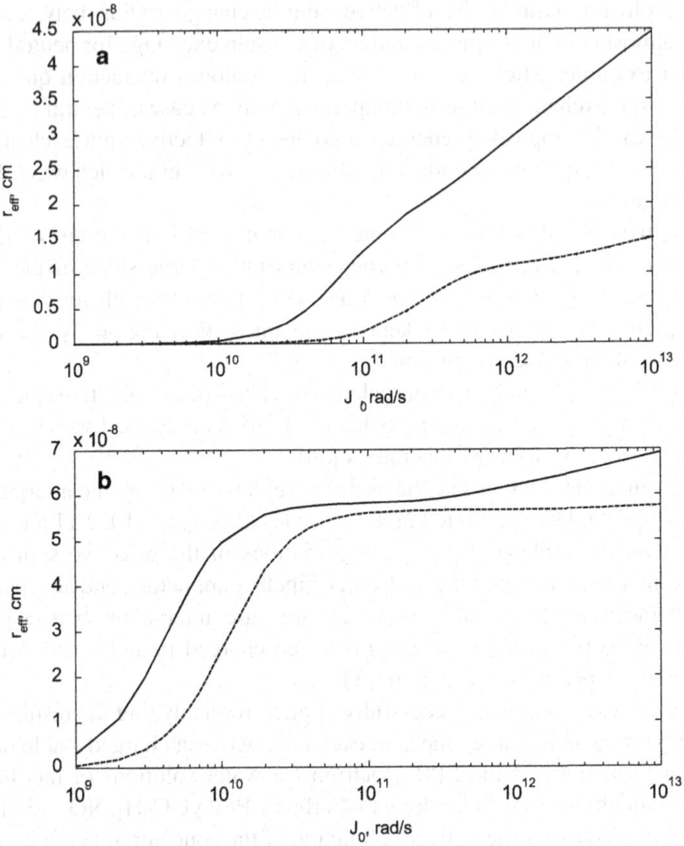

Fig. 2.11 Dependence of the effective radius of spin exchange between paramagnetic particles of like charged (**a**) and oppositely charged paramagnetic particles (**b**) for two values of the slope of the exchange integral decay with the increase in the distance between colliding paramagnetic particles: $\kappa = 10^8$ 1/cm (solid curves), $\kappa = 3\ 10^8$ 1/cm (dotted curve). Calculations were performed for the following parameter values: $r_0 = 6\ 10^{-8}$ cm, $\varepsilon=20$, $r_D = 10\ 10^{-8}$ cm, $D_{AB} = 1.38\ 10^{-5}$ cm^2/s. (Adapted with permission from [34])

from 0.4 to 1.3 cP. The radius of the closest approach of studied radicals was $r_0 = 0.7$ nm. The ionic strength is determined by the quantity

$$\sum C_i Z_i^2 = 2\left(C_{cp^-} + C_{NaCl} + C_{NaOH} + C_{HCl}\right) = 2C_{total},$$

in which C_k, Z_k are the concentration and charge number of the k-th ion. In the studied system all ions are singly charged, $Z_k = 1$.The Debye screening radius r_D is determined by equation $r_D = \{\varepsilon kT/(8\pi q^2 C_{total}\}^{-1/2}$. Here ε is the dielectric permittivity of the solution, q is the electron charge, κ is the Boltzmann constant.

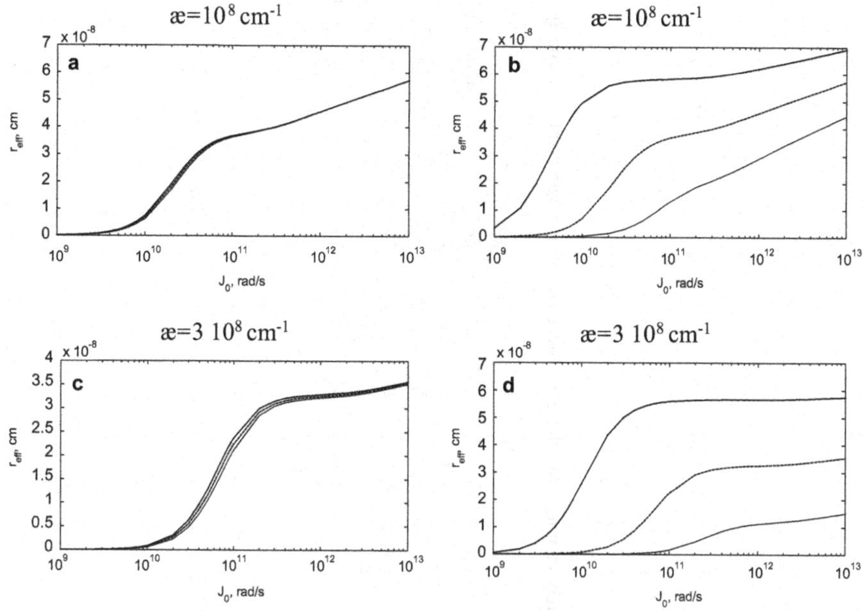

Fig. 2.12 Comparison of effective spin exchange radii for charged and neutral particles at different values of the Debye screening radius: (**a**, **c**) $r_D = 1.6\ 10^{-8}$ cm, (**b**, **d**) $r_D = 10\ 10^{-8}$ cm. Other calculation parameters are: $r_0 = 6\ 10^{-8}$ cm,, $\varepsilon = 20$, $D_{AB} = 10^{-5}$ cm^2/s. Abscissa for the exchange integral uses the log scale to present the variation of the effective spin exchange radius in the wide interval of J_0 {10^9 rad/s – 10^{13} rad/s}. Thin solid curves present the data for particles of like charge, thick dotted curves present data for neutral particles and thick solid curves are data for oppositely charged particles. (Adapted with permission from [34])

By analyzing concentration dependence of the EPR spectrum shape the spin exchange rate constant was determined. The radical was studied in the charged ($3CP^-$) and neutral (3CP) states. In the presence of NaOH the COOH group of the radical dissociates into COO^- and H^+, and the radical becomes charged. For all radicals in the solution to remain charged, it is necessary to fulfill the condition $\frac{C_{NaOH}}{C_{3CP}} = 2.5$.The maintenance of a certain value of the ionic strength of the solution is achieved by the adjustment of the NaCl content. To keep the radical neutral in the solution with NaCl, it is necessary to add counter ions HCl. Therefore, in this case it is necessary to fulfill the condition $\frac{C_{HCl}}{C_{3CP}} = 5$.

The spin exchange rate for charged radicals appears to be less than that for neutral ones. This should be expected since for the studied systems the Debye screening radius is larger than the radius of the closest approach of radicals and therefore the Coulomb repulsion of radicals of like charge decreases the collision frequency. In the studied systems with the ionic strength of 130 and 75 the Debye screening radius is 0.8 nm and 1.08 nm, respectively. Note that these Debye screening radii are larger

Table 2.7 Values of the effective spin exchange radius in Angstrem units between oppositely charged particles for a selected set of parameters of exchange and electrostatic interaction, and diffusion coefficient

J_0, rad/s		10^{10}				10^{11}				10^{12}				10^{13}			
κ, cm^{-1}		10^8		$3*10^8$		10^8		$3*10^8$		10^8		$3*10^8$		10^8		$3*10^8$	
D, cm^2/s		10^{-5}	10^{-6}	10^{-5}	10^{-6}	10^{-5}	10^{-6}	10^{-5}	10^{-6}	10^{-5}	10^{-6}	10^{-5}	10^{-6}	10^{-5}	10^{-6}	10^{-5}	10^{-6}
$\varepsilon = 20$	$r_D = 2A$	1.34	3.81	0.18	2.86	3.81	4.74	2.86	3.36	4.74	5.88	3.36	3.65	5.88	7.03	3.65	4.02
	$r_D = 10A$	5.33	5.84	3.5	5.63	5.84	6.30	5.63	5.67	6.30	7.04	5.67	5.78	7.04	7.89	5.78	5.94
$\varepsilon = 81$	$r_D = 2A$	1.88	3.78	0.14	2.69	3.78	4.73	2.69	3.3	4.73	5.88	3.3	3.61	5.88	7.03	3.61	3.99
	$r_D = 10A$	2.01	4.30	0.34	3.51	4.30	5.13	3.51	3.89	5.13	6.17	3.89	4.15	6.17	7.25	4.15	4.48

The distance of the closest approach $r_0 = 0.6$ nm [34]. (Adapted with permission from [34])

Table 2.8 Values of the effective spin exchange radius in Angstrem units between particles of like charge for a selected set of parameters of exchange and electrostatic interaction, and diffusion coefficient

J_0, rad/s		10^{10}				10^{11}				10^{12}				10^{13}			
κ, cm^{-1}		10^8		$3*10^8$		10^8		$3*10^8$		10^8		$3*10^8$		10^8		$3*10^8$	
D, cm^2/s		10^{-5}	10^{-6}	10^{-5}	10^{-6}	10^{-5}	10^{-6}	10^{-5}	10^{-6}	10^{-5}	10^{-6}	10^{-5}	10^{-6}	10^{-5}	10^{-6}	10^{-5}	10^{-6}
ε =20	$r_D = 2A$	0.95	3.72	0.09	2.39	3.72	4.72	2.39	3.18	4.72	5.88	3.18	3.53	5.88	7.03	3.53	3.94
	$r_D = 10A$	0.07	1.6	0.004	0.27	1.6	3.12	0.27	1.15	3.12	4.67	1.15	1.57	4.67	6.13	1.57	2.08
ε =81	$r_D = 2A$	1.08	3.75	0.12	2.57	3.75	4.73	2.58	3.25	4.73	5.88	3.25	3.58	5.88	7.03	3.58	3.97
	$r_D = 10A$	0.59	3.22	0.05	1.77	3.22	4.33	1.77	2.67	4.33	5.58	2.67	3.04	5.58	6.81	3.04	3.48

The distance of the closest approach $r_0 = 0.6$ nm (Adapted with permission from [34])

than the closest approach radius $r_0 = 0.7$ nm. It was found in [34] that for the system studied $r_{eff} = 2.4\ 10^{-8}$ cm for the neutral radical and $r_{eff} = 1.7\ 10^{-8}$ cm for the charged radical for both values of the ion force. The diffusion coefficient was calculated according to the Stokes formula. Fitting the exchange interaction parameters, it is possible to achieve that the effective radii of the spin exchange calculated theoretically are in satisfactory agreement with those obtained from the analysis of the concentration dependence of the EPR spectrum shape. Using parameters $r_0 = 0.7$ nm, $T = 298$ K, viscosity η=0.92 cP, $D = 1.38 \cdot 10^{-5}$ cm^2/s, $J_0 = 5.65 \cdot 10^{10}$ rad/s, æ $= 2.21 \cdot 10^8$ cm^{-1}, ε=74 (for water), $r_D = 0.8$ nm (for the ionic strengthof 130 mmole/L) and $r_D = 1.08$ nm (for ionic strength of 75 mmole/L), the calculations gave the following results: $r_{eff} = 2.46\ 10^{-8}$ cm for the neutral radical and $r_{eff} = 1.65\ 10^{-8}$ cm (at $r_D = 0.8$ nm), $r_{eff} = 1.54\ 10^{-8}$ cm (when $r_D = 1.08$ nm) for the charged radical. The Debye screening radius was calculated according to the formula $r_D = (\varepsilon kT/\Sigma(4\pi C_i q^2 Z_i^2))^{1/2}$, where q is the electric charge, Z_i is the charge number, C_i is the concentration of ions, T is temperature.

We note that in the studied experimental situation for $3CP^-$(3550) the Debye screening radius was smaller ($r_D = 0.16$ nm) than the collision radius, so that in this system the Coulomb interaction almost does not affect the efficiency of the spin exchange. However in the situation $3CP^-$(130), $3CP^-$(75), $3CP^-$(60) the Debye screening radius is larger than the collision radius of particles and the Coulomb interaction noticeably affects the manifestations of the exchange interaction and spin exchange in the EPR observed parameters.

In the systems studied experimentally in [34] the collision radius is 0.7 nm. Within the model of sudden collisions in the limit of strong collisions the effective spin exchange radius would reach the maximum value of 0.35 nm. The experimental values of the effective spin exchange radii were less than this value. Therefore in the studied systems the situation of the weak spin exchange is implemented. The theoretical calculations show that the experimental values of the effective spin exchange radius can be described using the exchange integral at the collision radius of about 6 10^{10} rad/s and κ=3 10^8 cm^{-1}.

Within the proposed theory the effective spin exchange radius of charged particles is determined by the parameters of the exchange interaction, $\{J_0, \kappa\}$, parameters of the electrostatic interaction, {dielectric permittivity ε, Debye screening radius r_D}, the collision radius r_0 and the coefficient of mutual diffusion of particles D_{AB}. The effective spin exchange radius is a function $r_{eff}(J_0,k,\varepsilon,r_D,r_0, D_{AB})$, which can be calculated, if values of all parameters are known. The spin exchange rate $Z_{ex} = 4\pi r_{eff} D_{AB} C_B$ is measured in EPR experiments. Thus, the spin exchange rate measured in EPR experiments accumulates information about the interactions between paramagnetic particles, kinematics of the mutual diffusion of colliding pairs of paramagnetic particles and the local concentration C_B of paramagnetic particles.

Let us assume that there is given a system, for which all parameters, which determine Z_{ex}, except for the exchange interaction parameters are known. In this case one can determine the effective spin exchange radius from the EPR data, and then to select such values of the exchange integral parameters, which lead to the coincidence of the calculated spin exchange radius with the value for this radius found from data of EPR spectroscopy. As a result, from data on the spin exchange in solutions one obtains information about the overlap of orbitals of unpaired electrons of two paramagnetic particles colliding in the solution, since in the acceptable approximation the exchange integral is proportional to the square of the overlap integral of these orbitals [3]. Data about the overlap of electron orbitals obtained in this manner are useful in the analysis of the kinetics of the electron transfer during the binary collisions in the solution [1, 2].

2.12 Approximate Estimates of the Effective Spin Exchange Radius for Charged Particles

It is of interest to compare the results of the exact calculations with approximate estimates. In the model of sudden switching of the exchange interaction the effective spin exchange radius between charged paramagnetic particles is conventionally estimated as (see Eqs. 2.6 and 2.62)

$$r_{eff} = f_D\, p_{ex}\, r_0 \tag{2.213}$$

In this expression f_D is given by Eq. (2.6), and the efficiency of the spin exchange p_{ex} is given by Eq. (2.62), in which the time τ is used as the effective collision time. In this estimate extended character of the exchange interaction is ignored, and it is considered that the exchange interaction is switched on suddenly only in a narrow layer between spheres with radii r_0 and $r_0 + \delta$. There was studied in [34] to what extent the considered approximation is applicable for calculation of the effective spin exchange radius between charged paramagnetic particles. Figure 2.13 shows the results of calculations according to formula (2.213) and calculations using the consistent theory of spin exchange for particles of like charge and oppositely charged particles developed above.

It follows from Fig. 2.13 that the approximation $r_{eff} = f_D\, p_{ex}\, r_0$ (Eq. (2.213) gives the satisfactory value of the effective spin exchange radius between oppositely charged paramagnetic particles in nonviscous solution. In the case of opposite charges of colliding partners, the approximate results deviate noticeably in viscous solutions with low diffusion coefficients and/or strong exchange integral (compare Fig. 2.13a, b, c). However, for particles of like charge it is seen (Fig. 2.13) that the approximation $r_{eff} = f_D\, p_{ex}\, r_0$ considerably underestimates the effective spin exchange radius. The last statement can be interpreted in the following way. When particles have like charges the Coulomb repulsion is strongly reducing the collisions with closest approach, $f_D \ll 1$ (see Eq. (2.6)). In this situation the exchange interaction at distances between particles more than the collision radius r_0 give a major contribution to the spin exchange efficiency. This contribution is taken into account in the calculations using the consistent theory

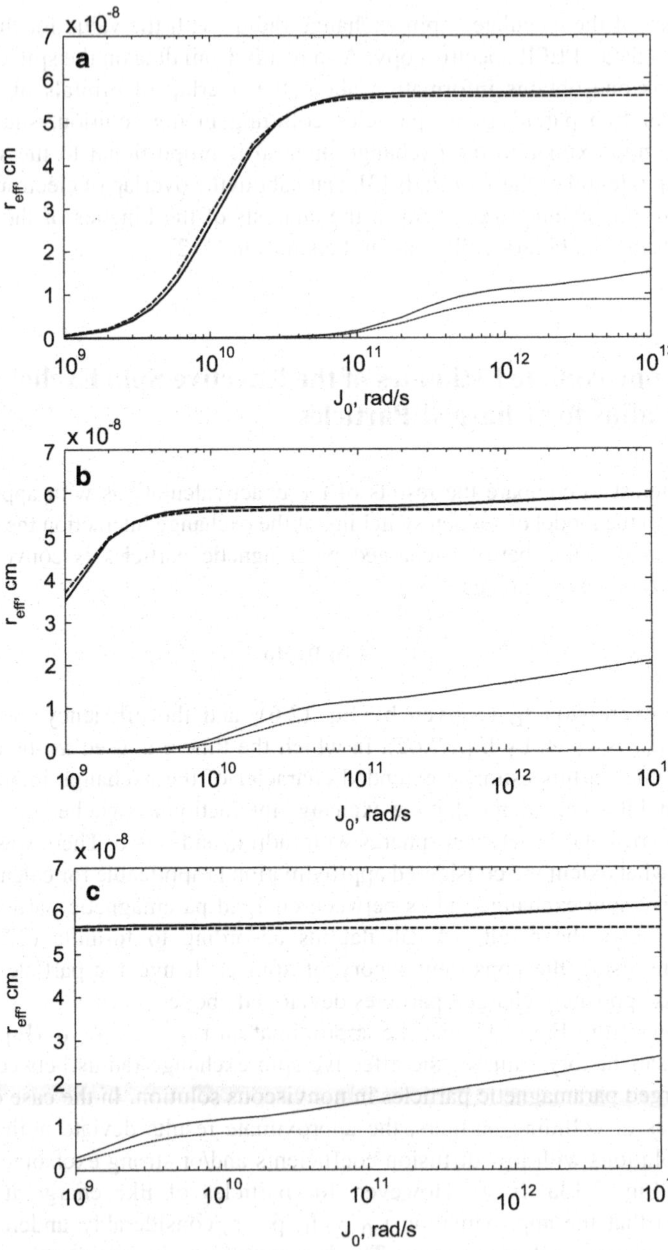

Fig. 2.13 Comparison of the effective spin exchange radius calculated for oppositely charged particles (thick curves) and particles of like charge (thin curves) within the theory described in previous section (solid curves) and in the approximation $r_{eff} = f_D\, p_{ex}\, r_0$ (dotted curves). Calculations were performed for the following parameter values: $r_0 = 6\ 10^{-8}$ cm,, ε=20, $r_D = 10^{-7}$ cm, κ=3 10^8 1/cm; D_{AB} = 1.38 10^{-5} cm^2/s (**a**), $D_{AB} = 10^{-6}$ cm^2/s (**b**), $D_{AB} = 10^{-7}$ cm^2/s (**c**). (Adapted with permission from [34])

developed in [34] and presented above, but it is not taken into account when the approximation $r_{eff} = f_D\, p_{ex}\, r_0$ is used.

Thus the discrepancy between the effective spin exchange radii calculated within the consistent theory and within the sudden collision approximation arises due to the extended character of the exchange interaction.

References

1. Zamaraev, K.I., Molin, Y.N., Salikhov, K.M.: Spin exchange. Nauka, Sibirian branch, Novosibirsk (1977)
2. Molin, Y.N., Salikhov, K.M., Zamaraev, K.I.: Spin exchange. Principles and applications in chemistry and biology. Springer, Heidelberg/Berlin (1980)
3. McWeeny, R., Sutcliffe, B.T.: Methods of molecular quantum mechanics. Academic, New York (1969). Chaps. 3, 4
4. Rabinovich, E., Wood, W.C.: The collision mechanism and the primary photochemical process in solutions. Trans. Faraday Soc. **32**, 1381–1387 (1936)
5. Adrian, F.J.: Role of diffusion-controlled reaction in chemically induced nuclear spin polarization. J. Chem. Phys. **53**, 3374 (1970)
6. Salikhov, K.M., Molin, Y.N., Sagdeev, R.Z., Buchachenko, A.L.: Spin polarization and magnetic effects in radical reactions. Elseivir/AkademikKiado, Amsterdam/Budapest (1984)
7. Salikhov, K.M.: Magnetic isotope effect in radical reactions. Springer, Wien/New York (1996)
8. Smoluchowski, M.V.: VersucheinerMathematischenTheorie der Koagulations Kinetic KolloiderLousungen. Z. Phys. Chem. **92**, 129–168 (1917)
9. Debye, P.: Reaction rates in ionic solutions. Trans. Electrochem. Soc. **82**, 265–272 (1942)
10. Heitler, W., London, F.: Wechselwirkung neutraler Atome und homopolare Bindung nach der Quantenmechanik. Zeitschrift für Physik. **44**, 455–472 (1927)
11. Dacre, P.D., McWeeny, R.: The interaction of two electronic systems at short and medium range. Proc. Roy. Soc. A. **317**, 435–454 (1970)
12. Gorkov, L.P., Pitaevsky, L.P.: The energy splitting of terms of the hydrogen molecule, DokladyAkademiiNauk. USSR. **151**, 822–825 (1963)
13. Smirnov, B.M.: Asimptotic methods in theory of collisions of atoms. Atomizdat, Moscow (1973)
14. Hirschfelder, J.O., Curtiss, C.F., Bird, R.B.: Molecular theory of gases and liquids. Wiley, New York (1954)
15. Schastnev, P.V., Salikhov, K.M.: Spin polarization and exchange interaction of multielectronic systems. Teor. Exper. Khim. **9**, 291–299 (1973)
16. Musin, R.N., Schastnev, P.V.: Calculation of exchange interactions of atoms in the pair orbital approximation. I, Zhur. Strukt. Khim. **17**(411–418) (1977)
17. Owen, J.: Spin resonance of ion pairs in crystal lattices. J. Appl. Phys. Suppl. **33**, 355–357 (1962)
18. Carrington, A., McLachlan, A.D.: Introduction to magnetic resonance with applications to chemistry and chemical physics. Harper and Row, New York (1967). Chap. 6
19. Neese, F.: Prediction of molecular properties and molecular spectroscopy with density functional theory: from fundamental theory to exchange-couling. Coord. Chem. Rev. **253**, 526–563 (2009)
20. Ernst, R., Bodenhausen, G., Wokaun, A.: Principles of nuclear magnetic resonance in one and two dimensions. Clarendon Press, London (1987)
21. Schweiger, A., Jeschke, G.: Principles of pulse electron paramagnetic resonance. Oxford University Press, Oxford (2001)
22. Kivelson, D.: Theory of ESR linewidths of free radicals. J. Chem. Phys. **33**, 1094 (1960)

23. Abragam, A.: The principles of nuclear magnetism. Clarendon Press, Oxford (1961)
24. Currin, J.D.: Theory of exchangerelaxation of hyperfine structure in electron spin resonance. Phys. Rev. **126**, 1995–2001 (1962)
25. Freed, J.H.: On Heisenberg spin exchange in liquids. J. Chem. Phys. **45**, 3452–3453 (1966)
26. Bales, B.L.: Inhomogeneously broadened spin-label spectra. In: Berliner, L.J., Reuben, J. (eds.) Biological magnetic resonance, vol. 8, pp. 77–130. Plenum Publishing Corporation, New York (1989)
27. Salikhov, K.M.: The contribution from exchange interaction to line shifts in ESR spectra of paramagnetic particles in solutions. J. Magn. Res. **63**, 271–279 (1985)
28. Salikhov, K.M.: Contributions of exchange and dipole–dipole interactions to the shape of EPR spectra of free radicals in diluted solutions. Appl. Magn. Reson. **38**, 237–256 (2010)
29. Salikhov, K.M.: Consistent paradigm of the spectra decomposition into independent resonance lines. Appl. Magn. Reson. **47**, 1207–1228 (2016)
30. Salikhov, K.M.: Peculiar features of the spectrum saturation effect when the spectral diffusion operates: system with two frequencies. Appl. Magn. Reson. **49**, 1417–1430 (2018)
31. Salikhov, K.M., Bakirov, M.M., Galeev, R.T.: Detailed analysis of manifestations of the spin coherence transfer in EPR spectra of ^{14}N nitroxide free radicals in non-viscous liquids. Appl. Magn. Reson. **47**, 1095–1122 (2016)
32. Bales, B.L., Bakirov, M.M., Galeev, R.T., Kirilyuk, I.A., Kokorin, A.I., Salikhov, K.M.: The current state of measuring bimolecular spin exchange rates by the epr spectral manifestations of the exchange and dipole-dipole interactions in dilute solutions of nitroxide free radicals with proton hyperfine structure. Appl. Magn. Reson. **48**, 1399–1447 (2017)
33. Salikhov, K.M., Khairuzhdinov, I.T.: Theoretical investigation of the effect of saturation of the EPR spectrum taking into account spectral diffusion in a system with a Gaussian distribution of the resonance frequencies of spins. ZhETP. **155**, 806–823 (2019)
34. Salikhov, K.M., Mambetov, A.E., MBakirov, M., Galeev, R.T., Khairuzhdinov, I.T., Zaripov, R.B., Bales, B.: Spin exchange between charged paramagnetic particles in dilute solutions. Appl. Magn. Reson. **45**, 911–940 (2014)
35. Lynden-Bell, R.-M.: The line shapes of ESR spectrum of a system of interacting triplets. Mol. Phys. **8**, 71–80 (1964)
36. Salikhov, K.M., Nikitaev, A.T., Senyukova, G.A., Zamaraev, K.I.: Exchange broadening of EPR lines in dilute solutions containing two varieties of paramagnetic particles. Teoret. Experim. Khim. **7**, 619–624 (1971)
37. Berkovich, M.L., Salikhov, K.M.: Effect of rotational diffusion on exchange broadening of EPR spectra. Teoret. Experim. Khim. **9**, 586–591 (1973)
38. Salikhov, K.M., Doktorov, A.B., Molin, Y.N.: Exchange broadening of ESR lines for solutions of free radicals and transition metal complexes. J. Magn. Reson. **5**, 189–205 (1971)
39. Rumyantsev, E.E., Salikhov, K.M.: Spin exchange between triplet excitons. Opt. Spectrosc. **35**, 570–573 (1973)
40. Salikhov, K.M.: Kinetics of processes caused by spin-spin interactions of particles magnetically dilute systems. Doctor of science theses, Kazan State University (1973)
41. Balling, L.C., Hanson, R.J., Pipkin, F.M.: Frequency shifts in spin exchange optical pumping experiments. Phys. Rev. **133**, A607–A626 (1964)
42. Balling, L.C., Pipkin, F.M.: Spin exchange in a cesium-electron system. Phys. Rev. **133**, A46–A53 (1964)
43. Bales, B.L., Peric, M.: EPR line shifts and line shapechangesdueto spin exchangeofnitroxide free radicals in liquids. J. Phys. Chem. B. **101**, 8707–8716 (1997)
44. Bales, B.L., Peric, M., Dragutan, I.: EPR line shifts and line shape changes due to spin exchange of nitroxide free radicals in liquids 2. Extension to high spin exchange frequencies and inhomogeneously broadened spectra. J. Phys. Chem. A. **106**, 4846–4854 (2002)
45. Bales, B.L., Peric, M., Dragutan, I.: EPR line shifts and line shapechangesdueto spin exchange of nitroxide free radicals in liquids 3. Extension to five hyperfinelines. Additional line shiftsduetore-encounters. J. Phys. Chem. A. **107**, 9086–9098 (2003)

46. Bales, B.L., Meyer, M., Smith, S., Peric, M.: EPR line shifts and line shape changes due to spin exchange of nitroxide free radicals in liquids 4. Test of a method to measurere-encounterrates in liquids employing^{15}N and ^{14}N nitroxide spin probes. J. Phys. Chem. A. **112**, 2177–2181 (2008)
47. Kurban, M.R., Peric, M., Bales, B.L.: Nitroxide spin exchange due to re-encounter collisions in a series of n-alkanes. J. Chem. Phys. **129**, 064501 (2008)
48. Stunzhas, P.A., Bendersky, V.A.: DEER for two frequencies spectrum considering spin exchange with sudden collisions model. Opt. spectroscopy. **30**, 1041–1046 (1971)
49. Hibma, T., Kommandeur, J.: Dynamics of triplet excitons in the simple tetracyanoquinodimethane (TCNQ) salts of rubidium, potassium, and tri-methyl-benzimidazol. Phys. Rev. **B12**, 2608–2618 (1975)
50. Skubnevskaya, G.I., Salikhov, K.M., Smirnova, L.M., Molin, Y.N.: Effect of electron spin relaxation on exchange broadening of EPR lines in dilute solutions. KinetikaiKataliz. **9**, 888–892 (1970)
51. Berdnikov, V.M., Makarshin, L.L., Doktorov, A.B., Kim, V.I., Kiprianov, A.A.: On the mechanism of the elementary act of spin exchange between ferricyanide ion and stable nitric radical 2,2,6,6-tetramethylpiperidone-N-oxyl in aqueous solution. Khim. Fisika. **1**, 70–78 (1982)
52. Turro, N.J.: Modern molecular photochemistry. University Science Books, Mill Valley (1991)
53. Berdnikov, V.M., Doktorov, A.B.: The steric factor in the elementary act in the liquid state. Theoret. Exper. Khim. **17**, 318–326 (1981)
54. Chestnut, D.B., Phillips, W.D.: EPR studies of spin correlation in some ion radical salts. J. Chem. Physics. **35**, 1002–1012 (1961)
55. McConnell, M.H., Pooley, D., Bradbury, A.: The paramagnetic resonance of wurster's blue perchlorate. Proc. Nat. Acad. Sci. USA. **48**, 1480–1482 (1962)
56. Jones, M.T., Chesnut, D.B.: Triplet spin exchange in some ion radical salts. J. Chem. Phys. **38**, 1311–1317 (1963)
57. Stunzhas, P.A., Bendersky, V.A., Blumenfeld, L.A., Sokolov, E.A.: Double electron –electron resonance of triplet excitons I. Opt. Spectr. **28**, 278–283 (1970)
58. Stunzhas, P.A., Bendersky, V.A., Sokolov, E.A.: Double electron –electron resonance of triplet excitons II. Opt. Spectrosc. **28**, 487–491 (1970)
59. Agostini, G., Corvaja, C., Giacometti, G., Pasimeni, L.: Optical, zero-field ODMR and EPR studies of the triplet-states from singlet fission in biphenyl-TCNQ and Biphenyl-Tetrafluoro-TCNQ charge-transfer crystals. Chem. Phys. **173**(2), 177–186 (1993)
60. Goldanski, V.I.: Physical chemistry of positron and positronium. Nauka, Moscow (1968)
61. Doktorov, A.B., Salikhov, K.M., Molin, Y.N., Zamaraev, K.I.: Exchange conversion of positronium in dilute solutions of paramagnetic complexes. Dokl. Akad. Nauk USSR. **205**, 1385–1388 (1972)
62. Goldanski, V.I., Zusman, R.I., Molin, Y.N., Shantarovich, V.P.: Interaction of positronium with Fe(3+) and Co(2+) ions in alcohol solutions. Dokl. Akad. Nauk USSR. **188**, 1079 (1969)
63. Berdnikov, V.M., Makarshin, L.L., Murza, L.I.: Exchange interaction between a stable nitroxide radical and the ferricyanide ion in aqueous solution. J. Struct. Khim. **19**, 319–321 (1978)
64. Vandenberg, A.D., Bales, B.L., Salikhov, K.M., Peric, M.: Bimolecular encounters and re-encounters (cage-effect) of a spin labeled analogue of cholestane in a series of n-Alkanes: effect of anisotropic exchange integral. J. Phys. Chem. A. **116**, 12460–12469 (2012)
65. Salikhov, K.M.: Diffusion theory of exchange broadening of EPR spectra of dilute solutions of paramagnetic particles. Teor. Eksperim. Khim. **10**(310–317) (1974)
66. Doktorov, A.B.: The impact approximation in the theory of bimolecular quasi-resonant processes. Physica. **90A**, 109 (1978)
67. Mambetov, A.E., Salikhov, K.M.: Calculations of spin exchange rate constants in inviscid liquids between paramagnetic particles with arbitrary spins for difffusion type of particle motion. ZhETP. **128**, 1013–1024 (2005)

68. Mambetov, A.E., Brenerman, M.H., Salikhov, K.M., Lapushkin, S.S.: Nonequilibrium polarization of electron spins of triplet excitons in a molecular crystal caused by their mutual annihilation. Matemat. Model. **20**, 44–54 (2008)
69. Bogolyubov, N.N.: Problems of dynamic theory in statistical physics. M-L, Fizmatgiz (1946)
70. Korst, N.N.: Semiclassical theory of spin relaxation in high viscosity media. Teoret. Matemat. Fiz. **6**, 265 (1971)
71. Salikhov, K.M., Sarvarov, F.S.: Theory of spin-dependent recombination of radicals in homogeneous solution. React Kinet Catal Lett. **4**, 33–41 (1975)
72. Salikhov, K.M., Sarvarov, F.S., Sagdeev, R.Z., Molin, Y.N.: Diffusion theory of recombination of radical pairs taking into account singlet-triplet transitions. Kinetika I Kataliz. **16**, 279–287 (1975)
73. Bales, B.L., Cadman, K.M., Peric, M., Schwartz, R.N., Peric, M.: Experimental method to measure the effect of charge on bimolecular collision rates in electrolyte solutions. J. Phys. Chem. A. **115**, 10903–10910 (2011)

Chapter 3
Paramagnetic Relaxation Caused by the Spin-Spin Dipole-Dipole Interaction of Paramagnetic Particles in a Liquid

Abstract The theory of paramagnetic relaxation of the electron spins of paramagnetic particles in solutions, caused by random modulation of the spin-spin dipole-dipole interaction by the mutual diffusion of particles, is presented. The theory of this process is well known in the literature. But it unreasonably uses the so-called secular approximation. As a result, the well-known theory does not take into account the transfer of spin coherence between spins with different resonant frequencies. Here are presented kinetic equations that take into account correctly the transfer of spin coherence caused by the dipole – dipole interaction.

The dynamics of unpaired electrons of paramagnetic particles in solutions is determined by spin-dependent interactions of individual particles, e.g., their interaction with external magnetic fields, spin-orbit coupling, hyperfine interaction of electrons with magnetic nuclei, and their exchange and dipole-dipole interaction with each other. In experiments at the maximum dilution allowed by the sensitivity of the experiment, it is possible to determine the magnetic resonance parameters of individual particles. In principle, in experiments at different concentrations of paramagnetic particles, it is possible to distinguish the contribution of the spin-spin interaction between particles to the spin dynamics. At the same time, the problem of the separation of the contributions of exchange and dipole-dipole interactions remains. This section presents a theory that describes the contribution of the dipole-dipole interaction in liquids to the spin dynamics of paramagnetic particles.

In contrast to the exchange interaction, the dipole-dipole interaction has a long-range character, and therefore, even in dilute solutions, the theoretical description of the contribution of the dipole-dipole interaction to the spin motion differs from the situation with the short-range exchange interaction. As a rule, the exact exchange integral values for the paramagnetic particles colliding in the solution and their dependence on the distance and mutual orientation of the particles are not known. It is necessary to calculate the exchange integrals or estimate their values using experimental EPR spectroscopy data [1, 2]. However, in many situations, the exchange interaction of particles in the dilute solutions can be considered taking into account only bimolecular collisions as described in the previous section. The situation with the dipole-dipole interaction is different. It is a long-range

K. M. Salikhov, *Fundamentals of Spin Exchange*,
https://doi.org/10.1007/978-3-030-26822-0_3

interaction and the model of molecular collisions is not appropriate when describing its contribution to the spin dynamics. A good message is that the spin Hamiltonian of the dipole-dipole interaction is known. At distances exceeding van der Waals radii, the approximation of point dipoles can be applied and calculations can be performed. The common thing between these interactions is that their contributions to the spin motion depend on the concentration of paramagnetic particles. In the experiment, the problem arises to separate these contributions to extract the contribution of bimolecular collisions, during which spin exchange occurs due to the exchange interaction.

Exchange and dipole-dipole interactions can cause mutual flip-flops of two spins. But there are important differences between these interactions. In contrast to the exchange interaction, the dipole-dipole interaction can induce not only mutual flip-flops of the spins of colliding particles but, e.g., it can cause a flip or flop of a spin of only one of the colliding particles, etc.

The following are the main results of the theory of paramagnetic relaxation in a non-viscous liquid caused by the dipole-dipole interaction of paramagnetic particles. In liquids, the dipole-dipole interaction between two paramagnetic particles changes randomly over time as a result of mutual diffusion of particles [3]. The mutual diffusion of particles can be characterized by the average value for a long time period t of the correlation function of the radius vector r(t) connecting two particles, $j(\tau) = < r(t)\, r(t + \tau)>$.

The theory of paramagnetic relaxation caused by the dipole-dipole interaction of spins in dilute solutions is repeatedly presented in the literature. However, the widely accepted theory is based on the unjustified approximation when the non-secular spin coherence transfer terms are omitted in the derivation of kinetic equations for the spin density matrix in this situation (see, e.g., [3], Eq. (VIII.36); [4], Eq. (2.19)). As a result, no terms are left in the kinetic equations that describe the spin coherence transfer between spins with different resonance frequencies. A fundamentally important drawback of the theory was the neglect of the reaction in "course" of the dipole-dipole interaction. It looks as if, in classical mechanics we have neglected Newton's third law.

I presented the correct equations in [5], pp. 197–209 and see also [6, 7]. These correct equations remained unnoticed for a long time. But the effect of this "reaction" was experimentally detected and studied in [8, 9]. The current state of the theory of paramagnetic relaxation in solutions caused by the dipole-dipole interaction [5–7] is summarized below.

3.1 Spin Hamiltonian of the Dipole-Dipole Interaction of Paramagnetic Particles

Usually, paramagnetic relaxation is studied by EPR spectroscopy, i.e., under conditions when the spins are placed in an external constant magnetic field. Therefore, in addition to the interaction between paramagnetic particles, we include the Zeeman energy of the interaction of electron spins with the external magnetic field. The direction of the external magnetic field is chosen along the z axis of the laboratory coordinate system. We consider a system whose spin Hamiltonian can be written as

$$\mathbf{H} = \hbar\Sigma\omega_k\mathbf{S}_{kz} + \hbar\Sigma\mathbf{H}_{d-d}(k, n), \tag{3.1}$$

where ω_k is the resonance frequency of paramagnetic particles. Note that the difference in frequencies can be associated not only with different g-factor values of paramagnetic particles, but also with the difference in local magnetic fields due to the hyperfine interaction with magnetic nuclei or spin-dependent interactions of individual paramagnetic particles.

The first term in (3.1) describes the Zeeman interaction of spins with the external and local magnetic fields. The second term describes the spin-spin dipole-dipole interaction between all possible pairs of particles. In the considered problem of paramagnetic relaxation in a liquid, the contribution of the interaction of all pairs to the relaxation of any chosen spin is additive. Therefore, we can consider relaxation for a pair of spins and then add the contributions of all pairs.

For two randomly selected spins, we assume that the spin Hamiltonian is (see [3])

$$\begin{aligned} \mathbf{H} &= \hbar(\mathbf{H}_0 + \mathbf{H}_{d-d}), \\ \mathbf{H}_0 &= \omega_1\mathbf{S}_{1z} + \omega_2\mathbf{S}_{2z}, \\ \mathbf{H}_{d-d} &= \Sigma F^q(t)\mathbf{A}^q, \end{aligned} \tag{3.2}$$

where the following notations are introduced

$$\begin{aligned} &\mathbf{A}^0 = \alpha(-(2/3)\mathbf{S}_{1z}\mathbf{S}_{2z} + (1/6)(\mathbf{S}_{1+}\mathbf{S}_{2-} + \mathbf{S}_{1-}\mathbf{S}_{2+})); \\ &\mathbf{A}^1 = \alpha(\mathbf{S}_{1z}\mathbf{S}_{2+} + \mathbf{S}_{1+}\mathbf{S}_{2z}); \; \mathbf{A}^{-1} = \alpha(\mathbf{S}_{1z}\mathbf{S}_{2-} + \mathbf{S}_{1-}\mathbf{S}_{2z}); \\ &\mathbf{A}^2 = \alpha(\mathbf{S}_{1+}\mathbf{S}_{2+})/2; \; \mathbf{A}^{-2} = \alpha(\mathbf{S}_{1-}\mathbf{S}_{2-})/2; \end{aligned} \tag{3.3}$$

where

$$\begin{aligned} &\alpha = -(3/2)\gamma_1\gamma_2\hbar; \\ &F^{(0)}(t) = (1 - 3\cos^2\theta)/r^3; \\ &F^{(+1)}(t) = \sin\theta\cos\theta\exp(-i\varphi)/r^3; \\ &F^{(-1)}(t) = \sin\theta\cos\theta\exp(i\varphi)/r^3; \\ &F^{(2)}(t) = \sin^2\theta\exp(-i2\varphi)/r^3; \\ &F^{(-2)}(t) = \sin^2\theta\exp(i2\varphi)/r^3. \end{aligned}$$

In these expressions, r is the distance between the spins in the pair, θ is the angle between the radius vector **r** connecting the spins of the pair and the direction of the external magnetic field (axis Z). As a result of translational diffusion, the relative position of spins in space changes. Therefore, Fq(t) changes randomly over time.

3.2 Kinetic Equations for the Description of Electron Paramagnetic Relaxation in Liquids Caused by the Dipole-Dipole Interaction Between Paramagnetic Particles with Arbitrary Spins

In the situation under discussion, the equation of motion for the density matrix of the selected spin pair becomes a differential equation with coefficients that are random processes:

$$\partial\rho/\partial t = (-i/\hbar)[H(t), \rho]. \tag{3.4}$$

Solving this equation up to quadratic terms on the dipole-dipole interaction, averaging over all possible trajectories of the relative motion of two spins, we obtain the following equation for the elements of the two-spin density matrix [3, 5].

$$\partial\rho_{\alpha\alpha'}/\partial t = -(i/\hbar)[H_0, \rho]_{\alpha\alpha'} + \Sigma R_{\alpha\alpha',\beta\beta'}\rho_{\beta\beta'}. \tag{3.5}$$

The widely used unjustified approximation is that in the right-hand side of Eq. (3.5), the sum over $\beta\beta'$ is limited only to states with degenerate energies of transitions $\hbar\beta$ and $\hbar\beta'$ satisfying the condition $\beta\text{-}\beta'=\alpha\text{-}\alpha'$. This approximation means neglecting of a set of non-secular terms which describe the "reaction" effect, that is the spin coherence transfer caused by the dipole-dipole interaction.

In Eq. (3.5), α, β – eigenstates of spin Hamiltonian $\mathbf{H}_0$ of the pair of spins under consideration, and kinetic coefficients $R_{\alpha\alpha',\beta\beta'}$ are expressed by relations

$$R_{\alpha\alpha',\beta\beta'} = \\ (1/2)\left[J_{\alpha\beta,\alpha'\beta'}(\alpha'-\beta') + J_{\alpha\beta,\alpha'\beta'}(\alpha-\beta) - \delta_{\alpha'\beta'}\Sigma J_{\gamma\beta\gamma\alpha}(\gamma-\beta) - \delta_{\alpha\beta}\Sigma J_{\gamma\alpha'\gamma\beta'}(\gamma-\beta')\right]; \tag{3.6}$$

$$J_{\alpha\alpha',\beta\beta'}(\omega) = \int_{-\infty}^{\infty} G(\tau)exp(-i\omega\tau)d\tau,$$

$$G(\tau)_{\alpha\alpha',\beta\beta'} = \ll \alpha \mid H_{dd}(t) \mid \alpha' >< \beta' \mid H_{dd}(t+\tau) \mid \beta \gg t.$$

In the right-hand side of the last equation the sign $<..>_t$ means averaging over all possible trajectories of a random walk of a pair of particles.

Thus, the kinetic parameters $R_{\alpha\alpha',\beta\beta'}$ (3.6) are expressed in terms of the spectral densities $J_{\alpha\alpha',\beta\beta'}(\omega)$ of the correlation functions of the randomly changing Hamiltonian of the dipole-dipole interaction. For a stationary process of the random thermal motion of paramagnetic particles, the correlation functions of the dipole-dipole interaction should not depend on the choice of the time moment t but should depend only on the difference between the two time moments.

The probability that the relative position of this pair of spins at the initial time is given by the radius vector in the interval ($\mathbf{r}_0$,$\mathbf{r}_0 + d\mathbf{r}_0$) is $(1/V)d^3\mathbf{r}_0$ (V is the volume of the solution). The conditional probability that during the time τ the radius vector of the relative position of the spins as a result of diffusion falls into the volume element $d^3\mathbf{r}$ with the radius vector $\mathbf{r}$ is denoted by $p(\mathbf{r}_0|\mathbf{r},\tau)$. The averaging procedure $<\ldots>t$ in Eq. (3.6) for all possible random walks of spins can be written using the conditional probability $p(\mathbf{r}_0|\mathbf{r},\tau)$ as follows

$$G(\tau)_{\alpha\alpha',\beta\beta'} = (1/V)\iint d^3\mathbf{r}_0 d^3\mathbf{r} p(\mathbf{r}_0|\mathbf{r},\tau) < \alpha \mid H_{dd}(t) \mid \alpha' >< \beta' \mid H_{dd}(t+\tau) \mid \beta > . \tag{3.7}$$

From Eq. (3.5) we can find the contribution of one of the spins in the solution to the change in the density matrix of a selected spin. The one-particle spin S_1 density matrix, $\rho(1)$, is obtained from the pair density matrix $\rho^{(2)}(1,2)$ by convolution over the states of spins S_2

$$\rho(1) = Tr_2\rho^{(2)}(1,2). \tag{3.8}$$

The one-particle spin density matrix for the S_2 is determined in the same way

$$\rho(2) = Tr_1\rho^{(2)}(1,2). \tag{3.9}$$

In order to find the equation for the one-particle density matrix, it is necessary to convolute (3.5) by the spin states of the interaction partner. To this end, we present the states of the pair α, β as a product of the eigenfunctions of the operators S_{1z} (functions $|m_1>$) and S_{2z} (functions $| m_2>$)

$$| \alpha >=| m_1m_2 > ; | \alpha' >=| m'_1m'_2 > ; | \beta >=| m''_1m''_2 > ; | \beta' >=| m'''_1 m'''_2 > .$$

Convoluting Eq. (3.5) over S_2 states, we obtain

$$\begin{aligned}\partial\rho(1)_{m1m1'}/\partial t &= -i\,\omega_1(m1 - m1')\,\rho(1)_{m1m1'} + \\ &+ \sum R_{m1m2;m1'm2,\,m1''m2'';m1'''m2'''}\,\rho^{(2)}(1,2)_{m1''m2'';m1'''m2'''} .\end{aligned} \tag{3.10}$$

In the right-hand side of Eq. (3.10), the pair spin density matrix can be approximated as a direct product of single spin density matrices

$$\rho(1,2) \cong \rho(1) \times \rho(2). \tag{3.11}$$

In this approximation, Eq. (3.10) becomes a system of nonlinear equations with respect to single spin density matrices

$$\partial\rho(1)_{m1m1'}/\partial t = -i\,\omega_1(m1 - m1')\,\rho(1)_{m1m1'} + \\ + \sum R_{m1m2;m1'm2,\,m1''m2'';m1'''m2'''}\,\rho(1)_{m1''m1'''}\,\rho(2)_{m2''m2'''}. \tag{3.12}$$

When considering relaxation under conditions close to equilibrium, single spin density matrices can be represented as the sum of the equilibrium density matrix ρ_0 and the additive σ, which describes small deviations of the spin ensemble state from equilibrium

$$\rho(1) = \rho_0(1) + \sigma(1), \rho(2) = \rho_0(2) + \sigma(2). \tag{3.13}$$

Substituting (3.13) in (3.12) and neglecting quadratic terms $\sigma(1)\,\sigma(2)$, we obtain linearized kinetic equations for single spin density matrices

$$\partial\sigma(1)_{m1m1'}/\partial t = -i\,\omega_1(m1 - m1')\,\sigma(1)_{m1m1'} + \\ + + \sum R_{m1m2;m1'm2,\,m1''m2'';m1'''m2'''}\{\rho_0(1)_{m1''m1'''}\,\sigma(2)_{m2''m2'''}\delta_{m1''m1'''} + \sigma(1)_{m1''m1'''} \\ \rho_0(2)_{m2''m2'''}\delta_{m2''m2'''}\}. \tag{3.14}$$

In typical EPR conditions of spin exchange studying experiments the Zeeman splitting energy is significantly less than the thermal energy kT. Therefore, the high-temperature approximation is applicable for the equilibrium spin density matrix, and it is possible to introduce into Eq. (3.14)

$$\rho_0(1)_{m1,m1} \approx 1/(2S_1 + 1); \rho_0(2)_{m2,m2} \approx 1/(2S_2 + 1). \tag{3.15}$$

3.3 Kinetic Equations to Describe Electron Paramagnetic Relaxation in Liquids Caused by the Dipole-Dipole Interaction between Free Radicals

In the case of free radicals with $S_1 = 1/2$, $S_2 = 1/2$, the linearized Eq. (3.14) are significantly simplified and give the following system of kinetic equations for the elements of single spin density matrices $\sigma(1)$ and $\sigma(2)$ [5].

$$\partial\sigma(1)_{1/2,-1/2}/\partial t = -i\,\omega_1\sigma(1)_{1/2,-1/2} - (1/V)\left[\sigma(1)_{1/2,-1/2}/\tau_2(1) + \sigma(2)_{1/2,-1/2}/\tau_2'\right];$$
$$\partial\sigma(2)_{1/2,-1/2}/\partial t = -i\,\omega_2\sigma(2)_{1/2,-1/2} - (1/V)\left[\sigma(2)_{1/2,-1/2}/\tau_2(2) + \sigma(1)_{1/2,-1/2}/\tau_2'\right];$$
$$\partial n(1)/\partial t = -(1/V)[n(1)/\tau_1(1) + n(2)/\tau_1'];$$
$$\partial n(2)/\partial t = -(1/V)[n(2)/\tau_1(2) + n(1)/\tau_1']. \quad (3.16)$$

Here, $n(k) = \sigma(k)_{1/2,1/2} - \sigma(1)_{-1/2,-1/2}$, $k = 1,2$, defines the difference between the populations of spin states with $m=\pm 1/2$. The parameters τ_1, τ_1' and τ_2, τ_2' in Eqs. (3.16) characterize relaxation times of the longitudinal and the transverse components of the two spins due to dipole—dipole interaction between them, which varies randomly during diffusion of paramagnetic particles [5].

$$1/\tau_2(1) = (1/32)\gamma_1^2\gamma_2^2\hbar^2[4J^{(0)}(0) + J^{(0)}(\omega_1 - \omega_2) + 18J^{(1)}(\omega_2) + 36J^{(1)}(\omega_1) + 9J^{(2)}(\omega_1 + \omega_2)];$$

$$1/\tau_2(2) = (1/32)\gamma_1^2\gamma_2^2\hbar^2[4J^{(0)}(0) + J^{(0)}(\omega_1 - \omega_2) + 18J^{(1)}(\omega_1) + 36J^{(1)}(\omega_2) + 9J^{(2)}(\omega_1 + \omega_2)];$$

$$1/\tau_2' = (1/16)\gamma_1^2\gamma_2^2\hbar^2[J^{(0)}(0) + J^{(0)}(\omega_1 - \omega_2) + 9J^{(1)}(\omega_1) + 9J^{(1)}(\omega_2)];$$

$$1/\tau_1(1) = (1/16)\gamma_1^2\gamma_2^2\hbar^2[J^{(0)}(\omega_1 - \omega_2) + 18J^{(1)}(\omega_1) + 9J^{(2)}(\omega_1 + \omega_2)];$$
$$1/\tau_1(2) = (1/16)\gamma_1^2\gamma_2^2\hbar^2[J^{(0)}(\omega_1 - \omega_2) + 18J^{(1)}(\omega_2) + 9J^{(2)}(\omega_1 + \omega_2)]; \quad (3.17)$$
$$1/\tau_1' = (1/16)\gamma_1^2\gamma_2^2\hbar^2[-J^{(0)}(\omega_1 - \omega_2) + 9J^{(2)}(\omega_1 + \omega_2)].$$

If resonance frequencies are equal, $\omega_1=\omega_2=\omega_0$, then Eqs. (3.16) give the following equations (see [5])

$$
\begin{aligned}
\partial\sigma_{1/2,-1/2}/\partial t &= -i\omega_0\sigma_{1/2,-1/2} - -(1/V)[\sigma_{1/2,-1/2}/\tau_2];\\
\partial n/\partial t &= -(1/V)n/T_{1d};\\
1/\tau_2 &= (9/32)\gamma_1{}^2\gamma_2{}^2\hbar^2[J^{(0)}(0) + 10J^{(1)}(\omega_0) + J^{(2)}(2\omega_0)];\\
1/T_{1d} &= (9/8)\gamma_1{}^2\gamma_2{}^2\hbar^2[J^{(1)}(\omega_0) + J^{(2)}(2\omega_0)].
\end{aligned}
\tag{3.18}
$$

In these equations

$$
\begin{aligned}
J^{(q)}(\omega) &= V\int d\tau e^{-i\omega\tau} < F(q)(t)F*(q)(t+t)>_t\\
&= \int d\tau e^{-i\omega\tau}\iint d^3r_0 d^3r p(r0|\mathbf{r},\tau)F^{(q)}(\mathbf{r}_0)F*^{(q)}(\mathbf{r}).
\end{aligned}
\tag{3.19}
$$

In the derivation of Eqs. (3.17) and (3.18), it is assumed that paramagnetic particles diffuse into an isotropic medium and therefore the condition is fulfilled [3].

$$< F(q)(t)F*(q')(t+\tau)>_t = 0, \text{ when } q \neq q'.$$

Note that in the book [3] when discussing the contributions of the dipole-dipole interaction between spins with different Zeeman frequencies to their paramagnetic relaxation, the terms containing characteristic times $1/\tau_2'$ and describing the spin coherence transfer were not taken into account. In all other respects, Eqs. (3.17) and (3.18) reproduce the results given in [3].

In Eqs. (3.16, 3.17 and 3.18), the terms containing times $1/\tau_2'$ describe the spin coherence transfer and interfere in a rather unexpected way with the contribution of the exchange interaction to the spin coherence transfer. In the kinetic Eqs. (2.53 and 2.55) that describe the contribution of the exchange interaction to electron paramagnetic relaxation, there are also terms that describe the spin coherence transfer analogous to the terms containing $1/\tau_2'$ in Eq. (3.16). It is surprising that the exchange interaction and the dipole-dipole interaction give contributions to the kinetic equations, which have opposite signs. We will return to this issue in connection with the discussion of the manifestations of spin-spin interactions in the shape of EPR spectra of paramagnetic particles.

It is necessary to add more terms into Eq. (3.13) that take into account the contribution of the dipole-dipole interaction of the spins which have the same resonance frequencies. To this end, we can use kinetic Eq. (3.18). The contribution of all spins in solution to relaxation is additive. Therefore, in a solution containing N_1, N_2 spins of the type S_1, S_2, the relaxation of the selected spins is described by the equations

$$\partial\sigma(1)_{1/2,-1/2}/\partial t = -i\,\omega_1\sigma(1)_{1/2,-1/2} - (N_2/V)\left[\sigma(1)_{1/2,-1/2}/\tau_2(1) + \sigma(2)_{1/2,-1/2}/\tau_2'\right]$$
$$- (N_1/V)\,\sigma(1)_{1/2,-1/2}/\tau_2;$$

$$\partial\sigma(2)_{1/2,-1/2}/\partial t = -i\,\omega_2\sigma(2)_{1/2,-1/2} - (N_1/V)\left[\sigma(2)_{1/2,-1/2}/\tau_2(2) + \sigma(1)_{1/2,-1/2}/\tau_2'\right]$$
$$- (N_2/V)\,\sigma(2)_{1/2,-1/2}/\tau_2;$$

$$\begin{aligned}\partial n(1)/\partial t &= -(N_2/V)[n(1)/\tau_1(1) + n(2)/\tau_1'] - (N_1/V)n(1)/T_{1d(1)};\\ \partial n(2)/\partial t &= -(N_1/V)[n(2)/\tau_1(2) + n(1)/\tau_1'] - (N_2/V)n(2)/T_{1d(2)}.\end{aligned} \tag{3.20}$$

As already was shown (see Eqs. 2.130 and 2.131), the kinetic equations for density matrices (3.20) can be used to find kinetic equations for the macroscopic magnetization components of spins S_1 and S_2

$$\begin{aligned}M_{1z} &= (N_1/V)g\beta Tr(S_{1z}\rho(1)); M_{1\pm} = (N_1/V)\,g\beta Tr(S_{1\pm}\rho(1));\\ M_{2z} &= (N_2/V)\,g\beta Tr(S_{2z}\rho(2)); M_{2\pm} = (N_2/V)\,g\beta Tr(S_{2\pm}\rho(2)),\end{aligned} \tag{3.21}$$

where Tr means convolution over spin variables. Here we assume that g-factors of radicals are the same.

Multiplying the first and third equations in (3.20) by N_1/V, and the second and fourth equations by N_2/V, passing to the thermodynamic limit N, V --> ∞ at the constant spin concentration $C_1 = N_1/V$, $C_2 = N_2/V$ and performing convolution on spin variables, we obtain

$$\begin{aligned}\partial M_{1-}/\partial t &= -i\omega_1 M_{1-} - (C_2/\tau_2(1))M_{1-} - (C_1/\tau_2')M_2 - (C_1/\tau_2)M_{1-};\\ \partial M_{2-}/\partial t &= -i\omega_2 M_{2-} - (C_2/\tau_2')M_{1-} - (C_1/\tau_2(2))M_{2-} - (C_2/\tau_2)M_{2-};\\ \partial M_{1z}/\partial t &= -(C_2/\tau_1(1))M_{1z} - (C_1/\tau_1')M_{2z} - (C_1/T_{1d})M_{1z};\\ \partial M_{2z}/\partial t &= -\left(C_2/\tau_1'\right)M_{1z} - (C_1/\tau_1(2))M_{2z} - (C_2/T_{1d})M_{2z}.\end{aligned} \tag{3.22}$$

As a rule, the paramagnetic particles in the solution have magnetic nuclei, which give a hyperfine structure of the EPR spectra. The kinetic equations for magnetization (3.22) can be generalized to the case of solutions of free radicals whose EPR spectra consist of a number of hyperfine components. To do this, we introduce the partial magnetization $\mathbf{M}_k$ of the electron spins, which belong to the k-th component of the spectrum, and denote the statistical weight of the k-th component by φ(k). Given the contribution of the dipole-dipole interaction between all spins, we obtain kinetic equations

$$\begin{aligned}\partial M_{k-}/\partial t &= -i\omega_k M_{k-} - M_{k-}/T_2 - M_{k-}\Sigma(\varphi_n/T_2(k,n)) - \varphi_k \Sigma M_{n-}/T_2'(k,n),\\ \partial M_{kz}/\partial t &= -(M_{kz} - M_{k0})/T_1 - M_{kz}\Sigma(\varphi_n/T_1(k,n)) - \varphi_k \Sigma M_{nz}/T_1'(k,n).\end{aligned} \tag{3.23}$$

In these equations, parameters are introduced that characterize the contribution of the dipole-dipole interaction to paramagnetic relaxation (C-spin concentration)

$$\begin{aligned}1/T_2(k,n) &= (1/8)\gamma^4\hbar^2 C\{J^{(0)}(0) + (1/4)J^{(0)}(\omega_k - \omega_n) +\\ &\qquad (9/2)J^{(1)}(\omega_n) + 9J^{(1)}(\omega_k) + (9/4)J^{(2)}(\omega_k + \omega_n)\};\\ 1/T_2'(k,n) &= (1/16)\gamma^4\hbar^2 C\{J^{(0)}(0) + J^{(0)}(\omega_k - \omega_n) + 9J^{(1)}(\omega_k) + 9J^{(1)}(\omega_n)\};\\ 1/T_1(k,n) &= (1/16)\gamma^4\hbar^2 C\{J^{(0)}(\omega_k - \omega_n) + 18J^{(1)}(\omega_k) + 9J^{(2)}(\omega_k + \omega_n)\};\\ 1/T_1'(k,n) &= (1/16)\gamma^4\hbar^2 C\{-J^{(0)}(\omega_k - \omega_n) + 9J^{(2)}(\omega_k + \omega_n)\}.\end{aligned} \tag{3.24}$$

If diffusion is described by a continuous diffusion model, we obtain (see [3], Eq. (VIII.109)–(VIII.114))

$$J(q) = \alpha^{(q)}(1/(r_0 D))\mathrm{Re}\int du[J_{3/2}(u)]^2/(u(u^2 - i\omega\tau_D)), \tag{3.25}$$

where D is the diffusion coefficient of the radical,

$$\alpha^{(0)} = (48/15)\pi;\ \alpha^{(1)} = (8/15)\pi;\ \alpha^{(2)} = (32/15)\pi;\ \tau_D = r_0{}^2/(2D).$$

Here $J_{3/2}$ is the Bessel function. The value of the integral in (3.25) is [10].

$$\begin{aligned}\int du[J_{3/2}(u)]^2/(u(u2 - i\omega\tau D)) &= (i/(3\omega\tau_D))\\ &\quad + (\pi/(2\omega\tau_D))J_{3/2}((i\omega\tau_D)^{1/2})H^{(1)}{}_{3/2}((i\omega\tau_D)^{1/2}),\end{aligned} \tag{3.26}$$

where $H^{(1)}{}_{3/2}$ is a Hankel function of the first kind.

At the closest approach of particles, the distance r_0 between the spins is on the order of several Angstroms. In liquids, the diffusion coefficient $D{\sim}10^{-5}$ cm^2s^{-1}, time $\tau_D{\sim}10^{-10}$ s. The splitting between the components of the EPR spectrum of free radicals caused by, e.g., the hyperfine interaction with magnetic nuclei, ω_k-ω_n, usually does not exceed 10^9 rad/s. Therefore, when calculating the spectral densities, we can assume that $\omega_k = \omega_n = \omega$, as $|\omega_k - \omega_n|\tau_D << 1$ (ω is the mean frequency). This circumstance makes it possible to significantly simplify the kinetic Eq. (3.23), since the kinetic coefficients can be considered independent of the indices k, n:

$$
\begin{aligned}
\partial M_{k-}/\partial t &= -i\omega_k M_{k-} - -M_{k-}/T_2 - M_{k-}/T_{2dd} - \varphi_k \Sigma M_{n-}/T_{2dd}';\\
\partial M_{kz}/\partial t &= - -(M_{kz} - M_{k0})/T_1 - M_{kz}/T_{1dd} - \varphi_k \Sigma M_{nz}/T_{1dd}';\\
1/T_{2dd} &= (1/8)\gamma^4\hbar^2 C\{(5/4)J^{(0)}(0) + (27/2)J^{(1)}(\omega) + (9/4)J^{(2)}(2\omega)\};\\
1/T_{2dd}' &= (1/8)\gamma^4\hbar^2 C\{J^{(0)}(0) + 9J^{(1)}(\omega)\};\\
1/T_{1dd} &= (1/16)\gamma^4\hbar^2 C\{J^{(0)}(0) + 18J^{(1)}(\omega) + 9J^{(2)}(2\omega)\};\\
1/T_{1dd}' &= (1/16)\gamma^4\hbar^2 C\{-J^{(0)}(0) + 9J^{(2)}(2\omega)\}.
\end{aligned}
\tag{3.27}
$$

The dipole-dipole interaction is manifested in two ways in the relaxation of the longitudinal magnetization of spins belonging to different components of the EPR spectrum. On the one hand, the redistribution of the longitudinal magnetization of the components occurs at the $1/T_{1dd}'$ rate, which leads to the establishment of the stationary distribution of the magnetization of the components in proportion to their equilibrium values. On the other hand, the dipole-dipole interaction contributes to the spin-lattice relaxation of the total longitudinal magnetization. In the derivation of kinetic equations (see, e.g., (3.27)) it was assumed that the populations of spin levels are the same (the infinite lattice temperature), so the longitudinal component of magnetization tends to zero as a result of the dipole-dipole interaction. At finite temperatures, this term should be written as (see also Eqs. (VIII.64, VIII.75, VIII.87) in [3]) $\{-(M_z\text{-}M_0)/T_{1d}\}$. Taking into account this note the kinetics of the spin-lattice relaxation of a total magnetization is described by the equation

$$
\begin{aligned}
\partial M_z/\partial t &= -(M_z - M_0)/T_1 - (M_z - M_0)/T_{1d};\\
1/T_{1d} &= 1/T_{1dd} + 1/T_{1dd}' = (9/8)\gamma^4\hbar^2 C\{J^{(1)}(\omega) + J^{(2)}(2\omega)\}.
\end{aligned}
\tag{3.28}
$$

Thus, $1/T_{1d}$ is the contribution of the dipole-dipole interaction to the spin-lattice relaxation rate, and $1/T'_{1dd}$ characterizes the spin cross-relaxation rate. The kinetic equations for the longitudinal projections of the k-th component magnetization (see Eq. (3.27)) can be written in the form

$$
\begin{aligned}
\partial M_{kz}/\partial t = &-(M_{kz} - M_{k0})(1/T_1 + 1/T_{1d}) + (M_{kz})/T_{1dd}'\\
&- \varphi_k \Sigma(M_{nz})/T_{1dd}'.
\end{aligned}
\tag{3.29}
$$

Equation (3.29) shows that the contribution of the dipole-dipole interaction to the kinetics of longitudinal magnetization is reduced to the fact that the spin-lattice relaxation rate of spins increases by $1/T_{1d}$ and the dipole-dipole interaction redistributes with the rate $1/T_{1dd}'$ the longitudinal magnetization of the system between spins of different components according to their statistical weights.

3.4 Kinetic Equations for Typical Conditions of EPR Experiments

For further applications, we write down kinetic Eq. (3.29) for typical EPR experiments.

From Eqs. (3.25 and 3.26) we have:

$$J^{(q)}(\omega) \cong \alpha^{(q)}\left(2\tau_D/r_0^3\right)\left[2/15 - \left(1/(9\sqrt{2})\right)(\omega\tau_D)^{1/2}\right], \qquad \text{when } \omega\tau_D < 1;$$
$$J^{(q)}(\omega) \cong \alpha^{(q)}\left(1/(\sqrt{2}r_0^3)\right)\left(1/\left(\omega(\omega\tau_D)^{1/2}\right)\right)[1 - 1/(\omega\tau_D)], \quad \text{when } \omega\tau_D > 1, \tag{3.30}$$

where ω is the average spin frequency. In EPR experiments, the resonance frequencies $\omega > 10^{11}$ rad/s are usually found and therefore the condition $\omega\tau_D > 1$is satisfied.

Substituting (3.30) into (3.27, 3.28), the kinetic coefficients in Eq. (3.29) can be expressed in terms of a single independent parameter T_{dd}:

$$\begin{aligned}
&T_{dd} = 1/[(2\pi/75)\gamma^4\hbar^2 C\tau_D/r_0^3];\\
&1/T_{2dd} \equiv 1/T_{ddsd} \cong 5/Tdd; 1/T_{2dd}' \equiv 1/T_{ddsct} \cong 4/T_{dd};\\
&1/T_{1dd} \equiv 1/T_{ddrp} \cong 2/T_{dd}; -1/T_{1dd}' \equiv 1/T_{ddpt} \cong 2/T_{dd};\\
&1/T_{1d} = (9/8)\gamma^4\hbar^2 C\{J^{(1)}(\omega) + J^{(2)}(2\omega)\}
\end{aligned} \tag{3.31}$$

Indices show which process characterizes the corresponding time. T_{ddrp} characterizes the decrease in the longitudinal magnetization (the population difference of spin levels) of each spin caused by the dipole-dipole interaction with all spins of the system. We can say that T_{ddrp} characterizes the relaxation of the longitudinal spin polarization; T_{ddpt} characterizes the transfer of longitudinal polarization from all spins of the system to the selected spin,; T_{ddsct} characterizes the spin coherence transfer from all spins to the selected spin; T_{ddsd} is the characteristic decoherence time of each spin induced by the dipole-dipole interaction with all spins of the system.

Thus, in the limiting situation $\omega\tau_D \gg 1$, kinetic Eq. (3.29) are reduced to simpler equations

$$\begin{aligned}
&\partial M_{k-}/\partial t = -i\omega_k M_{k-} - M_{k-}/T_2 - (5/T_{dd})M_{k-} - \boldsymbol{\varphi_k(4/T_{dd})M_-};\\
&\partial M_{kz}/\partial t = -(M_{kz} - M_{k0})(1/T_1 + 1/T_{1d}) - (2/T_{dd})M_{kz} + \boldsymbol{\varphi_k(2/T_{dd})M_z};\\
&M_- \equiv \Sigma M_{n-}, M_z \equiv \Sigma M_{nz}.
\end{aligned} \tag{3.32}$$

In these equations, the two terms, written in bold italics, deserve attention.

Firstly, in the equation for the z-component of magnetization, the selected term has the same sign as the corresponding term in the case of spin exchange (see, e.g., 2.56 and 2.59). It is well known that for nuclear spins the longitudinal polarization transfer due to the dipole-dipole and scalar interaction occurs with the opposite sign (see [3], Eqs. (VIII.88a, VIII.128). This difference between nuclear and electron spins is due to the fact that in the case of electron spins, the contribution of flip-flop transitions dominates in the longitudinal polarization transfer, and in the case of nuclear spins, the contributions of flip or flop transitions dominate.

Secondly, in the equation for the transverse magnetization components the term written in bold italics appears, which was mistakenly thrown out (see [3], Eq. (VIII.89), [4], Eq. (2.19). Thus, it was not taken into account that the dipole-dipole interaction causes the spin coherence transfer analogously to the spin coherence transfer due to the exchange interaction. The comparison of (3.32) with (2.59) shows that the contributions of exchange and dipole-dipole interactions to the rate of longitudinal spin polarization transfer add up, and their contributions to the rate of the spin coherence transfer are subtracted.

References

1. Zamaraev, K.I., Molin, Y.N., Salikhov, K.M.: Spin exchange. Nauka, Sibirian branch, Novosibirsk (1977)
2. Molin, Y.N., Salikhov, K.M., Zamaraev, K.I.: Spin exchange. Principles and applications in chemistry and biology. Springer, Heidelberg/Berlin (1980)
3. Abragam, A.: The principles of nuclear magnetism. (Oxford at the Clarendon Press 1961) Freed J H. J. Chem. Phys. **45**, 3452 (1966)
4. Freed, J.H., Fraenkel, G.K.: Theory of linewidths in electron spin resonance spectra. J. Chem. Phys. **39**, 326–348 (1963)
5. Salikhov, K.M., Semenov, A.G., Tsvetkov, Y.D.: Electron spin echo and its application. Nauka, Siberian branch, Novosibirsk (1976)
6. Galeev, R.T., Salikhov, K.M.: The theory of dipolar broadening of magnetic resonance lines in non-viscous liquids. Khim. Fizika. **15**, 48–64 (1996)
7. Salikhov, K.M.: Contributions of exchange and dipole–dipole interactions to the shape of EPR spectra of free radicals in diluted solutions. Appl. Magn. Reson. **38**, 237–256 (2010)
8. Bales, B.L., Meyer, M., Smith, S., Peric, M.: EPR line shifts and line shape changes due to spin exchange of nitroxide free radicals in liquids 6: separating line broadenning due to spin exchange and dipolar interactions. J. Phys. Chem. A. **113**, 4930–4940 (2009)
9. Peric, M., Bales, B.L., Peric, M.: EPR line shifts and line shape changes due to Heisenberg spin exchange and dipole-dipole interactions of nitroxide free radicals in liquids 8: further experimental and theoretical efforts to separate the effects of the two interactions. J Chem Phys A. **116**, 2855–2866 (2012)
10. Pfeifer, H.: Der Translationsanteil der Protonenrelaxation in wässrigen Lösungen paramagnetischer Ion. Ann. Physik, Bd. **8**(S), 1–8 (1961)

Chapter 4
Modified Bloch Equations for Dilute Solutions of Free Radicals Taking into Account Exchange and Dipole-Dipole Interactions

Abstract Kinetic equations are given for the magnetization of different sub-ensembles of isochromatic spins in solutions, taking into account the exchange and dipole – dipole interactions. It is assumed that the exchange and dipole – dipole interactions make an additive contributions to the paramagnetic relaxation of spins in solutions.

In the previous sections, kinetic equations for spin magnetizations were given separately, taking into account only the exchange or dipole-dipole interaction. In real systems, these interactions affect simultaneously the movement of spins, and hence the spin magnetization of the system. It is assumed that these interactions make an additive contribution to the kinetics of spin relaxation processes, as well as the transfer of spin excitation and spin coherence. Under this assumption, using the above results for spin exchange (see 2.59) and the dipole-dipole interaction (see 3.32), we can write the generalized Bloch equation for magnetization of electron spins. These equations will be needed in the future when discussing the manifestation of spin exchange in EPR spectroscopy. We present these equations for typical situations of the EPR experiments (3.31).

EPR spectra of free radicals usually contain a number of hyperfine components. In this situation, when considering the effect of the exchange and dipole-dipole interaction on the paramagnetic relaxation, it should be borne in mind that the effect of these interactions in general can vary greatly depending on which hyperfine components the resonance frequencies of the interacting spin pairs belong to. In Sect. 2.8.1 (Fig. 2.5) the spin exchange was considered in detail in a model situation when colliding spins had different resonance frequencies. It has been shown that the effect of the exchange interaction is reduced to the equivalent exchange of magnetizations between the colliding partners if the particles collide with equal resonance frequencies or if the frequency difference can be neglected during the collision. In this case of equivalent exchange, the kinetic equations for the magnetizations of $\mathbf{M}_k$ electron spins, which refer to the k-th hyperfine component, can be written as [1, Eq. (4.75)]

K. M. Salikhov, *Fundamentals of Spin Exchange*,
https://doi.org/10.1007/978-3-030-26822-0_4

$$\partial M_{k-}/\partial t = -i\omega_k M_{k-} - M_{k-}/T_{2k} - (K_{ex}C + 5/T_{dd})\, M_{k-} + \varphi_k\, (K_{ex}C - 4/T_{dd})M_-;$$
$$\partial M_{kz}/\partial t = -(M_{kz} - M_{k0})(1/T_{1k} + 1/T_{1d}) - (K_{ex}C + 2/T_{dd})\, M_{kz} +$$
$$\varphi_k (K_{ex}C + 2/T_{dd})M_z;$$
$$M_- \equiv \Sigma M_{n-}, M_z \equiv \Sigma M_{nz}. \tag{4.1}$$

In these equations, the following parameters are introduced: ω_k is the resonance frequency of the isolated spins in the k-th component, φ_k is the statistical weight of the spins in the k-th component. Characteristic times of T_{1k} and T_{2K} describe the relaxation of the longitudinal and transverse magnetization components of isolated radicals. Other terms characterize the contribution of the exchange and dipole-dipole interaction to paramagnetic relaxation (C-concentration of spins): K_{ex} is the rate constant of equivalent spin exchange between radicals, T_{1d}, T_{dd} characterize the contribution of dipole-dipole interaction to relaxation.

For further analysis, it is helpful to write these kinetic equations in matrix form. Let us consider the situation with two components. Then the coefficient matrix of kinetic equations, e.g., for the transverse component of the magnetization has the form

$$\mathbf{L} = \begin{Bmatrix} -i\omega_1 - 1/T_{21} - (1-\varphi_1)K_{ex}C - (1/T_{dd})(5+4\varphi_1) & \varphi_1(K_{ex}C - 4/T_{dd}) \\ \varphi_2(K_{ex}C - 4/T_{dd}) & -i\omega_2 - 1/T_{22} - (1-\varphi_2)K_{ex}C - (1/T_{dd})(5+4\varphi_2) \end{Bmatrix} \tag{4.2}$$

It is well known from the theory of linear equations with constant coefficients that in the first approximation in the matrix **L** nondiagonal elements can be neglected if their absolute value is less than the absolute value of the difference of the corresponding diagonal elements. When this condition is met, the effect of the exchange and dipole-dipole interaction is reduced to the fact that instead of $1/T_{21}$ and $1/T_{22}$ relaxation rates of isolated particles, it is possible to introduce new, faster relaxation rates which can be characterized with $T_{21}*$ and $T_{22}*$ decoherence times of spins in the 1st and 2nd hyperfine components

$$\begin{aligned} 1/T_{21}* &= 1/T_{21} + (1-\varphi_1)K_{ex}C + (1/T_{dd})(5+4\varphi_1); \\ 1/T_{22}* &= 1/T_{22} + (1-\varphi_2)K_{ex}C + (1/T_{dd})(5+4\varphi_2). \end{aligned} \tag{4.3}$$

Nondiagonal elements of the matrix **L** (3.35) describe the spin coherence transfer to the considered spin from the interaction partners. If we talk about the spin exchange, the terms $\varphi_k K_{ex}C$ describe a reaction in a collision. In the next section, we discuss in detail how this quantum coherence transfer ("reaction" in a sense of the third law of Newton) is manifested itself in the shape of EPR spectra.

We have already noted that in most cases stable nitroxide radicals are used as spin probes [2–4]. For these radicals, the necessary condition of equivalent spin exchange is commonly satisfied that the exchange integral J is much larger than the splitting of

Δ hyperfine spectrum components, which is determined by the interaction with magnetic nitrogen nuclei and protons. But the condition $J \gg \Delta$ is not enough to get the equivalent spin exchange case. It needs one more condition to be satisfied: $\Delta\tau_c \ll 1$, where τ_c is the duration of the collision of two particles in the exchange interaction region. If the condition $\Delta\tau_c > 1$ is met, the exchange becomes not equivalent (see Fig. 2.5), the transfer of coherence ("reaction" effect) in the collision becomes negligible (see Fig. 2.5). In the case of nonequivalent spin exchange, Eq. (4.1) can no longer be applied. Under last conditions the kinetic equations take a form

$$\begin{aligned}
&\partial M_{k-}/\partial t = -i(\omega_k + \delta_k)M_{k-} - M_{k-}/T_{2k} - (KC + 5/T_{dd})\, M_{k-} - \varphi_k\, (4/T_{dd})M_-;\\
&\partial M_{kz}/\partial t = -(M_{kz} - M_{k0})(1/T_{1k} + 1/T_{1d}) - (K_{ex}C + 2/T_{dd})\, M_{k-} +\\
&\varphi_k\, (K_{ex}C + 2/T_{dd})M_z.
\end{aligned} \tag{4.4}$$

There are no terms that define the spin coherence transfer induced by the exchange interaction (the exchange "reaction" effect). It was shown in Sect. 2.8.1 (see Eq. 2.80, 2.81, 2.82, 2.83, 2.84, 28.5 and 2.86) that in the equations for the transverse magnetization components, the spin decoherence rate KC appears instead of $K_{ex}C$, and it was shown in Sect. 2.8.1 (see Fig. 2.5), $K > K_{ex}$.

In the conditions of nonequivalent exchange, when the spin coherence transfer induced by the exchange interaction is negligible, the transverse components of magnetization M_{k-} are coupled to each other only due to the dipole-dipole interaction.

There is an intermediate region, where kinetic equations for the magnetization (4.1) are applicable with a small modification in a quasi-equivalent spin exchange case when $J\tau_c > 1$, $\Delta\tau_c < 1$. It was shown in Sect. 2.8.1 (see, e.g., Fig. 2.5) that under the condition $\Delta\tau_c < 1$, the efficiency of spin coherence transfer between radicals almost coincides with the efficiency of equivalent spin exchange, but there is an additional shift in the resonance frequencies of spins. In this case, the kinetic equations for the magnetization of radicals can be written as

$$\begin{aligned}
&\partial M_{k-}/\partial t = -i(\omega_k + \delta_k)M_{k-} - M_{k-}/T_{2k} - (K_{ex}C + 5/T_{dd})\, M_{k-} + \varphi_k\, (K_{ex}C - 4/T_{dd})M_-;\\
&\partial M_{kz}/\partial t = -(M_{kz} - M_{k0})(1/T_{1k} + 1/T_{1d}) - (K_{ex}C + 2/T_{dd})\, M_{k-} +\\
&\varphi_k\, (K_{ex}C + 2/T_{dd})M_z.
\end{aligned} \tag{4.5}$$

In comparison with (4.1), a shift in the spin frequency is added here, which occurs due to the fact that at the moment of collision, along with the exchange interaction, it is necessary to take into account the movement of spins with their own frequencies inherent in isolated radicals [5]. Because of the difference in the spin frequencies of colliding particles, the total spin moment of the particles is not preserved in the elementary act of spin exchange.

In the general case, kinetic Eqs. (4.1, 4.2, 4.3, 4.4 and 4.5) can be presented in the form

$$\partial M_{k-}/\partial t = -i(\omega_k + \delta_k)M_{k-} - M_{k-}/T_{2k} - (KC + 5/T_{dd})\,M_{k-} + \varphi_k\left(K'C - 4/T_{dd}\right)M_-;$$
$$\partial M_{kz}/\partial t = -(M_{kz} - M_{k0})(1/T_{1k} + 1/T_{1d}) - (K_{ex}C + 2/T_{dd})\,M_{k-} + \varphi_k\,(K_{ex}C + 2/T_{dd})M_z. \quad (4.6)$$

When equivalent spin exchange occurs, then $K{=}K' = K_{ex}$ (4.1). When the spin exchange is non-equivalent, then in Eq. (4.6) K > Kex, $K' = 0$.

Note that the kinetic Eq. (4.1), (4.3, 4.4, 4.5 and 4.6) for the spin magnetization are given for electron spins under the conditions of EPR experiments. This is explicitly taken into account in the terms that describe the spin coherence transfer, M_{k-}, energy and spin excitation, M_{kz}. We note the well-known fact that in the case of nuclear spins, the transfer of the longitudinal magnetization due to the scalar and dipole-dipole interactions occurs with different signs (see [6], Eq. (VIII.128)), while in the case of electron spins, these contributions have the same sign.

In the following sections, the kinetic equations for the magnetization of stable radical solutions will be used to analyze the shape of the EPR spectra.

In conclusion of this section once again we will pay attention to terminology which is used in connection with the problem of spin exchange. On the one hand, the kinetic Eqs. (4.1, 4.2, 4.3, 4.4, 4.5 and 4.6) show that in general, the exchange interaction in bimolecular collisions is manifested in three aspects:

(a) spin decoherence occurs, spins “lose” phase (quantum coherence), and the rate of this process is $K_{ex}C$ or KC for equivalent or non-equivalent spin exchange case, respectively;
(b) spins “get” quantum coherence from collision partners (reaction in collision), the rate of this process is $K_{ex}C$ or K’C, and $K'C \leq K_{ex}C$;
(c) the excitation energy is transferred between the spins with different resonance frequencies (see the equation for M_{kz} (4.6)), and the rate of this process at $J \gg \Delta$ is given by $K_{ex}C$.

The spin excitation transfer between spins with different resonance frequencies was called cross-relaxation [7]. It is seen from the kinetic Eq. (4.6) that the exchange and dipole-dipole interactions can give cross-relaxation. An elementary act of cross-relaxation is a mutual flip-flop of spins.

Often in magnetic resonance, the paramagnetic relaxation of spins is caused by random changes (fluctuations) of the local magnetic field at the location of the spin. The fluctuating local field can cause an adiabatic shift of resonant frequencies. This process is called the random frequency migration. The random frequency migration causes spin phase relaxation (decoherence), but it does not change the population difference of spin states. This means that the frequency migration does not contribute to the change in the value of the z-projection of the total spin magnetization.

In the case of the small B_1 field, the manifestation of equivalent spin exchange in the relaxation of transverse magnetization components (i.e., in the relaxation of quantum coherence) can be described by a frequency migration model. The random frequency modulation is also very often called spectral diffusion.

The nonadiabatic component of the fluctuating local magnetic field can alter quantum coherence but also the population of spin states. The term spectral diffusion is often used for describing two different processes: adiabatic random change of resonance frequencies of spins and random migration of excitation between spins with different resonance frequencies.

References

1. Salikhov, K.M., Semenov, A.G., Tsvetkov, Y.D.: Electron spin echo and its application. Nauka, Siberian branch, Novosibirsk (1976)
2. Zamaraev, K.I., Molin, Y.N., Salikhov, K.M.: Spin exchange. Nauka, Sibirian branch, Novosibirsk (1977)
3. Molin, Y.N., Salikhov, K.M., Zamaraev, K.I.: Spin exchange. Principles and applications in chemistry and biology. Springer, Heidelberg/Berlin (1980)
4. Bales, B.L.: Inhomogeneously broadened spin-label spectra. In: Berliner, L.J., Reuben, J. (eds.) Biological magnetic resonance, vol. 8, pp. 77–130. Plenum Publishing Corporation, New York (1989)
5. Salikhov, K.M.: The contribution from exchange interaction to line shifts in ESR spectra of paramagnetic particles in solutions. J. Magn. Res. **63**, 271–279 (1985)
6. Abragam, A.: The principles of nuclear magnetism. Clarendon Press, Oxford (1961)
7. Bloembergen, H.N., Shapiro, S., Pershan, P.S., Artman, J.O.: Cross-relaxation in spin systems. Phys. Rev. **114**, 445–459 (1959)

Chapter 5
Manifestation of Exchange and Dipole-Dipole Interaction in the Form of EPR Spectra of Paramagnetic Particles in Solutions in Linear Response Case

Abstract A general formula is given for the EPR spectrum in solutions in the region of linear response. The differences in understanding the manifestations of spin exchange in the early and new paradigms are discussed in detail. A common feature of these paradigms is that the components of the spectrum broaden and shift to the center of gravity of the EPR spectrum with an increase in the rate of transfer of spin coherence in the region of slow exchange, while in the region of fast transfer of spin coherence, the entire spectrum turns into one narrow homogeneous line in the center of gravity. Within both paradigms, the spectrum is represented as the sum of the resonance lines. The fundamental differences between the two paradigms relate to the nature and form of each resonance line. The establishment of the fact that each resonance line has a mixed absorption + dispersion form has become the cornerstone of the new paradigm. In the new paradigm, each line of the spectrum is associated with a specific mode of coordinated (collective) motion of the coherence of spins. The exchange narrowing of the spectrum at a high spin exchange rate is explained by the fact that in this situation only one collective mode is excited by the microwave field. The remaining modes are dark states and are not observed in the EPR experiment.

First, we will follow the logic of early works on spin exchange [1–3], in which the widely spread paradigm of spin exchange in solutions and its study using EPR spectroscopy was formulated. Let us focus on the theory of EPR spectra in the case of a linear response, when the alternating magnetic field is weak enough and the saturation effect is not manifested. This means that when the energy of the external field is absorbed, there is no noticeable accumulation of energy in the spin system, since it is quickly withdrawn to the surrounding molecules, to the thermostat. In magnetic resonance spectroscopy, a thermostat is commonly called a lattice. The linear response of the spin system occurs when the condition is satisfied (see [4], Eq. (II.37, III.15))

$$(\gamma B_1)^2 T_2 < 1/T_1. \tag{5.1}$$

K. M. Salikhov, *Fundamentals of Spin Exchange*,
https://doi.org/10.1007/978-3-030-26822-0_5

Given that the most common spin probes are nitroxide radicals, below we will focus more on the manifestations of spin exchange in the EPR spectra by the example of nitroxide radicals. Naturally, these arguments can be used in the study of bimolecular collisions of other paramagnetic particles.

The hyperfine interaction with magnetic nuclei splits spin energy levels and therefore the EPR spectra of radicals have a hyperfine structure, hfs. Each of the spectrum hfs component corresponds to a certain "configuration" of spins of magnetic nuclei. In nitroxide radical due to the hyperfine interaction of unpaired electron with nitrogen nucleus, the spectrum has two nitrogen hfs components for the isotope ^{15}N and three nitrogen components of hfs for ^{14}N, since for nitrogen isotopes the projection of the nuclear spin on the direction of the external constant magnetic field can have two values + 1/2 or − 1/2 for ^{15}N (nuclear spin I = 1/2), and three values 1, 0, −1 for ^{14}N (nuclear spin I = 1). Each nitrogen hfs component of the spectrum consists of a number of lines due to hyperfine interaction with protons/deuterons. EPR experiments are carried out in sufficiently strong magnetic fields, so that the hyperfine interaction of electron, S, and nuclear, I, spins can be described by spin-Hamiltonian

$$\mathbf{H}_{\mathrm{hfi}} = \hbar a \mathbf{S}_Z \mathbf{I}_Z, \tag{5.2}$$

where *a* is the constant of hyperfine interaction. This means that the whole ensemble of radicals can be divided into sub-ensembles with a different set of projections of nuclear spins, which form the hfs of the EPR spectrum.

For describing the manifestations of spin exchange in EPR spectra, it is useful to consider a simple model in which the radical has a single magnetic nucleus with spin 1/2, for example, ^{15}N.

5.1 Features of the EPR Spectrum of the Solution of Radicals, Which Have a Single Magnetic Nucleus I = 1/2, in the Presence of Equivalent Spin Exchange

To highlight manifestations of the equivalent spin exchange let us simplify the situation as much as possible and assume that the contribution of the dipole-dipole interaction to the relaxation of spins can be neglected in comparison with the contribution of the exchange interaction. This situation is realized, for example, at sufficiently low viscosity of the solution.

In EPR experiments there is usually a constant magnetic field B_0 and a linearly polarized alternating magnetic field with frequency ω and amplitude $2B_1$, $B_x = 2B_1\cos(\omega t) = B_1(\exp(-i\omega t) + \exp(i\omega t))$ is supplied. Only one of the rotating components of this field causes a resonant response. Leaving only one component of the field B_x and moving to a rotating with frequency ω coordinate system, we obtain the kinetic equations for the partial magnetizations of spins belonging to different hyperfine components of the EPR spectrum. In the considered model situation $\varphi_1 = \varphi_2 = 1/2$, and T_1, T_2 times of paramagnetic relaxation of isolated spins in both subensembles are considered to be equal. As a result, we have

$$\begin{aligned}\partial M_{1-}/\partial t = &- i(\omega_0 + a/2 - \omega)M_{1-} - M_{1-}/T_2 - (1/2)K_{ex}CM_{1-} + (1/2)K_{ex}CM_{2-}\\ &- i\gamma B_1(1/2)M_0;\\ \partial M_{2-}/\partial t = &- i(\omega_0 - a/2 - \omega)M_{2-} - M_{2-}/T_2 - (1/2)K_{ex}CM_{2-} + (1/2)K_{ex}CM_{1-}\\ &- i\gamma B_1(1/2)M_0.\end{aligned} \tag{5.3}$$

In the region of linear response in these equations, it is allowed to put $M_{kz} = \varphi_k M_0 = (1/2)M_0$, where M_0 is the equilibrium longitudinal magnetization of electron spins.

The intensity J_{EPR} of the EPR signal is proportional to the y-component of magnetization ([4], Eq. (III.21))

$$J_{EPR} \sim M_{1y} + M_{2y} = -\mathrm{Im}\{M_{1-} + M_{2-}\} \tag{5.4}$$

To find it, it is necessary to solve the Eq. (5.2) in steady state condition

$$\begin{aligned}-i(\omega_0 + a/2 - \omega)M_{1-} - M_{1-}/T_2 - (1/2)K_{ex}CM_{1-} + (1/2)K_{ex}CM_{2-} = i\gamma B_1(1/2)M_0;\\ -i(\omega_0 - a/2 - \omega)M_{2-} - M_{2-}/T_2 - (1/2)K_{ex}CM_{2-} + (1/2)K_{ex}CM_{1-} = i\gamma B_1(1/2)M_0.\end{aligned} \tag{5.5}$$

Having solved (5.5)), we obtain the following expression for the shape of the spectrum [5].

$$\begin{aligned}J_{EPR}(\omega) = \gamma B_1 T_2 M_0 Im\Big\{&4iT_2(1 + K_{ex}CT_2 + iT_2(\omega_0 - \omega))\\ &/\Big(4 + T_2\Big(4K_{ex}C + a^2T_2 + 4i(2 + K_{ex}CT_2)(\omega_0 - \omega) - 4T_2^2(\omega_0 - \omega)^2\Big)\Big)\Big\}.\end{aligned} \tag{5.6}$$

The roots of the denominator in (5.6) determine the values of ω for which resonance is expected. In principle, the roots of the denominator are complex numbers. Equating the denominator to zero, we get the roots

$$\begin{aligned}&\omega(1) = \omega_0 + R/2 - i/T_2 - (i/2)K_{ex}C; \omega(2) = \omega_0 - R/2 - i/T_2 - (i/2)K_{ex}C;\\ &R = \left(a^2 - (K_{ex}C)^2\right)^{1/2}.\end{aligned} \tag{5.7}$$

The real parts of the roots give the resonance frequencies, and the imaginary parts give the resonance widths. The behavior of the roots is determined by the ratio between the splitting of the spectrum lines, a, and the rate of spin exchange, $K_{ex}C$. In the slow exchange region when $K_{ex}C < a$, $R \approx a-(K_{ex}C)^2/(2a)$. As a result, we obtain that in the region of slow exchange the resonance frequencies are equal to the values

$$\omega(1) = \omega_0 + a/2 - (K_{ex}C)^2/(4a), \omega(2) = \omega_0 - a/2 + (K_{ex}C)^2/(4a). \quad (5.8)$$

It is seen that the frequencies of the resonances become closer to each other. This frequency shift is quadratically dependent on the spin concentration. In this case, both resonances have the same width

$$\Delta\omega_{1/2} = 1/T_2 + (1/2)K_{ex}C. \quad (5.9)$$

Spin exchange additionally broadens both resonances. This exchange broadening of the resonance lines increases linearly in spin concentration (see (5.9)).

While achieving fast exchange, when $K_{ex}C \geq a$, there is only a resonance at a medium frequency $\omega = \omega_0$. Although the resonance frequencies coincide in the $K_{ex}C \geq a$ region, the widths of these resonances are different. In this situation, the resonance widths are equal to the values (see imaginary parts of (5.7))

$$\Delta\omega_{1/2}(1) = 1/T_2 + a^2/(4K_{ex}C); \quad \Delta\omega_{1/2}(2) = 1/T_2 + K_{ex}C. \quad (5.10)$$

It is seen that the width of one of the resonances, $\Delta\omega_{1/2}(1)$, decreases with the growth of the spin exchange rate, and the width of the other resonance $\Delta\omega_{1/2}(2)$ increases.

In the experiment, under the condition of fast spin exchange, an exchange narrowed line is observed (see Fig. 2.4). It could be assumed that in the experiment the second resonance is not observed because this resonance is very wide, and therefore it is difficult to register. But then, near the critical spin exchange rate, when $K_{ex}C$ is greater than the splitting a of the resonance frequencies of the isolated spins, but not much more, the spectrum would have to represent the sum of two Lorentzian lines with different widths. In addition, the integral intensity of the observed narrow line would be equal to half of the integral intensity of the entire spectrum. This is not observed in the experiment. This paradox was resolved in [5]. The results of this work will be presented below. Here is given only the main result. The transfer of coherence between spins induced by the exchange interaction forms the collective modes of motion of the quantum coherence of all spins. The external alternating magnetic field excites only certain collective modes, others are practically not excited. In the model system under consideration, a narrow collective resonance is excited. It turns out that the above wide resonance with a width of $\Delta\omega_{1/2}(2)$ (see (5.10)) is “dark” in EPR spectroscopy.

5.2 General Solution for the Shape of the EPR Spectrum

The structure of kinetic equations for spin magnetization in solutions allows us to find an explicit expression for the shape of the spectrum of radicals with an arbitrary hfs of the EPR spectrum. In EPR experiments in the field of linear response of a system on a microwave field, when saturation effects can be neglected and $M_{kz} \approx M_{k0} = \varphi_k M_0$, the measurement results can be analyzed using kinetic equations, which were discussed in the Sect. 5.4, (see Eqs. 4.1 and 4.4)

$$\begin{aligned} \partial M_{k-}/\partial t &= -i(\omega_k + \delta_k - \omega)M_{k-} - M_{k-}/T_{2k} - (K_{ex}C + 1/T_{ddsd})\, M_{k-} \\ &+\varphi_k(\boldsymbol{K_{ex}C - 1/T_{ddsct}})M_- - i\gamma B_1 M_{k0}; \end{aligned} \tag{5.11}$$

$$M_- \equiv M_x - iM_y \equiv \Sigma M_{n-}.$$

Note that in these equations an additional frequency shift δ_k is associated with a possible realization of the non-equivalent spin exchange during the elementary collision act. This shift was discussed in detail in Sect. 2.8.2, (see also [6–8]). Experimentally, this shift has been comprehensively investigated in a number of works (see, for example. [9–15]).

The transfer rate of spin coherence $V_{sct} = (K_{ex}C - 1/T_{ddsct})$ at a certain diffusion coefficient of the spin probe can vanish. From the equality to zero of the total rate of spin coherence transfer, $V_{sct} = (K_{ex}C - 1/T_{ddsct}) = 0$, we can estimate that this can happen when the coefficient of diffusion of the spin probes $D \approx 10^{-6}$ cm^2/s. i.e. in the case of the viscosities of about 10–50 cP.

Measured on the experiment, the EPR spectrum is proportional to the stationary value of the projection of the magnetization on the y-axis, M_y. It can be found by equating Eq. (5.11) to zero. The solution is obtained in two steps. First, we express all M_{k-} through the components of the total magnetization of spins. Assuming steady state condition $\partial M_{k-}/\partial t = 0$, we obtain

$$M_{k-} = (-\varphi_k(\boldsymbol{K_{ex}C - 1/T_{ddsct}})M_- + i\gamma B_1\varphi_k M_0)/(-i(\omega_k + \delta_k - \omega) - 1/T_{2k} - (K_{ex}C + 1/T_{ddsd})). \tag{5.12}$$

We sum the left and right parts of the Eq. (5.12) and obtain the equation for the steady state value of the total transverse magnetization of the system M_-

$$M_- = -(K_{ex}C - 1/T_{ddsct})G(\omega)M_- + i\gamma B_1 M_0 G(\omega); \tag{5.13}$$

$$G(\omega) = \sum_k \frac{\varphi_k}{\{-i(\omega_k + \delta_k - \omega) - 1/T_{2k} - (K_{ex}C + 1/T_{ddsd})\}}.$$

This equation gives the total transverse magnetization of the system [2, 7, 8, 14, 15]

$$M_- = i\gamma B_1 M_0 G(\omega)/[1 + (K_{ex}C - 1/T_{ddsct})G(\omega)]. \tag{5.14}$$

Equations (5.11, 5.12, 5.13 and 5.14) are valid in the case of equivalent spin exchange. When nonequivalent spin exchange is realized then we have to substitute in Eqs. (5.11, 5.12, 5.13 and 5.14) kinetic parameters ($K_{ex}C + 1/T_{ddsd}$) and ($K_{ex}C-1/T_{ddsct}$) by ($KC + 1/T_{ddsd}$) and ($K'_{ex}C-1/T_{ddsct}$), respectively (see (4.1 and 4.6)). Thus in the nonequivalent spin exchange case the steady state EPR spectrum in the linear response region ig given by

$$M_- = i\gamma B_1 M_0 G_{ne}(\omega)/\left[1 + \left(K'_{ex}C - 1/T_{ddsct}\right)G_{ne}(\omega)\right];$$
$$G_{ne}(\omega) = \sum_k \frac{\varphi_k}{\{-i(\omega_k + \delta_k - \omega) - 1/T_{2k} - (KC + 1/T_{ddsd})\}}. \tag{5.15}$$

M_y–projection of magnetization is equal to $M_y = -\,Im[M_-]$. Note that the steady state EPR spectrum is proportional no M_y.

The fact that the exchange interaction and dipole-dipole interaction lead to the "reaction" coherence transfer with opposite signs has one remarkable consequence if the condition ($K_{ex}C-1/T_{ddsct}$) = 0 is satisfied. Note that $1/T_{ddsct} = K_dC$, where K_d does not depend on concentration C of spins. If the condition ($K_{ex}C-1/T_{ddsct}$) = 0 is satisfied at some definite viscosity, then it will satisfy as well at all concentrations. Under this particular situation we can vary the rate of the spin exchange $K_{ex}C$ by changing spin concentration but never there will appear the spin exchange induced shift of the resonance frequencies and collaps of the spectrum into one narrow homogeneously broadened spectral line. At the same time, exchange broadening of lines will occur. Indeed, from the general solution (5.14 and 5.15) we can see that in the absence of spin coherence transfer, i.e. when ($K_{ex}C-1/T_{ddsct}$) = 0, the spectrum is a superposition of individual Lorentzian lines with resonance frequencies $\omega_k + \delta_k$, and their widths are given by the values

$$\Delta\omega_{1/2}(k) = 1/T_{2k} + (K_{ex}C + 1/T_{ddsd}), \tag{5.16}$$

when ($K_{ex}C - 1/T_{ddsct}$) = 0.

At first glance it may seem that (5.16) contradicts to the well-known result that in the slow spin exchange region the contribution of the exchange interaction to the line broadening is given by the value of $(1-\varphi_k)K_{ex}C$ rather than $K_{ex}C$ as in Eq. (5.16). This is because, in this case, the coherence transfer caused by the dipole-dipole interaction fully compensates the contribution of the coherence transfer caused by the exchange interaction between spins at the same frequency, i.e. between spins belonging to the same component of the spectrum. Since the exchange and dipole-dipole interactions transfer coherence in the antiphase, with different signs, there is a negative interference of the contributions of the two mechanisms in the transfer of spin coherence. It would be interesting to implement an experiment in which the viscosity of the solution makes it possible to have the equal contributions of the exchange and dipole-dipole interaction to the transfer of spin coherence.

To illustrate the impact of the transfer of spin coherence on the shape of the spectrum in Fig. 5.1 the EPR spectra of a model nitroxide radical calculated by formulas (5.14 and 5.15) at different concentrations of spin probes are presented. In EPR experiments, the derivative of the spectrum is detected, so in Fig. 5.1 spectrum derivatives are also given. It is assumed that an unpaired electron interacts with one magnetic nucleus ^{14}N with spin $I = 1$, which gives three nitrogen hyperfine components in the EPR spectrum at low radical concentrations. We also assume that each nitrogen hyperfine component has an unresolved hyperfine structure due to interaction with protons and that the inhomogeneous broadening of each nitrogen component of the spectrum can be described by a Gaussian distribution with dispersion σ.

Figure 5.1 shows that the shape of the spectrum is highly dependent on the concentration of radicals. At low concentrations, all spectral lines are broadening and the extreme components shift to the center of the spectrum. At high concentrations of radicals the whole spectrum "collapses" into one homogeneously broadened line in the center of gravity of the spectrum. If the dipole-dipole interaction makes a negligible contribution, then at high concentrations of radicals the spectrum turns into one narrow line (see the behavior of dotted curves when the concentration changes in Fig. 5.1). This effect is called exchange narrowing of the spectrum. But if, along with the spin exchange, the dipole-dipole interaction also contributes to the kinetics of quantum coherence, then at high concentrations also the EPR spectrum is broadened by the dipole-dipole interaction of spins, although the collapse of all spectral lines into one homogeneously broadened line occurs (cf. the width of two spectra at a concentration of $C = 500$ mM/L, the lower series of curves in Fig. 5.1). The collapse of the spectrum occurs when the rate of spin coherence transfer becomes equal to the "nitrogen" splitting of the spectrum. In this case, the EPR spectrum is described by the Lorentzian line with a width equal to (see, for example, [7, 8, 14, 16])

$$1/T_{2\mathrm{eff}} \equiv 1/T_{\mathrm{ddsd}} + 1/T_{\mathrm{ddsct}} + 2a_N{}^2/(3(K_{ex}C - 1/T_{\mathrm{ddsct}})) + 1/T_2. \qquad (5.17)$$

In the slow exchange region the dipole-dipole interaction contribution to the line width is equal to $1/T_{\mathrm{ddsd}} + 1/(3T_{\mathrm{ddsct}})$, while under the exchange narrowing conditions the dipolar contribution to the width is equal to $1/T_{\mathrm{ddsd}} + 1/T_{\mathrm{ddsct}}$ (see 5.17). Note that in the region of exchange narrowing the dipole-dipole interaction broadens the spectrum $(27/19) \approx 1.4$ times more than in the region of slow spin exchange. The latter observation correlates with the result known in the theory of magnetic resonance. In magnetically concentrated systems, the second moment of dipole-dipole interaction is greater if the resonance frequencies of all spins are the same [4, 17]. In the conditions of exchange narrowing, a situation is realized when

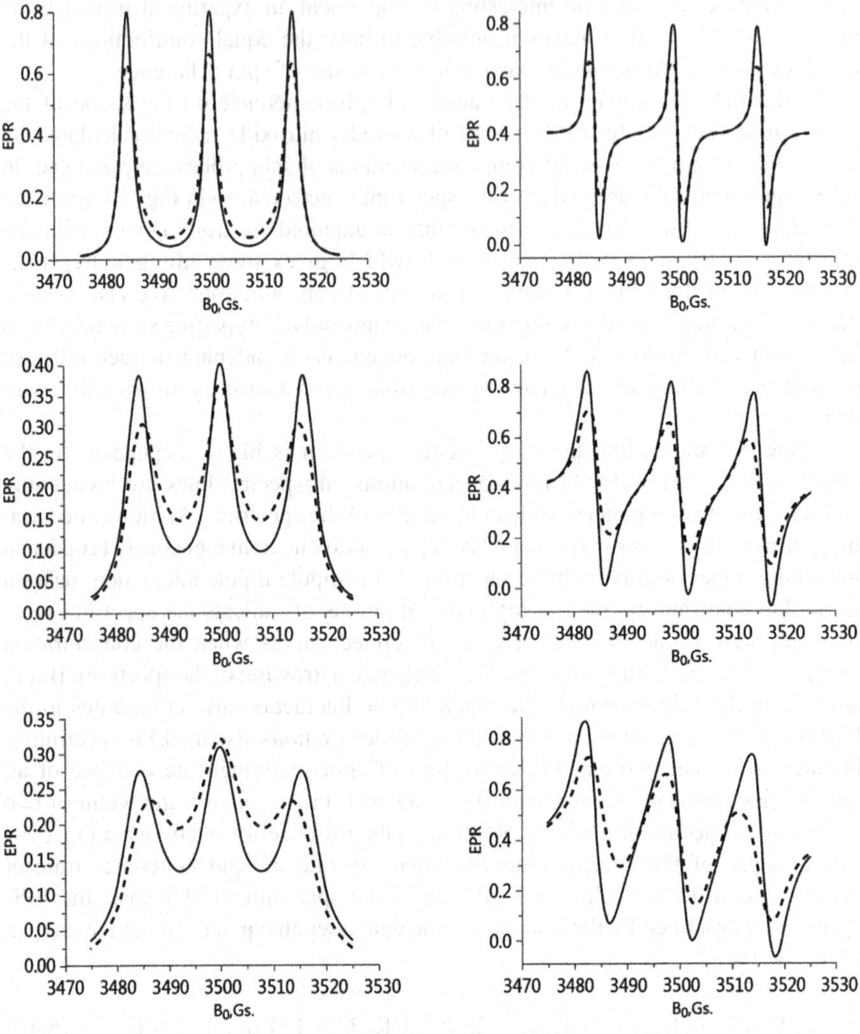

Fig. 5.1 Numerical simulations of the EPR spectrum (left) and its derivative (right) of ^{14}N nitroxide radicals for several concentrations of spins. The following parameters were used during these calculations: $a_N = 16$ Gs, g = 2, $\sigma = 0.12\,\text{Gs}^2$, $\Gamma_{0k} = 0.2$ Gs, W = V = K_{ex} C = 0.1 C Gs (dashed lines), $K_{ex} = 0.05$ Gs L/mM, $K_{dsd} = 0.011$ Gs L/mM, $K_{dsct} = 0.009$ Gs L/mM (solid lines), C is concentration of radicals in mM/L units, B is external magnetic field strength. In EPR spectroscopy, frequency is usually measured in Gauss, so the rate of the spin processes are given in Gauss. One Gauss corresponds to a angular frequency of 1.76 10^7 rad/s. Curves on rows from the top to bottom correspond to concentrations C = 20 mM/L, 50 mM/L, 75 mM/L, 200 mM/L, 500 mM/L, respectively. (Adapted with permission from Ref. [14])

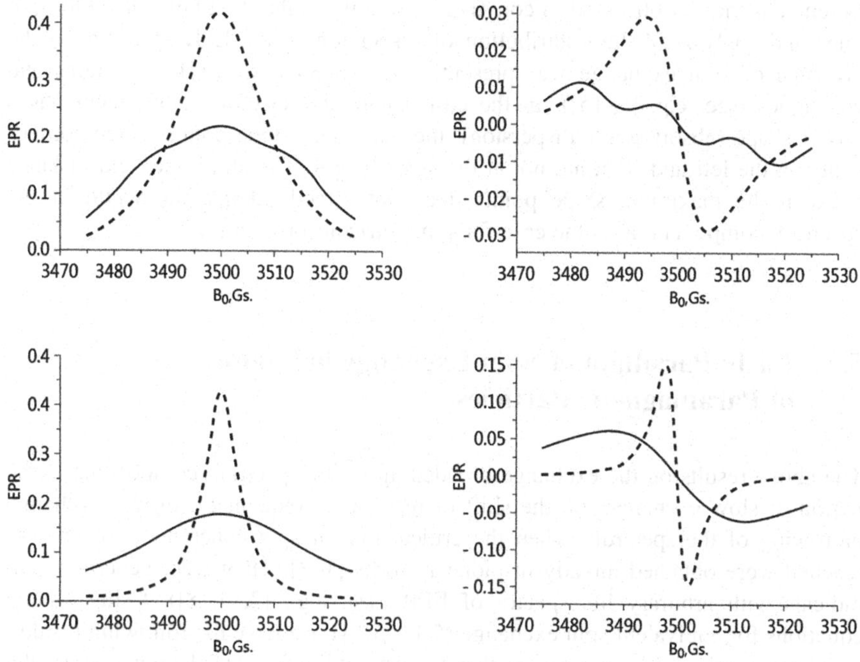

Fig. 5.1 (continued)

phenomenologically the frequency of all spins becomes effectively the same in contrast to the situation of slow exchange, when each spin has its resonance frequency.

The curves given in the second and third rows in Fig. 5.1, show that in the region of slow spin exchange (at intermediate values of the concentration of spins) the extreme components of the spectrum are asymmetric and are the sum of the symmetric absorption line and the asymmetric dispersion line. Proof of this conclusion is presented in the next section. The asymmetry of the extreme lines may be indicated by the fact that in the derivative of the spectrum for the extreme components of the spectrum, the values of the derivatives at the points of maximum slope do not coincide (see the curves on the right in the three upper rows in Fig. 5.1). In fact, these values of the derivatives of components of the spectrum at the maximum slope points may also differ due to the overlap of the wings of the extreme components of the spectrum with the central component. However, it turns out that the asymmetry of the extreme components is larger than could be expected as a result of simply overlapping of the adjacent components of the spectrum. The mixed form of EPR spectrum lines due to spin coherence transfer I theoretically predicted [7, 8]. But in the early years of studying spin exchange, the manifestation of a mixed form of resonance lines in the region of slow spin exchange was not studied in the experiment. Apparently, this was due to the sensitivity of EPR equipment. However, it should be noted that in [18] for an aqueous solution of $K_2(SO_3)_2NO$, it turns out that already in 1963 a deviation of the line shape from the

Lorentzian one was observed. A convincing experimental proof of the mixed form of lines and analysis of the contribution of dispersion to the EPR spectrum in the presence of spin exchange was presented in a number of works by Bales and colleagues (see, e.g. [9–15]). In the case where the spectrum component has a mixed shape (absorption + dispersion), the maximum slope points of the components on the left and right are not at the same height. Therefore, the field distance between the maximum slope points does not directly define the width of the spectrum component, as is the case for symmetric absorption lines.

5.3 Early Paradigm of Spin Exchange in Solutions of Paramagnetic Particles

The above results on the exchange broadening of the spectrum components in the region of slow exchange, on the shift of resonance frequencies, on the exchange narrowing of the spectrum when the critical rate of spin coherence transfer was reached were obtained already in pioneer works [1–3]. They were generalized to radicals with arbitrary hfs spectra of EPR (see. e.g., [2, 7, 8]). Using kinetic equations for equivalent spin exchange (5.11, 5.14), we have the following results:

in the region of slow spin exchange, the additional broadening due to the exchange interaction is given by the value

$$\Delta\omega_{1/2}(\mathrm{k}) = (1 - \varphi_\mathrm{k})\mathrm{K_{ex}C}. \tag{5.18}$$

If we also consider the contribution of the dipole-dipole interaction, the concentration broadening of the k-th component of the spectrum is given by

$$\Delta\omega_{1/2}(\mathrm{k}) = (1 - \varphi_\mathrm{k})\mathrm{K_{ex}C} + 1/\mathrm{T_{ddsd}} + \varphi_\mathrm{k}/\mathrm{T_{ddsct}}. \tag{5.19}$$

In the situation of slow equivalent spin exchange, the total shift of the resonance frequency of the k-th component is given

$$\Delta\omega_\mathrm{k} = \omega_\mathrm{k} - \omega_\mathrm{0k} = \delta_\mathrm{k} - \varphi_\mathrm{k}(\mathrm{K_{ex}C} - 1/\mathrm{T_{ddsct}})^2 \sum_{m \neq k} \varphi_m/(\omega_{0k} - \omega_{0m}). \tag{5.20}$$

Here ($\mathrm{K_{ex}C} - 1/\mathrm{T_{ddsct}}$) is the rate of the spin coherence transfer caused by the exchange and the dipole-dipole interactions, ω_0k–resonance frequency of the k-th component of the spectrum of the isolated radical, ω_k is the expected frequency in the presence of the spin coherence transfer, φ_k is the statistical weight of the k-th component of the EPR spectrum, δ_k -additional shift of the resonance frequencies induced by the exchange interaction in a course of a meeting of paramagnetic particles (in a course of the spin exchange elementary act). Note that $1/\mathrm{T_{ddsd}}$, $1/\mathrm{T_{ddsct}}$ are proportional to the spin concentration, and therefore the broadening of the components (5.19) due to the exchange and dipole-dipole interactions is proportional to C, and two terms of the frequency shifts in (5.20) are proportional to C and C^2, respectively.

Note that the formulas (5.19 and 5.20) were obtained [7, 8] using the perturbation theory for the small parameter $(K_{ex}C-1/T_{ddsct})/(<\Delta^2>)^{1/2}<1$, where $<\Delta^2>$ is the dispersion of the resonance frequencies of spins relative to the center of gravity of the spectrum, which characterizes the spread of the resonance frequencies of spins in different components of the steady state EPR spectrum. If the frequency spread is caused by the hyperfine interaction of an unpaired electron of radical with its magnetic nuclei,

$$< \Delta^2 >= (1/3)\sum_n I_n(I_n+1)a_n^2. \tag{5.21}$$

Here the summation is carried out on all the magnetic nuclei of the radical. For example, for radicals having one nuclear with spin $I = 1/2$, $<\Delta^2> = a^2/4$. As already noted, in early theories of paramagnetic relaxation of spins in solutions, it was believed that the dipole-dipole interaction contributes only to the dephasing (decoherence) of spins, but does not contribute to the transfer of coherence, namely, reaction of coherence in spin-spin dipole-dipole interaction). Therefore, in early theories it was believed that the frequency shift in the case of slow transfer of spin coherence is given by an expression in which there is no effect of the dipole-dipole interaction

$$\Delta\omega_k = \omega_k - \omega_{0k} = -\varphi_k(K_{ex}C)^2\sum_{m\neq k}\varphi_m/(\omega_{0k}-\omega_{0m}). \tag{5.22}$$

On the example of a radical with one magnetic nucleus above, it was shown that at a sufficiently high rate of equivalent spin exchange, the entire spectrum turns into one narrow line with a width of $\Delta\omega_{1/2}(1) = 1/T_2 + a^2/(4K_{ex}C)$. It can also be expressed through the dispersion of the distribution of resonant frequencies

$$\Delta\omega_{1/2}(1) = 1/T_2 + < \Delta^2 > /(K_{ex}C) \tag{5.23}$$

In a situation of fast equivalent spin exchange when

$$K_{ex}C > \left(< \Delta^2 >\right)^{1/2} \tag{5.24}$$

the expression (5.23) describes the contribution of spin exchange to the width of the EPR spectrum for an arbitrary hfs of the EPR spectrum of the radical.

Thus, already in the first studies of spin exchange using EPR spectroscopy the following observations were made:

1. In the situation of slow spin exchange, when $K_{ex}C < (<\Delta\omega^2>)^{1/2}$, the components of the resolved hfs components of a spectrum are broadening by the value of $(1 - \varphi_k)K_{ex}C$. This exchange broadening increases linearly with increasing concentration of the radicals C. The exchange broadening of the more intense hfs components of the spectrum is less than the broadening of the less intense component of the spectrum.
2. In the situation of slow spin exchange, the resonance frequencies of the spectrum components are shifted to the center of gravity of the spectrum in accordance with Eq. (5.21). The contribution of the dipole-dipole interaction to the discussed frequency shift was not taken into account. The frequency shift (5.21) increases in proportion to the square of the spin concentration.
3. In the situation of fast spin exchange, $K_{ex}C > (<\Delta^2>)^{1/2}$, the whole inhomogeneously broadened spectrum collapses into one homogeneously broadened Lorentzian line at the center of gravity of the frequency spectrum of isolated radicals. In this region of the exchange narrowing of the spectrum, the contribution of the exchange interaction to the width is inversely proportional to the concentration of spins (5.23).

These observations formed the theoretical basis for the paradigm of determining the rate of spin exchange using EPR spectroscopy. It is proposed to measure the concentration dependence of the line width and/or the shift of the resonant frequencies of the EPR spectrum components in the region of low concentrations of radicals (the situation of slow exchange) and to find the rate of spin exchange. This rate can also be found by studying the concentration dependence of the line width in the region of sufficiently high concentrations of radicals, when the situation of fast exchange and collapse of the spectrum into one line is realized.

It may seem that all three methods of measuring the rate of spin exchange look quite equivalent, at least for the case of equivalent exchange should give the same value of the rate of spin exchange. But the experiment mainly uses the broadening of lines in the area of slow spin exchange. To measure the rate of exchange in a situation of exchange narrowing of the EPR spectrum, it is necessary to have a sufficiently high concentration of paramagnetic particles, usually free radicals. Often it is simply not possible to obtain solutions with the necessary high concentrations of spins. As for the shift of lines, the situation is complicated due to the fact that there are several mechanisms for the shift of lines. Therefore, the concentration-dependent frequency shift observed in the experiment is not limited to the value given by the expression (5.23). Below we will return to this issue.

Note that in the case of non-equivalent spin exchange, the broadening of the lines in the slow exchange, on the one hand, and the frequency shift in the slow exchange and the broadening of the lines in the fast spin exchange case (fast transfer of coherence), on the other hand, can lead to *different values of the rate* of change of the spin state, since they are associated with different elementary spin processes: decoherence of spins and transfer of spin coherence, respectively. Concentration changes in the shape of the spectrum can also be associated with the dipole-dipole spin-spin interaction, which also gives a spin-dependent contribution to paramagnetic relaxation. All this was well understood always. However, this paradigm has a drawback: it was commonly accepted that the dipole-dipole interaction does not contribute to the transfer of spin coherence, i.e. does not cause the effect of "reaction" of quantum coherence from interaction with the spin-partner (see, for example, [4], Eqs. (VIII.35, VIII.36, VIII.89). Thus, in the widespread paradigm of spin exchange in solutions it was accepted that the dipole-dipole interaction causes only decoherence of spins. This means that the dipole-dipole interaction does not contribute to the shift of the resonant frequencies of the spectrum components and does not lead to the collapse of all components of the STS spectrum in one line, as in the case of the exchange interaction. Only the contribution of the dipole-dipole interaction to the broadening of the lines was taken into account, both in the case of slow spin exchange and in the conditions of the exchange narrowing of the spectrum.

If we assume that the dipole-dipole interaction does not contribute to the transfer of spin coherence, the problem of separating the contributions of the exchange and dipole-dipole interactions is greatly simplified. Indeed, then the frequency shift should be determined only by the exchange interaction in collisions (5.19), and from the frequency shift it would be possible to directly find the rate of the spin exchange. Note also that in the conditions of the exchange narrowing of the spectrum, the dipole-dipole interaction gives a contribution to the width of the line, which grows linearly in the concentration of spins, and the transfer of coherence due to the exchange interaction gives a contribution to the width of the line, which is inversely proportional to the concentration of spins (see Eq. 5.16). If we neglect the transfer of spin coherence due to the dipole-dipole interaction, the concentration dependence of the spectrum width in the region of exchange narrowing, in principle, makes it possible to separate the contributions of the exchange and dipole-dipole interactions. To do this, we need to present the width of the spectrum with a formula of the form $\Delta\omega_{1/2} = 1/T_2 + K_{dd}C + <\Delta^2>/(K_{ex}C)$ and choose unknown parameters so that this formula best describes the experimental data obtained at different spin concentrations.

In the region of exchange broadening of the spectrum components, both interactions, exchange and dipole-dipole, give contributions that grow linearly with the concentration of spins. In this situation, a difference in their dependence on the coefficient of mutual diffusion of radicals may be useful for the separation of these contributions. But here may be problems associated with the fact that the rate of spin exchange in general case is controversially depending on the diffusion coefficient of the particles, this dependence is actually the opposite for weak and strong spin exchange cases (see discussion of this issue in Sect. 5.2).

5.4 A New View on the Manifestation of Spin Exchange in EPR Spectra. Collective Modes of Evolution of Quantum Coherence of Spins

Calculations of EPR spectra with spin exchange is reduced to solving a system of non homogeneous linear equations for spin magnetizations with constant coefficients. The numerical solution of these equations using computers is not a problem. The implementation of such spectra calculations will be given below. But it should be noted that this "gross" approach does not allow to highlight the physical nature of the observed transformations of the spectrum (see, e.g., Fig. 5.1).

There is another approach that allows us to better highlight the physics of what is happening, based on finding collective independent modes of quantum coherence kinetics [5, 7, 8, 14–16]. To do this, we write the system of Eq. (5.11) in another form. We introduce vectors **M** $_-$ and $\mathbf{M}_0$ with projections $\{M_{k-}\}$ and $\{\varphi_k M_0\}$, respectively. From the coefficients of the system of Eq. (5.11) we make a matrix L. Equations (5.11) for the steady state solution can be written in the form

$$\boldsymbol{LM}_- - i\gamma\, B_1 \boldsymbol{M}_{eq} = 0. \tag{5.25}$$

One can find the eigenvalues $\{\lambda_k\}$ and eigenvectors X_k of the evolution operator **L** by solving the equation

$$LX_k = \lambda_k X_k. \tag{5.26}$$

Eigenvectors of **L** are a superposition of transverse magnetizations of spins M_{k-} belonging to different components of the EPR spectrum. In fact, these eigenvectors are independent collective modes of evolution of quantum coherence of the system. Each collective mode, in principle, can give the effect of resonance response to external influences. The observed absorption spectrum is the sum of the independent contributions of the resonant responses of collective modes to the external variable field. It remains to find how the external field excites each collective mode [5, 7, 8]. To do this, we need to find a matrix **U** that performs a similarity transformation from the matrix **L** to the diagonal matrix $\mathbf{ULU}^{-1}$ [20]. Accordingly, it is necessary to transform the vector **F**, which describes the action of an external perturbation $\mathbf{F} \equiv i\gamma B_1 M_0$ (see right part (5.25)). After conversion, this vector is set as

$F* = i\gamma B_1 \mathbf{U} \mathbf{M}_0$. Similarly, the vector $\mathbf{M}_-$ is also transformed to collective modes, $\mathbf{M}_-* = \mathbf{UM}_-$. As a result, the contribution of collective modes to the observed spectrum is given by the values:

$$M*_{k-} = F*_k/\lambda_k. \tag{5.27}$$

The resonance effect is observed when $|\lambda_k|$ takes the minimum value. Note that λ_k have the form $\lambda_k = (-i(\Omega_k - \omega) - \Delta\Omega_k)$. The imaginary part of λ_k determines the resonance frequency Ω_k, and the real part sets the "width" of the resonance. The contribution of collective modes to the spectrum is determined by the components of the vector **F**∗. It was shown [5, 7, 8] that in the general case $F*_k = c_k + if_k$. As a result, we obtain that the contribution of a single collective mode to the absorption spectrum is given as

$$J_k* = -ImM*_{k-} = f_k(\Delta\Omega_k + p_k(\Omega_k - \omega))/\left((\Omega_k - \omega)^2 + \Delta\Omega_k{}^2\right). \quad (5.28)$$

It can be seen that in the general case the collective mode resonance is a mixture of the Lorentzian absorption line and the Lorentzian dispersion line. The spectrum observed in the experiment is determined by

$$M_y = -ImM_- = -ImUM_-* \quad (5.29)$$

and is a superposition of lines J_k* (5.28). Thus, we have two identical representations of the spectrum observed in the experiment: one as the sum of the lines of independent collective modes (5.28), and the other is given by Eqs. (5.14 and 5.15).

In general, when the EPR spectrum consists of many components, the problem can be solved analytically only for the case of slow transfer of quantum coherence, when it is possible to apply the perturbation theory for the small parameter $|K_{exsct}C - \varphi_k(1/T_{2ddsct})|/\Delta\omega < 1$, where $\Delta\omega$ is the inhomogeneous width of the spectrum. For the first time I noted this mixed shape of the EPR lines in [7, 8]. Correct kinetic equations for spin relaxation in solutions taking into account the exchange and dipole-dipole interaction was presented in [21] and further developed in refs. [14–16, 22].

From Eq. (5.11) in the first order of perturbation theory we obtain the contribution of spin exchange and dipole-dipole interaction to the homogeneous broadening of the k-th collective resonance. For example, in the case of equivalent spin exchange, when $\Delta\omega\tau_c < 1$, the line width of the collective resonance is

$$\begin{aligned} \Delta\Omega_k &= 1/T_{2ddsd} + \varphi_k(1/T_{2ddsct}) + (1 - \varphi_k)K_{ex}C, \quad \text{(equivalent spin exchange)}; \\ \Delta\Omega_k &= 1/T_{2ddsd} + \varphi_k(1/T_{2ddsct}) + KC - \varphi_k K'_{ex}C, \quad \text{(non-equivalent spin exchange)}. \end{aligned} \quad (5.30)$$

In the same situation, in the second order of the perturbation theory, we obtain a frequency shift of the k-th collective resonance due to spin exchange and dipole-dipole interaction:

$$\delta\omega_k = \varphi_k\left(K'_{ex}C - (1/T_{2ddsct})\right)^2 \Sigma'\varphi_n/(\omega_n - \omega_k), \quad (5.31)$$

where the summation in the right part is under the condition $n \neq k$.

The contribution of a dispersion term for each collective mode is given by the parameter ([5, 7, 8, 14])

$$p_k = \left(K'_{ex}C - (1/T_{2ddsct})\right)\sum_{n\neq k} 2\varphi_n/(\omega_k - \omega_n) \quad (5.32)$$

Hence, we have that for low-field components of the spectrum $p_k > 0$, and for high-field components $p_k < 0$, provided that the exchange interaction contributes more to the transfer of coherence than the dipole-dipole interaction.

In the case of slow equivalent spin exchange the Eqs. (5.30, 5.31 and 5.32) allow, in principle, to find the rate constant of spin exchange, if there are known dependences on the concentration of spins of a broadening of the resonances, of a frequency shift of the resonances, and of a contribution of dispersion to the form of the resonance lines. But the realization of this potential possibility is not a straightforward matter. From the experiment, in any case, we obtain the total contribution of the exchange and dipole-dipole interaction (see Eqs. 5.30, 5.31 and 5.32). Therefore, there is a problem of separation of these contributions. In addition, the mixed shape of the lines makes it difficult to find the width and resonance frequency from the EPR spectrum (see below).

5.5 Collective Modes of Motion of the Spin Coherence for Model Systems

Two-Frequency Model To illustrate the formation of collective modes in the evolution of spin coherence in solutions of paramagnetic particles, we consider a simple system [5]. Consider a dilute solution of stable free radicals in which an unpaired electron interacts with only one ^{15}N magnetic nucleus. The hyperfine interaction constant is a. The whole ensemble of radicals can be divided into two sub-ensembles with different projections of the nuclear spin into the direction of the external constant magnetic field. The transfer of coherence is induced by spin exchange and dipole-dipole interaction. To calculate the shape of the EPR spectrum under conditions where saturation effects do not occur, it is necessary to find a steady state solution of (5.11). For the considered two-frequency model kinetic equations have already been given (5.3). Let's bring them again for convenience of consideration in case when the equivalent spin exchange operates and makes dominant contribution so that dipole-dipole contribution can be neglected

$$\begin{aligned} \partial M_{1-}/\partial t &= (-i(\omega_0 + a/2 - \omega) - \Gamma)M_{1-} - VM_{1-} + \mathbf{VM_{2-}} - i(1/2)\omega_1 M_0, \\ \partial M_{2-}/\partial t &= (-i(\omega_0 - a/2 - \omega) - \Gamma)M_{2-} - VM_{2-} + \mathbf{VM_{1-}} - i(1/2)\omega_1 M_0. \end{aligned} \quad (5.33)$$

Here ω_0 is the Zeeman frequency of the electron spins of radicals, $\Gamma = 1/T_2$ is the decoherence rate of the spins of isolated radicals, ω and $\omega_1 = \gamma B_1$ are the carrier frequency and the Rabi frequency of the microwave field, respectively. Note that Eq. (5.33) are recorded in a coordinate system rotating at a frequency ω. In this situation, $V = (1/2)K_{ex}C$, where C is the concentration of all radicals and (1/2) C is the concentration of radicals in each of the two sub-assemblies with different nuclear spin projections. For the considered model system (5.33) we have a matrix of coefficients

$$\boldsymbol{L} = \{\{-\mathrm{i}\,(\omega_0 + a/2-\omega)-\Gamma-V, V\}, \{V, -\mathrm{i}(\omega_0-a/2-\omega)-\Gamma-V\}\}. \quad (5.34)$$

The eigenvalues of **L** and the corresponding eigenstates X_k are

$$\begin{aligned} &\lambda_1 = -\mathrm{i}(\omega_0-\omega)-R/2-\Gamma-V, \quad \lambda_2 = -\mathrm{i}(\omega_0-\omega) + R/2-\Gamma-V; \\ &X_1 = c_1\{-i(a+R)/(2V), 1\}; \quad X_2 = c_2\{(-ia+R)/(2V), 1\}. \end{aligned} \quad (5.35)$$

Here $R \equiv (-a^2 + 4V^2)^{1/2}$, c_1 and c_2 are normalizing factors.

If the rate of spin coherence transfer is negligibly small, $V \approx 0$, the distance between the two lines in the spectrum is equal to the constant of hyperfine interaction a. In region $2\,|V| < |a|$ line splitting in the spectrum is equal to $Im\{R\} = (a^2 - 4V^2)^{1/2} \approx a - 2\,V^2/a$, i.e. with increasing rate of the spin coherence transfer the splitting of the lines decreases, the lines move towards each other. In this case, both resonances have the same width equal to $V + \Gamma$. Two frequencies (imaginary parts of λ_1 and λ_2) coincide when the critical rate of coherence transfer $|V_c| = |a|/2$ is reached. There are also two resonances in the $|V| \geq |a|/2$ region. The frequencies of these resonances coincide, but they have different widths $\Delta\Omega_{\pm} \approx V + \Gamma \pm (1/2)R$. When $2|V| \gg |a|$, the line widths can be approximated as $\Delta\Omega_{\pm} \approx V + \Gamma \pm (V-a^2/(8V))$. One of the resonances turns out to be narrow, $\Delta\Omega- \approx \Gamma + a^2/(8V)$, and as V grows, its width decreases with increasing spin concentration. This effect is well known as the exchange narrowing of the spectra. The second resonance has a large width $\Delta\Omega+ \approx 2V + \Gamma-a^2/(8V) \rightarrow 2V + \Gamma$. Thus, under the conditions of exchange narrowing, the spectrum should consist of two lines with the same resonance frequency, but one of them is a narrow line and the other is wide. The eigenvalues of the operator **L** give information about the frequency and width of possible resonant excitations of the system. But the shape of the resonance lines and the integral intensity of the resonance response of different collective modes depends on how the external field excites these collective modes, i.e. on the values of $F*_k$. Calculations carried out according to the algorithm described above lead to the result

$$\mathbf{F}^* = \mathrm{i}\omega_1 \mathbf{X}\mathbf{M}_{eq} = (\mathrm{i}\omega_1/2)M_0\{(2V-ia - R)/(2R), (2V - ia + R)/(2R)\}. \quad (5.36)$$

In the conditions of exchange narrowing, when the rate of coherence transfer $V > a/2$,

$$F^* \approx \{(i\omega_1/2)M_0\{(-ia/(4V), 2 + ia/(4V)\} \rightarrow i\omega_1/2)M_0\{0, 2\}. \quad (5.37)$$

It can be seen that the external microwave field B_1 effectively excites both collective modes when spin exchange is slow, $V < a/2$, and only one of collective

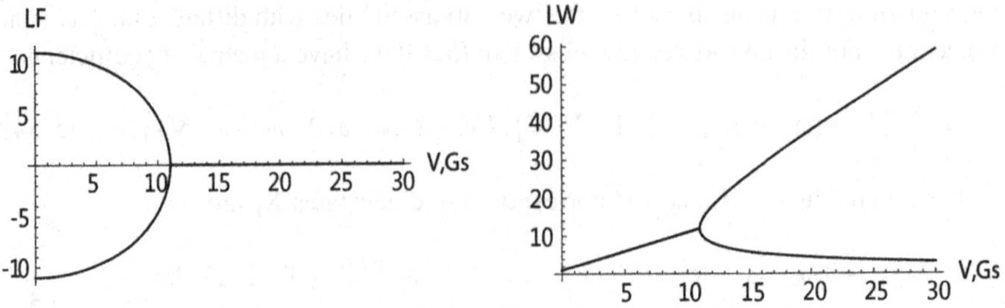

Fig. 5.2 The dependence of the resonance frequencies (figure on the left) of the collective modes of spin coherence and widths (figure on the right) of these collective resonances on the spin coherence transfer rate V for the two-frequency model. Calculations are carried out for the values of the parameters: $a = 22$ Gs, $\Gamma = 1$ Gs. (Adapted with permission from Ref. [5])

modes, which is given by the eigenvector X_2 when spin exchange is fast, $V > a/2$. Let us analyze the eigenstates, X_k, $k = 1, 2$ (5.35), the generalized magnetization vectors of the system, which give independent collective modes of evolution. In the region of slow spin exchange, when $2V \ll a$, the eigenvectors of the system (5.35) are equal to $X_1 \approx (M_{1-} - i(V/a)M_{2-})$; $X_2 \approx (-i(V/a)M_{1-} + M_{2-})$. In this case, the collective eigenvectors consist of the vector of the magnetization of a single component and contain a small contribution of the vector of magnetization of the other components (sub-ensembles of spins with different resonance frequency).

In another limiting situation of fast spin exchange, $2V > a$, i.e. in the situation of exchange narrowing of the spectrum, the eigenvectors tend to

$$X_1 \approx (-M_{1-} + M_{2-})/2^{1/2};\ X_2 \approx (M_{1-} + M_{2-})/2^{1/2}. \tag{5.38}$$

So at $V > a/2$ the collective motion mode with X_2 corresponds to in-phase mode motion of the magnetization components. In the EPR experiment, the microwave field excites practically only the symmetric eigenvector X_2.

To visualize the marked changes in the shape of the EPR spectra with a change in the spin exchange rate, we present several figures. Note that in EPR spectroscopy hyperfine constants of the interaction, the splitting in the spectra, the frequencies of the resonances and the broadening of the lines is measured in Gauss. Therefore, we express the parameter of the coherence transfer rate V in Gauss. To obtain the numerical value of V in rad/s units, one has to multiply the numerical value of V in Gauss by $1.76\ 10^7$.

Figure 5.2 shows the behavior of the resonance frequencies and resonance widths for collective modes of quantum coherence evolution with increasing spin exchange rate. It is seen that at $2V_c = \text{KexC} = a$, there is a sharp transition in the behavior of the frequencies and resonances widths of collective modes. It is interesting that the change in the nature of the motion of the magnetization vectors near the critical point of the exchange narrowing has the character of a phase transition. But such a sharp transition was obtained under the assumption that the decoherence times of spins for

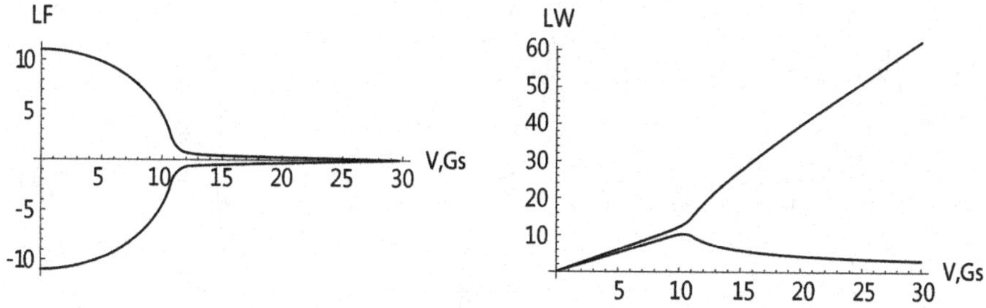

Fig. 5.3 The dependence of the resonance frequencies (figure on the left) of the collective modes of spin coherence evolution and the widths (figure on the right) of these collective resonances on the velocity V of the spin coherence transfer for the two-frequency model. Calculations are carried out for the values of the parameters: a = 22 Gs, $\Gamma_1 = 0.3$ Gs, $\Gamma_2 = 1$ Gs. (Adapted with permission from Ref. [16])

isolated particles in both sub-ensembles are the same. If these times are different, then the collapse of the entire spectrum in one exchange narrowed line ceases to be very contrast, and becomes more blurred (see Fig. 5.3).

Figure 5.3 shows the case when, in the absence of a spin exchange between two subunits of spins, the widths of the resonance lines of spins from different subensembles are not equal. It turns out that in this situation there is no sharp collapse of the spectrum at $V = a/2$, as was the case in $\Gamma_1 = \Gamma_2$ (see Fig. 5.2).

According to the general theory (see Eq. 5.28), the resonances of collective modes are described by the mixed form of the line as a mixture of the absorption and dispersion line: $J_k = J_{kabs} + J_{kdis}$. The exact decomposition of the total spectrum into contributions of two collective modes is found for the two-frequency model in general case when both exchange and dipole-dipole interactions operate [5]. In the case of slow transfer of coherence $a > 2$ V, $R = i(a^2 - 4V^2)^{1/2} \equiv i R_0$, we have [5]

$$\begin{aligned} J_1 - (1/2)\omega_1 M_0(\Gamma + W + (2\,V/R_0)(\omega_0 + R_0/2 - \omega))/((\omega_0 + R_0/2 - \omega)^2 \\ + (\Gamma + W)^2) = J_{1abs} + J_{1dis} \equiv J_{1absL} + pJ_{disL1}, \\ J_2 = (1/2)\omega_1 M_0(\Gamma + W - (2\,V/R_0)(\omega_0 - R_0/2 - \omega))/((\omega_0 - R_0/2 - \omega)^2 \\ + (\Gamma + W)^2) = J_{2abs} + J_{2dis} \equiv J_{2absL} - p\,J_{disL2}. \end{aligned} \tag{5.39}$$

In Eq. (5.39) W is the total rate of the spin decoherence induced by the exchange and dipole-dipole interactions and V is the total rate of the spin coherence transfer caused by both these interactions. In case of dominant equivalent spin exchange for the two frequency situation, $W = V = (1/2)K_{ex}C$.

Both components of the spectrum contribute to absorption and dispersion. The contribution of the absorption is described by Lorentzian curve of the form $J_{absL} = = (\Gamma + W)/(x^2 + (\Gamma + W)^2)$. The dispersion contribution can be represented as the product of a factor $p = \pm 2\,V/R_0$ on the Lorentzian dispersion curve of the

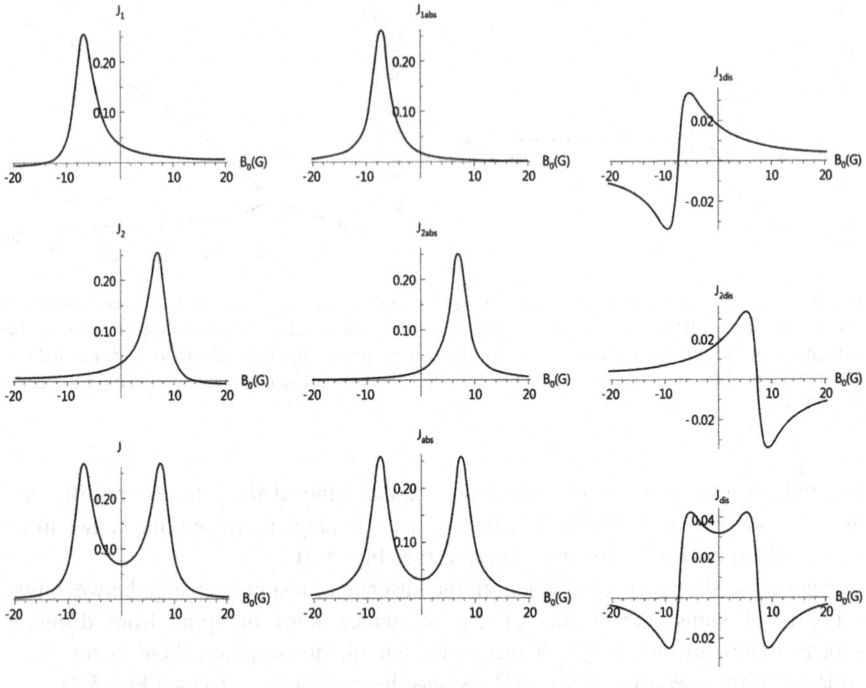

Fig. 5.4 Contributions of the collective mode resonances with the resonance frequencies $\omega_0 + R_0/2$ (top row) and $\omega_0 - R_0/2$ (central row) to the EPR spectrum, and the contribution of both collective independent modes, $J = J_1 + J_2$, (bottom row). Total contributions are on the left, absorption contributions are in the middle and dispersion contributions are on the right. Parameters are $a = 15$ Gs, $V = 2$ Gs, $W = V$, $\Gamma = 0$. (Adapted with permission from Ref. [5])

form $J_{disL} = x/(x^2 + (\Gamma + W)^2)$. When $V < a$, the parameter p approximately equals $2V/a$. The spectra calculated for the case of slow transfer of coherence and assuming that the exchange interaction dominates and the contribution of the dipole-dipole interaction can be neglected, i.e. $W = V$, are shown in Figs. 5.4 and 5.5.

Figure 5.4 shows that the collective modes of the spin coherence (transverse magnetization) manifest the EPR spectra of the mixed shape: they contain the absorption and dispersion terms. For parameters chosen in Fig. 5.4 the amplitude of the dispersion contribution is by the order of magnitude less than the amplitude of the absorption contribution.

Figure 5.5 shows spectra simulated in the case when the spin coherence exchange rate approaches the critical value $V \rightarrow a/2$, $R_0 \rightarrow 0$.

Figure 5.5 shows that close to the critical spin coherence transfer rate $V = a/2$ the dispersion contribution to the signal is pronounced in the response of the collective modes J_1 and J_2. Figure 5.5 illustrates that the dispersion contributions (the right-

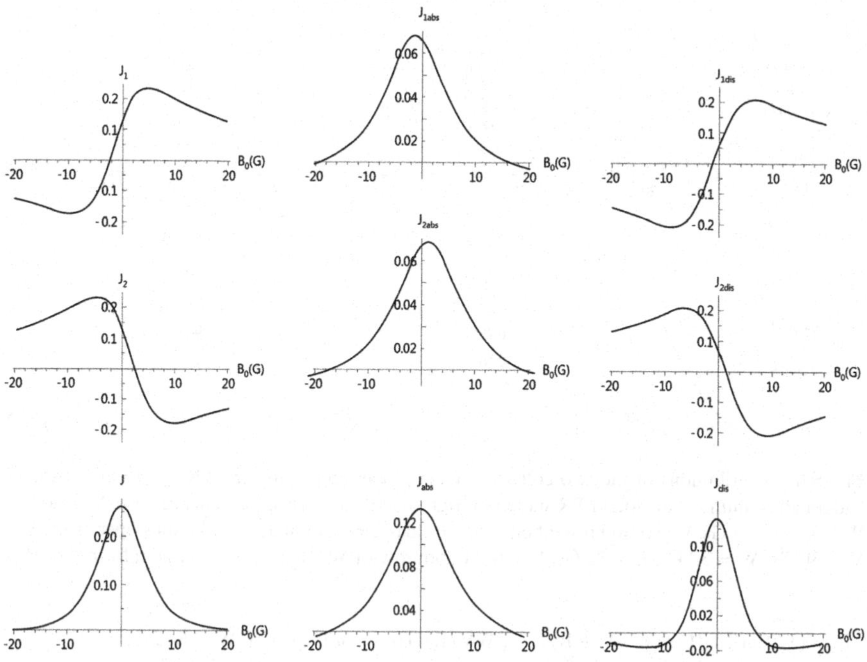

Fig. 5.5 Contributions of the collective mode resonances with the resonance frequencies $\omega_0 + R_0/2$ (top row) and $w_0 - R_0/2$ (middle row) to the EPR spectrum, and the contribution of both collective independent modes, $J = J_1 + J_2$, (bottom row). Total contributions are on the left, absorption contributions are in the middle and dispersion contributions are on the right. Parameters are $a = 15$ Gs, $V = 7.4$ Gs, $W = V$, $\Gamma = 0$. (Adapted with permission from Ref. [5])

hand curve in upper and middle rows) are broad curves and have amplitudes three times larger than the absorption contributions to the individual signals of the collective modes of the system. However, the total dispersion contributions of both collective modes, J_{dis} (the right-hand curve in the bottom row) has the amplitude that is almost equal to the amplitude of the total absorption contribution, $J_{abs} = J_{1abs} + J_{2abs}$ (the middle curve in the bottom row)

The analysis of the curves in Fig. 5.5 shows that the total spectrum observed in experiment has almost equal contributions from the absorption terms of both collective modes, $J_{1abs} + J_{2abs}$, and the dispersion terms of these modes, $J_{dis} = J_{1dis} + J_{2dis}$.

In the case of the fast coherence transfer, when $a < 2$ V, $R = (4\,V^2 - a^2)^{1/2}$, the contributions of the collective modes to the spectrum are given by Eq. (5.40)

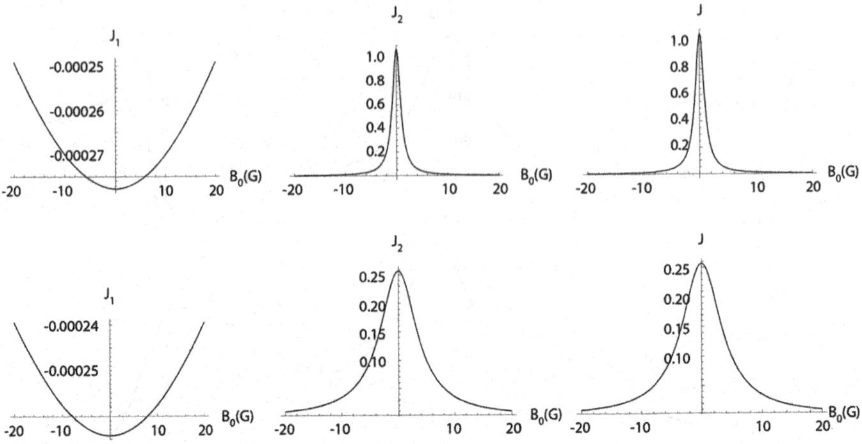

Fig. 5.6 Contributions of the two collective independent modes to the EPR spectrum (left-hand and middle columns), the total EPR spectrum (right-hand column): spin exchange case, V = 30 Gs, W = V, *a* = 15 Gs, Γ = 0 (top row); both, spin exchange and dipolar, contributions operating case, V = 30 Gs, W = 33 Gs, *a* = 15 Gs, Γ = 0 (bottom row). (Adapted with permission from Ref. [5])

$$
\begin{aligned}
J_1 &= (1/2)\omega_1 M_0(\Gamma + W - V + R/2 - 2WV/R)/((\omega_0 - \omega)^2 + (\Gamma + W + R/2)^2) \\
&= (1/2)\omega_1 M_0\{\Gamma + W - V - (a^2 + 4\,V(W - V))/(2\,R)\}/((\omega_0 - \omega)^2 + (\Gamma + W + R/2)^2), \\
J_2 &= (1/2)\omega_1 M_0(\Gamma + W - V - R/2 + 2\,W\,V/R)/((\omega_0 - \omega)^2 + (\Gamma + W - R/2)^2) \\
&= (1/2)\omega_1 M_0\{\Gamma + W - V + (a^2 + 4V(W - V))/(2\,R)\}/((\omega_0 - \omega)^2 + (\Gamma + W - R/2)^2
\end{aligned}
\tag{5.40}
$$

As it was already noted, in this case both collective mode resonance frequencies are the same but the widths of those resonances are different. According to Eq. (5.40), the J_1 resonance is broader than the J_2 resonance. It follows from Eq. (5.40) that under the resonance condition $\omega_0 = \omega$ the amplitude of the broader component J_1 is less than the narrower J_2 component. For example, when the contribution of the dipole-dipole interaction is negligible, i.e., W = V, asymptotic values of J_1 and J_2 are $lim_{V \to \infty} J_1(\omega_0 = \omega) = 0$, $lim_{V \to \infty} J_2(\omega_0 = \omega) = 1$. These features of the J_1 and J_2 collective mode resonance lines directly correlate with their excitation by the alternating magnetic field B_1.

Figure 5.6 shows spectra simulated in the case of the fast spin coherence transfer.

Figure 5.6 shows that in the fast spin coherence transfer case the exchange narrowing of the spectrum occurs. The dipole-dipole contribution to the spin relaxation leads to the concentration-dependent broadening of the exchange narrowed spectrum (compare the top and bottom spectra in Fig. 5.6). Note that only one of the two collective modes is manifested in the observed spectrum. Other mode appears as a signal with the negligible intensity and the negative sign (see left-hand spectra in Fig. 5.6).

The intensity of J_1 is small for two reasons. On one hand, this resonance is much broader than the J_2 resonance (see Fig. 5.6). This is an expected observation. Indeed, for the case $a = 15$ Gs, $V = W = 30$ Gs the eigenvalues of $\boldsymbol{L}$ are $\lambda_1 = -59.04 - i(\omega_0 - \omega)$, $\lambda_2 = -0.95 - i(\omega_0 - \omega)$. Thus, the width of the resonance which corresponds to λ_1 should be about 60 times larger than that corresponding to λ_2. On the other hand, the symmetries of the eigenvectors of the system corresponding to λ_1 and λ_2 are different. They are, respectively, $X_1 = \{-0.97 + i\,0.25, 1\}$ and $X_2 = \{0.97 + i\,0.25, 1\}$. When the coherence transfer rate increases, then $X_1 \to \{-1, 1\}$ asymptotically and $X_2 \to \{1, 1\}$. The external magnetic field is acting most efficiently on the collective evolution mode, which is described by the symmetric (in-phase) vector X_2. The asymmetric (anti-phase) mode X_1 is not excited and will not be observed.

Three-Frequency Model In [5] the spectrum transformation for the three-frequency model is analyzed in detail. In EPR, this situation is realized for ^{14}N nitroxide radicals. In general, the effect of coherence transfer on the shape of the observed spectrum is similar to what occurs in a two-frequency situation. But there are differences. Suppose that in the absence of spin coherence transfer resonances appear at frequencies $\omega_1 = \omega_0 + a$, $\omega_2 = \omega_0$, $\omega_3 = \omega_0 - a$. It could be expected that under the conditions of fast transfer of coherence (exchange narrowing of the spectrum) all three collective modes will give resonance at the average frequency ω_0. But it turns out that this does not actually happen (see below).

We present the results obtained for the three-frequency model system [5]. It is assumed that the total contribution of the exchange and dipole-dipole interactions to the decoherence of spins of each component is given by the rate W, and the transfer of spin coherence is given by the rate V. If the situation of equivalent spin exchange is realized and the exchange interaction makes a dominant contribution, then $W = (2/3)K_{ex}C$ and $V = (1/3)K_{ex}C$ (see 4.1, 5.11). It is shown [5] that at an arbitrary transfer rate of spin coherence, the frequencies of collective resonances can be written as a splitting of the hyperfine interaction with the effective "hyperfine interaction constant". For example, in case that the exchange interaction leads to the equivalent spin exchange and the dipole-dipole interaction contribution to the spin dynamics can be neglected we obtain the following results

$$
\begin{aligned}
&\omega_1 = \omega_0 + a_{ef}, \omega_2 = \omega_0, \omega_3 = \omega_0 - a_{ef}, \\
&a_{ef} = (1/6)\left(3^{7/6}\left(a^2 - 3V^2\right)/Q + 3^{5/6}Q\right), \\
&Q = \left(9V^3 + 3^{1/2}\left(a^6 - 9a^4V^2 + 27a^2V^4\right)^{1/2}\right)^{1/3}.
\end{aligned}
\tag{5.41}
$$

Q is always a real positive number

The behavior of a_{ef} with increasing coherence transfer rate V is shown in Fig. 5.7. In limit cases of slow spin coherence transfer, $V \to 0$, $a_{ef} = a$, as expected. But in the limit of the fast coherence transfer, $V \to \infty$, $a_{ef} \to a/3^{1/2}$(!) The result was unexpected. In a similar situation in the two-frequency model, the splitting in the spectrum disappears in the region of exchange narrowing. In contrast to the

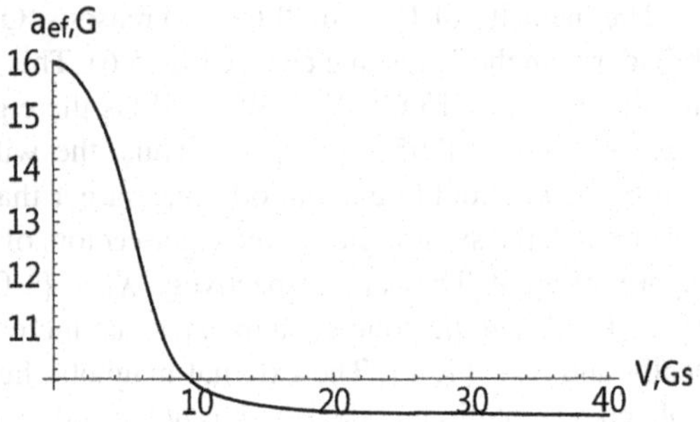

Fig. 5.7 The dependence of the effective “constants” a_{ef} (5.41) which gives splitting of spectral lines, on the rate V of the spin coherence transfer. Calculations are done for $a = 16$ Gs, W = 2 V. (Adapted with permission from Ref. [5]).

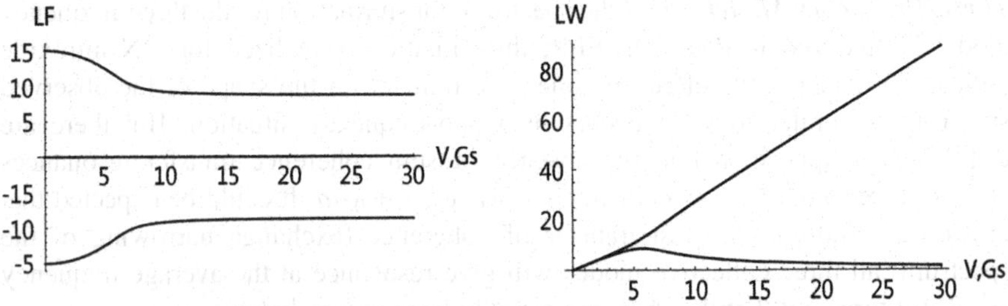

Fig. 5.8 The dependence of the resonance frequencies (left figure) of collective modes of the evolution of spin coherence and widths (right figure) of these collective resonances on the rate V of the transfer of spin coherence for three-frequency models: $a = 15$ Gs, $\Gamma = 0$. (Adapted with permission from Ref. [5])

two-frequency model, in the three-frequency system, at all rates of spin coherence transfer, three frequencies of collective modes remain. Unlike the two-frequency model (see Fig. 5.2) in the three-frequency model there is no collapse of all frequencies of collective modes at the frequency ω_0, in the center of gravity of the spectrum. The behavior of resonance frequencies with an increase in the coherence transfer rate is shown in Fig. 5.8 (curves on the left).

This figure shows that in the three-frequency model the resonance frequencies of the three collective modes do not become the same at any coherence transfer rates.

Figure 5.8 (curves on the right) shows how the widths of the resonances of collective modes change with the growth of the quantum coherence transfer rate.

The widths of the resonances are given by expressions

$$\begin{aligned}
\Delta\omega(1) &= \mathrm{W} + \Gamma - (1/6)3^{2/3}\left(a^2 - 3\mathrm{V}^2\right)/\mathrm{Q} - 3^{1/3}\mathrm{Q};\\
\Delta\omega(2) &= \mathrm{W} + \Gamma + \left(a^2 - 3\mathrm{V}^2\right)/\left(3^{1/3}\mathrm{Q}\right) - \mathrm{Q}/3^{2/3};\\
\Delta\omega(3) &= \Delta\omega(1).
\end{aligned} \tag{5.42}$$

The two extreme resonances (1,3) have the same width, it grows almost linearly with the growth of the coherence transfer rate. The width of the central spectrum line, at the average frequency, passes through the maximum with the growth of the spin coherence transfer rate. The resonance width of the central line in the region of slow spin coherence transfer coincides with the width of the extreme resonance lines, but in the region of fast coherence transfer the width of the central line, $\Delta\omega(2)$, decreases with the increase in the coherence transfer rate (see Fig. 5.8, right, curve with maximum). We emphasize that in Fig. 5.8 calculations are presented for the case when the exchange interaction dominates in the spin coherence transport and the contribution of the dipole-dipole interaction can be neglected, so that $W = 2V$.

But in the EPR experiment at high rates of spin exchange collective resonances at frequencies ω_1 and ω_3 (5.41) are not observed. And just as in the case of the two frequencies discussed above, these resonances are not observed not because of their large width, but because they are not excited by microwave pulses, they are "dark" states for EPR spectroscopy.

In [5] the EPR spectrum is decomposed into contributions of three independent collective modes of coherence motion

$$
\begin{aligned}
&J = J_1 + J_2 + J_3,\\
&J_1 = c(d_1\Delta\omega_1 - a_1(\omega_0 + a_{ef} - \omega))/\left((\omega_0 + a_{ef} - \omega)^2 + \Delta\omega_1{}^2\right),\\
&J_2 = c\,(d_2\Delta\omega_2)/\left((\omega_0 - \omega)^2 + \Delta\omega_2{}^2\right)\\
&J_3 = c\,(d_1\Delta\omega_1 + a_1(\omega_0 - a_{ef} - \omega))/\left((\omega_0 - a_{ef} - \omega)^2 + \Delta\omega_1{}^2\right).
\end{aligned}
\tag{5.43}
$$

Here are introduced the notations:

$$
\begin{aligned}
&c = \gamma B_1 M_0/3;\\
&a_1 = \left(a^2(-\Delta\omega_2 + \Delta\omega_1) + 3\left((-\Delta\omega_2 + \Delta\omega_1)(-\Delta\omega_1 + V + W)^2\right.\right.\\
&\left.\left.+\, a_{ef}{}^2\,(\Delta\omega_2 + \Delta\omega_1 - 2\,(V + W))\right)\right)/\left(2a_{ef}\left(a_{ef}{}^2 + (\Delta\omega_2 - \Delta\omega_1)^2\right)\right);\\
&d_1 = -\left((a^2 - 3\,(a_{ef}{}^2 - (2\,\Delta\omega_2 - \Delta\omega_1 - V - W)(\Delta\omega_1 - V - W)))/\left(2\left(a_{ef}{}^2 + (\Delta\omega_2 - \Delta\omega_1)^2\right)\right)\right);\\
&a_3 = -a_1;\\
&a_2 = 0;\\
&d_3 = d_1;\\
&d_2 = 3 - 2\,d_1.
\end{aligned}
$$

Resonance J_2 at the frequency of $\omega_2 = \omega_0$ describes a Lorentzian absorption curve. Resonances J_1 and J_3 at the frequency ω_1 and ω_3 (5.41) are decribed by the sum of Lorentzian absorption curve, J_{absL}, and Lorentzian dispersion curve, J_{disL},

$$
J_1 = d_1 J_{1absL} + p\,J_{1disL}, J_3 = d_1 J_{3absL} - p\,J_{3disL}, p = -a_1/d_1. \tag{5.44}
$$

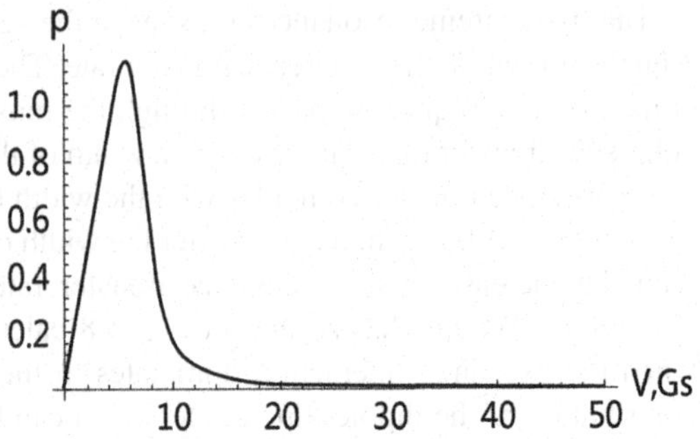

Fig. 5.9 Dependence of the dispersion contribution to the resonance line J_1 (5.44) on the coherence transfer rate. Calculations are carried out for the parameters $a = 16$ Gs, $W = 2V$, $\Gamma = 0$. (Adapted with permission from Ref. [5])

In the case of slow spin exchange we have $p \approx 3V/a$ [5]. If the exchange interaction gives the main contribution to the transfer of spin coherence, then

$$V = (1/3)\, K_{ex}C/a \text{ and } p = K_{ex}C/a. \tag{5.45}$$

This result for the situation of slow exchange was obtained already in [7, 8] in the framework of the perturbation theory for the small parameter $K_{ex}C/a$. Figure 5.9 shows the dependence of the dispersion contribution to the observed EPR spectrum on the coherence transfer rate calculated from the above formulas. The calculation is carried out for the case when the contribution of the dipole-dipole interaction can be neglected.

Figure 5.9 shows that the contribution of the dispersion first grows linearly on the coherence transfer rate, with this linear relationship performing well up to $K_{ex}C \sim a$. It turns out that the result obtained for p in the framework of perturbation theory can be used not only for $K_{ex}C/a \ll 1$, but also for $K_{ex}C/a \sim 1$. Figure 5.9 shows that in the area of $K_{ex}C/a \sim 1$ the curve p(V) passes through the maximum.

B. Bales and M. Perik conducted fundamental studies of the effect of the spin concentration on the shape of the EPR spectrum of Fremy salt solutions and of stable nitroxide radical 4-oxo-2,2,6,6-tetramethylpiperidine-d17 (deuterated Tempone) [9], see also [10–13]. The aim of this work was to confirm the predictions of the theory ([7, 8]) on the mixed form of the spectrum lines due to the transfer of spin coherence, on the dependence of the dispersion contribution on the rate of spin coherence transfer. The system chosen in [9] is very appropriate for testing the theory, since in the studied system three nitrogen hyperfine components of the EPR spectrum are almost homogeneously broadened Lorentzian lines. Note that in this work all experimental data were interpreted very well without taking into account the dipole-dipole contribution. The experiments confirmed all predictions of the theory. Special attention deserves the fact that the experiment confirmed the prediction of a linear dependence of the contribution of the dispersion on the rate of the coherence transfer and this linearity was performed in a wide interval of concentrations of spins. This opens up a new possibility to determine the rate of spin coherence transfer from EPR data. The authors of [9] made the statement that the measurement of the contribution of the dispersion into a spectrum allows to measure directly the

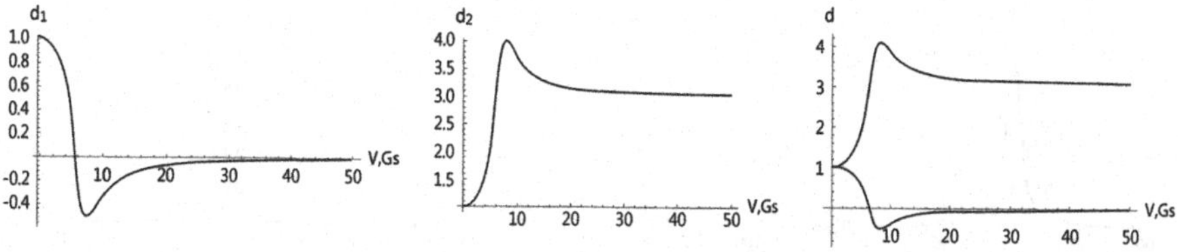

Fig. 5.10 Dependence on the rate of spin coherence transfer of the contribution of the absorption of collective modes at frequencies ω_1, ω_3 (parameter d_1) and at frequency ω_2 (parameter d_2). The figure on the right combines d_1 and d_2 curves. Calculations are carried out for the parameters $a = 16$ Gs, $W = 2V$. (Adapted with permission from Ref. [5])

contribution of the exchange interaction and that it seems that in this case there is no problem of separation of contributions of exchange and dipole-dipole interactions in the transfer of spin coherence. Of course, it may well be a situation in which the dipole-dipole interaction makes a negligible contribution to the transfer of spin coherence in comparison with the exchange interaction. And Fremy salt seems to be a system of that type. But in general, we need to take into account the contribution of both interactions to the spin coherence transfer and solve the problem of separation of their contributions if we want to obtain information about bimolecular collisions of molecules using EPR spectroscopy. A real achievement of Bales and Peric is that they not only confirmed experimentally the manifestation of dispersion in the observed signal, but also demonstrated the ability to determine the rate of spin exchange with great accuracy from experimental data on the contribution of dispersion to the shape of the EPR spectrum.

The contribution of absorption to the observed resonance lines of collective modes is determined by the parameters $d_1 = d_3$ and d_2 (see. (5.43)). Their behavior with the growth of the coherence transfer rate is shown in Fig. 5.10.

Figure 5.10 shows that the contribution of the absorption of collective modes with resonant frequencies ω_1, ω_3 (parameter d_1) does not change monotonically with the increase of the coherence transfer rate (see the curve on the left in Fig. 5.10). It even changes sign when the spectrum collapse condition is reached. With a very fast transfer of spin coherence, the absorption contribution on these collective modes tends to zero. The collective mode with resonance frequency $\omega_2 = \omega_0$ under $V > a$ conditions gives a narrow line and the contribution of this mode (parameter d_2) to the observed spectrum reaches the plateau (see the middle Fig. 5.10). Under the conditions of the exchange narrowing of the spectrum, the resonance at the frequency $\omega_2 = \omega_0$ corresponds to the in-phase motion of the transverse magnetizations of all spins of the system, so that this collective mode is most effectively excited by the external B_1 field. For example, for $a = 16$ Gauss, $V = 50$ Gauss, $W = 2V$, the collective mode with frequency $\omega_2 = \omega_0$ is a vector $X_2 = \{0.98 - i0.21, 0.999 - i0.11, 1\}$ in the basis M_{1-}, M_{2-}, M_{3-}. This vector practically coincides with the vector of in-phase-mode motion of magnetizations of all components $\{1, 1, 1\}$.

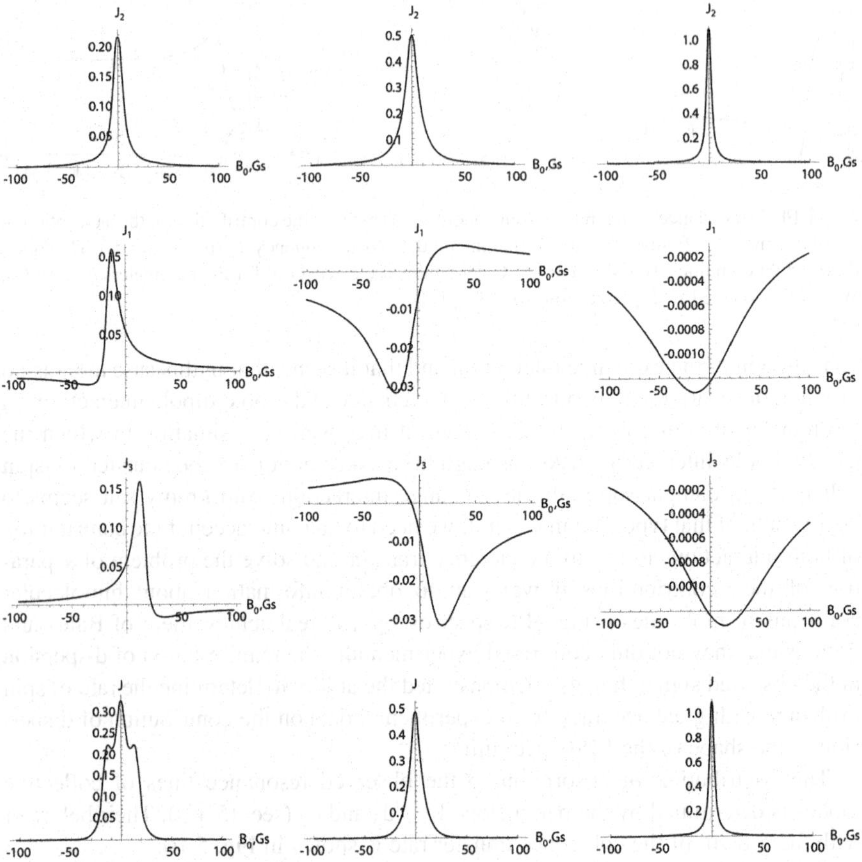

Fig. 5.11 Contributions of collective modes with the resonance frequencies $\omega_2 = \omega_0$ (top row), $\omega_1 = \omega_0 + a_{ef}$ (second row), $\omega_3 = \omega_0 - a_{ef}$ (third row) and the sum of all three contributions (bottom row). Curves on the left are obtained for V = 3 G; curves in the middle are obtained for V = 8 Gs; curves on the right correspond to V = 20 Gs. For all curves a = 16 Gs, W = 2V (the dipole-dipole interaction is not included here). (Adapted with permission from Ref. [5])

To illustrate the contributions of different collective spin modes to the observed EPR spectrum in Fig. 5.11 the contributions of each independent mode and the total spectrum for the three values of the coherence transfer rates are presented.

The curves in Fig. 5.11 clearly show that in the limit of the slow spin coherence transfer two collective modes $J_1$and J_3 have the asymmetric shapes (see left-hand column curves in Fig. 5.11). They are mixtures of the absorption and dispersion terms (see Eq. (5.43). The absorption terms dominate in the contribution of J_1 and J_3 to the total spectrum J. In the case of the intermediate spin coherence transfer rates (see central column curves in Fig. 5.11) the dispersion terms dominate in the contribution of J_1 and J_3 to the total spectrum J. But the large dispersion terms of J_1 and J_3 resonances interfere destructively, they practically cancel each other.

Figure 5.11 shows that even under the condition of the spectrum exchange narrowing (see right-hand column curves in Fig. 5.11) there are three resonances of three collective modes. The frequencies of these modes differ (!), they never collapse into one average frequency. However, in the case of the fast spin exchange one can detect in EPR spectra only the contribution of one of the collective mode resonances (see Fig. 5.11, curves J_2) since other two modes (see J_1, J_3 on Fig. 5.11) lead to broad lines with very small integral intensities.

5.6 What Happens When Dipole-Dipole Interaction Is Giving a Dominant Contribution to the Spin Coherence Transfer?

Typically, the spin exchange is studied in non-viscous solutions and the exchange interaction is giving a major contribution to the spin coherence transfer, so that $V_{sct} = K_{ex}C - V_{ddsct} > 0$. In all previous figures, there were presented numerical simulation under the condition that the spin coherence transfer rate $V_{sct} > 0$. But there might be situations when the dipole –dipole interaction dominates and $V_{sct} < 0$.

In the non-typical $V_{sct} < 0$ case the manifestations of the spin coherence transfer in EPR spectra are not the same as they are in the $V_{sct} > 0$ case [16, 22]. To illustrate, let us consider transformations of the EPR spectrum for a two frequency model. Solution of a problem of the EPR spectrum and contributions of collective modes to the spectrum are given in Eqs. (5.39) and (5.40) for cases of slow and fast spin coherence transfer, respectively. These solutions are valid for the $V_{sct} < 0$ situation as well as for the $V_{sct} > 0$ case.

Using the Eqs. (5.39 and 5.40), contributions of two collective modes to the EPR spectrum and the total EPR spectrum were calculated under condition $V_{sct} < 0$. The results are presented on Figs. 5.12, 5.13 and 5.14 (compare with Figs. 5.4, 5.5 and 5.6).

Figures 5.12 and 5.13 correspond to the cases when the spin coherence transfer induced by the dipole-dipole interaction is rather slow (Fig. 5.12, $|V_{sct}| \ll a$) and has an intermediate value (Fig. 5.13). One can see that the resonance frequencies of the system are shifting to the center of the spectrum due to the spin coherence transfer if we consider the absorption contributions of the collective modes (compare absorption contributions on Figs. 5.12 and 5.13, curves on the central columns). However, positions of maxima of lines which are observed in EPR spectrum (curves on the left columns in Figs. 5.12 and 5.13) do not practically shift from their positions in the absence of any spin coherence transfer, i.e., at $V_{sct} = 0$. This happens since spin coherence transfer with $V_{sct} < 0$ leads to a contribution of the dispersion to the observable in EPR experiments which is resulting in a repulsion of maxima in the EPR spectrum considered.

When $|V_{sct}| \geq a/2$ both collective modes of the spin system have the same resonance frequency. It is similar to the situation of the exchange merging of the

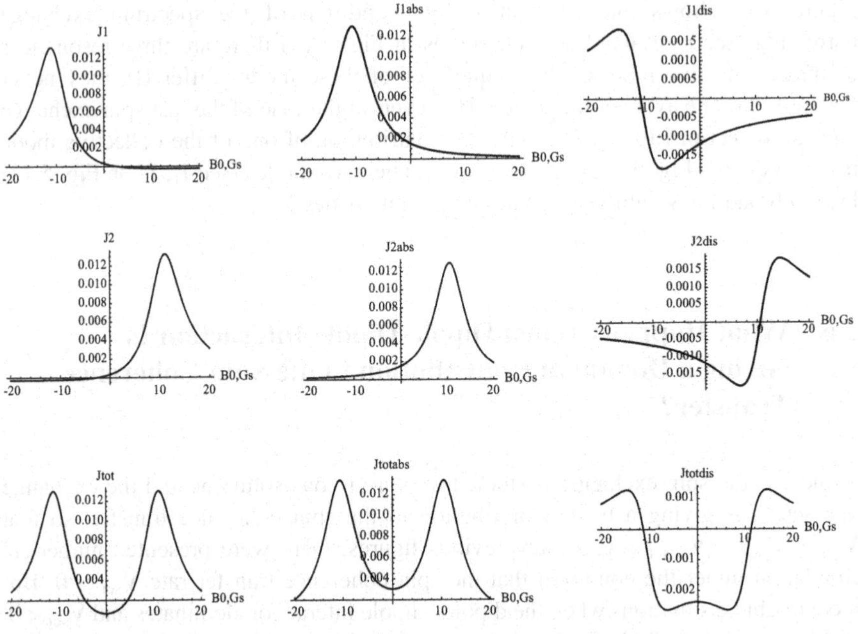

Fig. 5.12 Contributions of the collective mode resonances with the resonance frequencies $\omega_0 + R_0/2$ (top row) and $\omega_0 - R_0/2$ (central row) to the EPR spectrum, and the contribution of both collective independent modes, $J = J_1 + J_2$, (bottom row). Total contributions are on the left, absorption contributions are in the middle and dispersion contributions are on the right. Parameters are $a = 22$ Gs, $V = -3$ Gs, $W = 3.75$ Gs, $\Gamma = 0.1$ Gs, $\omega_1 = 0.1$ Gs

two frequency spectrum into one line (see Fig. 5.6). Thus, two resonance lines are merging into situation when there is only one resonance frequency independent on a sign of the V_{sct}, since only absolute value of V_{sct} is responsible for this frequencies merging effect. But sign of V_{sct} appears to be of great importance, since it determines a sign of the dispersion contribution to a resonance line.

When $V_{sct} > 0$, under conditions of a merging of two resonance frequencies two collective modes have different widths of resonances and they contribute to the observed spectrum with different signs. It was described above (see, e.g., Fig. 5.6) that when exchange interaction gives main contribution to the spin coherence transfer then in a case of fast spin coherence transfer in the EPR experiment we see practically only narrow collective mode resonance. Other collective mode is practically not detected since it gives a broad resonance, it is not excited by the microwave field, and its contribution to the observable corresponds to an irradiation not absorption.

When $V_{sct} < 0$, again under conditions of a merging of two resonance frequencies two collective modes have different widths of resonances (both widths are large) and they contribute to the observed spectrum with different signs. But now it appears that in EPR spectra we detect a resonance with larger width, negative contribution of the less width resonance, in principle, is disturbing a Lorentzian shape of the spectrum

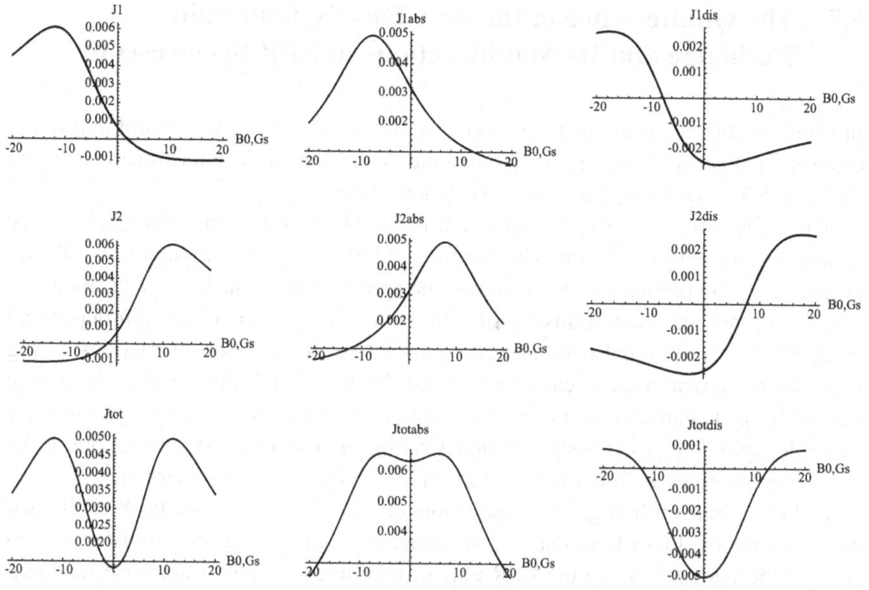

Fig. 5.13 Contributions of the collective mode resonances with the resonance frequencies $\omega_0 + R_0/2$ (top row) and $w_0 - R_0/2$ (central row) to the EPR spectrum, and the contribution of both collective independent modes, $J = J_1 + J_2$, (bottom row). Total contributions are on the left, absorption contributions are in the middle and dispersion contributions are on the right. Parameters are $a = 22$ Gs, $V = -8$ Gs, $W = 10$ Gs, $\Gamma = 0.1$ Gs, $\omega_1 = 0.1$ Gs.

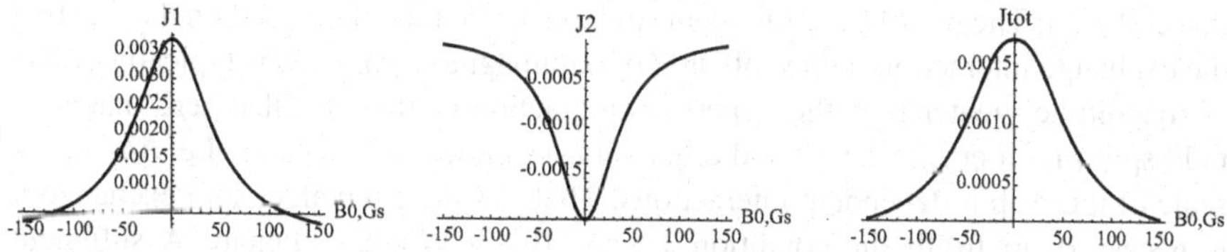

Fig. 5.14 Contributions of the two collective independent modes to the EPR spectrum (left-hand and middle columns), the total EPR spectrum (right-hand column): $V = -12$ Gs, $W = -(7/2)V$, $a = 22$ Gs, $\Gamma = 0.1$Gs

near the center of the spectrum. This effect is seen on Fig. 5.14 since the total spectrum (right curve in Fig. 5.14) is broader than the resonance line of the collective mode (see left curve in Fig. 5.14). The negative contribution of the second collective mode (see curve in the center in Fig. 5.14) gives a "depression" in the center of the spectrum (see also discussion of this effect in [16, 22]. In fact, there can manifest itself a shallow hole in the center of a spectrum (examples of numerical calculations which demonstrate this shallow hole are presented in [16]).

5.7 The Quintessence of the New Paradigm of Spin Exchange and Its Manifestations in EPR Spectroscopy

In order to formulate the principal novelty of today's view of the problem under discussion, it is necessary to briefly recall the main ideas of the widespread paradigm (in Sect. 5.3 it was designated as an early paradigm).

The early spin exchange paradigm was based on the results obtained for the equivalent exchange situation. The exchange interaction at the moment of collision of paramagnetic particles is described by the spin-Hamiltonian $\mathbf{V}_{ex} = \hbar J \mathbf{S}_A \mathbf{S}_B$ and it is assumed that the scale of this interaction is much larger than other spin-dependent interactions of individual paramagnetic particles, so these latter interactions during the collision, it would seem, can be neglected. Then the total spin $\mathbf{S} = \mathbf{S}_A + \mathbf{S}_B$ during the collision of particles A and B is preserved. This means that in a collision, for example, spin A gives (loses) its magnetization, its quantum coherence, but at the same time receives an equivalent "reaction" from the interaction partner.

In fact, when colliding, from the point of view of each particle A or B, two processes occur: loss of spin state (magnetization) and "reaction" (transfer of spin state of the partner). Since the EPR experiment detects dipole transverse magnetization, it is important in the region of linear response in the EPR experiment that the particle loses its coherence (phase) when colliding, but at the same time the coherence of the collision partner can be transferred as well. It is fundamentally important to distinguish these two aspects of the collision, as they lead to completely different effects, appear quite differently in EPR spectroscopy. In the early paradigm of spin exchange, we consider the situation in which decoherence (loss of coherence) and transfer of coherence (reaction) are reduced to the equivalent exchange of spin states in a collision. The use of the equivalent spin exchange model is based on the fact that the exchange interaction in the collision of paramagnetic particles is typically orders of magnitude greater than the hyperfine interaction or the so-called zero magnetic field splitting of energy levels and other spin-dependent interactions. Let $\hbar\Delta$ set the scale of these spin-dependent interactions. Then for the equivalent spin exchange it is necessary to fulfill the condition $J > \Delta$. But it is not sufficient. A sufficient condition is $\Delta\tau_c < 1$, where τ_c is the characteristic collision time. In the opposite case $\Delta\tau_c > 1$ non equivalent spin exchange is realized: during collisions spins lose their coherence (decoherence of the spins occurs), but the "reaction" of the spin coherence decreases and may even become zero.

Thus, a new and fundamentally important aspect of today's spin exchange paradigm is that two different effects are clearly distinguished: spin decoherence (spin phase relaxation) and spin coherence transfer from the collision partner (collision "reaction" effect).

In the early paradigm, the exchange broadening in the region of slow spin exchange is interpreted by the uncertainty relation: spin-exchanging collisions are shortening "life time" of spin states of particles. This collision induced dephasing of spins describes the broadening of the lines quite satisfactorily. But dephasing of spins in collisions can not explain the shift of the resonance frequency of spins which is observed.

In the new paradigm, the broadening and shift of resonance frequencies observed in the EPR experiment are interpreted in a completely different way. The transfer ("reaction") of spin coherence between spins creates a correlation in the spin motion and forms collective modes of spin motion. These collective modes evolve independently. Each collective mode resonates with its own frequency and width. It is important to emphasize that the integral intensity of the resonance lines of different collective modes is different (see Fig. 5.10). For example, in a situation of exchange narrowing, almost all the integral intensity is accumulated with only one of the collective resonances. Thus, in the new paradigm of spin exchange, the EPR spectrum is a superposition of independent resonance lines of collective spin modes. This approach allows us to describe the frequency (and hence the frequency shifts), the width and the observed resonances, and their intensity.

The new approach provides a completely new interpretation of the effect of the exchange narrowing of the spectrum. But for the sake of fairness it should be emphasized that in the field of slow spin exchange, concentration broadening of collective mode resonance lines is quite satisfyingly consistent with the rate of decoherence of spins, which gives a common paradigm. Thus, in terms of resonance width, the new paradigm inherits much of the old paradigm. Therefore, the determination of the spin decoherence rate within the framework of the widespread paradigm (exchange broadening) remains one of the quite working approaches. Below we will return to this issue in connection with the discussion of protocols for the practical determination of the spin exchange rate using EPR spectroscopy.

For the modern development of spin exchange studies it turned out to be important my theoretical prediction that in the region of slow spin exchange the independent collective modes of spin coherence motion are formed which manifest themselves in EPR experiments as a resonance lines of mixed form: each line in the EPR spectrum contains the contribution of absorption and dispersion (see [7, 8], Sect. 2.4). Like the formation of collective modes, the appearance of the dispersion contribution is due solely to the spin coherence transfer between the spins (due to "reaction" effect in collisions). The theoretical prediction of the mixed form of resonance lines in the EPR spectra has long remained out of the field of view of experimenters. So it was, until the work of Bales and Peric [9]. In this work, very careful measurements of the EPR spectrum of radicals in solutions were carried out and it was convincingly shown that in the region of slow spin exchange the components of the spectrum have a mixed form. The observed spectrum was presented by them as a sum of absorption and dispersion lines and it was shown that the proportion of dispersion in the EPR spectrum opens a new practical method for determining the rate of spin exchange using EPR spectra. Later in numerous works extensive studies of the dispersion contribution to the EPR spectra of nitroxide radical in solutions caused by equivalent spin exchange were carried out [10–15].

The analysis of dispersion contribution to the observed EPR spectrum gives a completely new protocol for measuring the spin coherence transfer rate.

The dispersion contribution to the experimental EPR spectrum has become another and potentially very promising way to find the rate of spin coherence transfer between paramagnetic particles in the solution.

Note that a similar situation might occur for chemical exchange. For example, the appearance of the dispersion contribution was observed in [23, Sect. 4.6.3] in the phase of the proton free induction signal in the presence of a reversible chemical reaction.

A common paradigm in the study of spin exchange involves measuring the shift of position of the maxima of the EPR spectrum lines. It is taken into account that the position of this maximum is determined by two factors: the real shift of the resonance frequency and the overlap of the line with other lines. But the common paradigm of spin exchange studying does not include one more mechanism of a shift of position of the maxima of the EPR spectrum lines. This mechanism comes out from the mixed shape of the resonance lines in the presence of the spin coherence transfer. Note that this shift does not mean that the frequency of the collective spin mode changes. Figures 5.5 or 5.11 illustrate that this form-induced shift of a position of maximum might be really big. Without taking into account the contribution of this mechanism, the interpretation of observed line shifts may lead to completely incorrect conclusions.

Concentration changes in the shape of EPR spectra are due not only to the exchange interaction, but also to the dipole-dipole interaction of spins. Therefore, the contribution of dipole-dipole interaction should be taken into account to find the spin exchange rate from the experimental data. It has already been noted above that the widespread paradigm proceeds from the fact that, unlike the exchange interaction, the dipole-dipole interaction can only cause decoherence of spins and cannot cause the transfer of coherence to the spin from its interaction partner (see, for example, [4], Eqs. (VIII.35, VIII.36, VIII.89); [19] Eq. (2.19)). Meanwhile, in the book [21] I have shown that the dipole-dipole interaction in solutions of paramagnetic particles also causes the transfer of coherence ("reaction" of spin coherence) similar to the exchange interaction. But it is very interesting that the spin-to-partner coherence transfer in the case of dipole-dipole interaction occurs with the phase change by 180°, so in the kinetic Eq. (5.11) the terms corresponding to the spin coherence transfer have different signs for the exchange and dipole-dipole interactions. According to (5.11) the total rate of spin coherence transfer is $V_{sct} = K_{ex}C - 1/T_{ddsct}$. The theory predicts [7, 8, 16] that the dispersion contribution appears in the EPR spectrum with a sign that has V_{sct}. In principle, by changing the viscosity of the solution, one can change the sign of V_{sct}. At low viscosity values, the contribution of the exchange interaction should dominate and $V_{sct} > 0$ is expected, but at high viscosity values, the contribution of the dipole-dipole interaction to the transfer of spin coherence might dominate and $V_{sct} < 0$ is expected. Accordingly, when the viscosity in the EPR spectrum changes, the sign with which the dispersion contribution enters the observed EPR signal will change. As a result, when the contribution of the dipole-dipole interaction dominates the contribution of the exchange interaction to the transfer of spin coherence, i.e. when $V_{sct} < 0$, the observed in the experiment spectral lines "break apart" (resonance lines are repulsed!) with an increase in the rate of transfer of spin coherence, while the predominant contribution from the exchange interaction (when $V_{sct} > 0$) leads to the convergence of the observed resonance lines in the EPR spectrum.

A new paradigm for determining the spin exchange rate from the EPR spectra of spin probes is based on the consistent theory of paramagnetic relaxation in solutions due to the dipole-dipole interaction of spin probes (see, [5, 14, 15]).

Thus, according to modern concepts in the presence of spin coherence transfer between spins, in the situation of linear response of the system to the microwave field, the EPR spectrum can be represented as the sum of resonance lines of independent collective modes in the form of

$$I_{exp} \sim \sum J_k(\Delta\Omega_k + p_k(\omega_k - \omega))/\left((\omega_k - \omega)^2 + \Delta\Omega_k{}^2\right), \quad (5.46)$$

where we have introduced the parameters ω_k and $\Delta\Omega_k$ –frequency and width of resonance of the k-th collective mode, and p_k is the contribution of dispersion. The intensity J_k of each resonance is proportional to the square of the transition matrix element of the spin-Hamiltonian of the interaction of the spin system with the microwave field. All these parameters depend on the rate of spin coherence transfer. For collective resonance frequencies, their widths and J_k analytical solutions are known only for the model with two and three frequencies [5]. In general, we are to obtain these parameters numerically.

Note that formally (5.46) is a generalization of the $T_1 - T_2$ Portis model [24] in the approximation of the linear response of the spin system on the microwave field: the spectrum is a superposition of lines, each of which has its own resonance frequency. As it will be presented in Chap. 7 the EPR spectrum can be presented as a sum of independent lines at any power of the microwave field including spectra under saturation conditions [25]. Resonance frequencies of those lines depend on microwave field amplitude B_1.

But unlike the situation that Portis considered, in this case, each resonance line is not described by the Lorentzian absorption curve, but is a mixture of absorption and dispersion. These resonances do not refer to individual spins, but reflect the modes of collective, correlated, motion of all spins and MW field. And a very important difference from the Portis model is that each collective mode has its own coefficient J_k, i.e. different collective modes give a different contribution to the spectrum. When the exchange interaction gives a major contribution to the spin coherence transfer, $V_{sct} > 0$, in the conditions of fast coherence transfer all J_k but one tend to zero. In this limit case only in-phase motion of all spins of the system is observed in the spectrum. Other modes are "dark", non visible for EPR spectroscopy.

References

1. Kivelson, D.J.: Theory of ESR linewidths of free radicals. J. Chem. Phys. **33**, 1094–1106 (1960)
2. Currin, J.D.: Theory of exchange relaxation of hyperfine structure in electron spin resonance. Phys. Rev. **126**, 1995–2001 (1962)
3. Freed, J.H.: On Heisenberg spin exchange in liquids. J. Chem. Phys. **45**, 3452–3453 (1966)

4. Abragam, A.: Nuclear magnetism. Principles of nuclear magnetism. Oxford University Press, London (1961)
5. Salikhov, K.M.: Consistent paradigm of the spectra decomposition into independent resonance lines. Appl. Magn. Reson. **47**, 1207–1228 (2016)
6. Salikhov, K.M.: The contribution from exchange interaction to the line shifts in ESR spectra of paramagnetic particles in solutions. J. Magn. Res. **63**, 271–279 (1985)
7. Zamaraev, K.I., Molin, Y.N., Salikhov, K.M.: Spin exchange. Nauka, Sibirien branch, Novosibirsk (1977). in Russian
8. Molin, Y.N., Salikhov, K.M., Zamaraev, K.I., Exchange, S.: Principles and applications in chemistry and biology. Springer, Berlin/Heidelberg/New York (1980)
9. Bales, B.L., Peric, M.: EPR line shifts and line shape changes due to spin exchange of nitroxide free radicals in liquids. J. Phys. Chem. B. **101**, 8707–8716 (1997)
10. Bales, B.L., Peric, M.: EPR line shifts and line shape changes due to spin exchange of nitroxide free radicals in liquids 2. Extension to high spin Exchange frequencies and inhomogeneously broadened spectra. J. Phys. Chem. A. **106**, 4846–4854 (2002)
11. Peric, M., Bales, B.L.: Lineshapes of spin exchange broadened EPR spectra. J. Magn. Reson. **169**, 27–29 (2004)
12. Bales, B.L., Peric, M., Dragutan, I.: EPR line shifts and line shape changes due to spin exchange of nitroxide free radicals in liquids 3. Extension to five hyperfine lines. Additional line shifts due to re-encounters. J. Phys. Chem. A. **107**, 9086–9098 (2003)
13. Bales, B.L., Meyer, M., Smith, S., Peric, M., Bales, B.L., Meyer, M., Smith, S., Peric, M.: EPR line shifts and line shape changes due to spin exchange of nitroxide free radicals in liquids 4. Test of a method to measure re-encounter rates in liquids employing ^{15}N and ^{14}N nitroxide spin probes. J. Phys. Chem. A. **112**, 2177–2181 (2008)
14. Salikhov, K.M., Bakirov, M.M., Galeev, R.T.: Detailed analysis of manifestations of the spin coherence transfer in EPR spectra of ^{14}N nitroxide free radicals in non-viscous liquids. Appl. Magn. Reson. **47**, 1095–1122 (2016)
15. Bales, B.L., Bakirov, M.M., Galeev, R.T., Kirilyuk, I.A., Kokorin, A.I., Salikhov, K.M.: The current state of measuring bimolecular spin exchange rates by the epr spectral manifestations of the exchange and dipole-dipole interactions in dilute solutions of nitroxide free radicals with proton hyperfine structure. Appl. Magn. Reson. **48**, 1399–1447 (2017)
16. Salikhov, K.M.: The state of the theory of spin exchange in dilute solutions of paramagnetic particles. Physics-Uspekhi **189**, 1017–1043 (2019). https://doi.org/10.3367/UFNr.2018.08.038421
17. Salikhov, K.M.: Contributions of exchange and dipole–dipole interactions to the shape of EPR spectra of free radicals in diluted solutions. Appl. Magn. Reson. **38**, 237–256 (2010)
18. Altschuler, S.A., Kozyrev, B.M.: Electron paramagnetic resonance of compounds of elements of intermediate groups. Fizika-Mathemat. Literat. Moscow (1972)
19. Jones, M.T.: Electron spin exchange in aqueous solutions of $K_2(SO_3)_2NO$. J. Chem. Phys. **38**, 2892–2895 (1963)
20. Smirnov, V.I.: Course of high mathematics, vol. 3, part I (M., Publishing Hoouse of Fisiko-Math. Literature, 1958)
21. Salikhov, K.M., Semenov, A.G., Tsvetkov, Y.D.: Electron spin echo and its application, p. 342. Nauka, Siberian branch, Novosibirsk (1976)
22. Galeev, R.T., Salikhov, K.M.: The theory of dipolar broadening of magnetic resonance lines in non-viscous liquids. Khim. Fizika. **15**, 48–64 (1996)
23. Ernst, R.R., Bodenhausen, G., Wokaun, A.: Principles of nuclear magnetic resonance in one and two dimensions. Clarendon Press, Oxford (1987)
24. Portis, A.M.: Electronic structure of F centers: saturation of the electron spin resonance. Phys. Rev. **91**, 1071–1078 (1953)
25. Salikhov, K.M.: Peculiar features of the spectrum saturation effect when the spectral diffusion operates: system with two frequencies. Appl. Magn. Reson. **49**, 1417–1430 (2018)

Chapter 6
Experimental Determination of the Spin Exchange Rate from the Analysis of the EPR Spectrum Shape

Abstract It is shown which parameters of paramagnetic spin relaxation can be found using a simulation of the EPR spectra in a linear response. Much attention is paid to the discussion of the characteristic features of the derivative of the EPR spectrum, which is observed experimentally. In the framework of the early paradigm, it is from the special points of the spectrum that the frequency and width of the resonances are obtained. But one can make a big mistake, since the resonance lines have a mixed shape. Therefore, the new paradigm proceeds from the fact that a fraction of the contribution of dispersion is distinguished in the experimental spectrum. It is this fraction of the dispersion that allows one to directly determine the rate of coherence transfer.

The spin exchange rate can be found using EPR spectroscopy techniques. The most theoretically developed, accessible and widely used method is the experimental study of steady-state EPR spectra in the case of the linear response of a spin system to the microwave field B_1. Therefore, this approach is discussed here in some detail.

Widely used methods of finding the spin exchange rate are based on the measurement of the concentration broadening of the EPR spectrum lines and the shift of the position of the maxima of these lines.

The new spin exchange paradigm has introduced another method based on the measurement of the dispersion contribution to the observed spectrum. Here special attention is paid to this issue.

Recently, progress has been made in the study of steady-state EPR spectra in the presence of the spin coherence transfer between spins in strong microwave fields, when the effect of saturation of the microwave energy absorption by the spin system is manifested. This method may be very useful as well for the study of spin exchange. The steady-state double electron-electron resonance and pulse EPR methods are also useful when studying spin exchange. These methods will be considered in the next chapter.

In the previous sections, it was shown that, in general, the exchange interaction of two particles in a collision can lead to quite different results and, accordingly, the spectroscopic manifestations of spin exchange in the shape of the EPR spectrum can be quite different. For example, there may be a nonequivalent spin exchange

K. M. Salikhov, *Fundamentals of Spin Exchange*,
https://doi.org/10.1007/978-3-030-26822-0_6

situation, when the exchange interaction in bimolecular collisions causes only decoherence of spins and does not transfer spin coherence from the collision partner, i.e., the "reaction" coherence transfer does not occur. In this situation, there is exchange broadening, but there is no exchange shift of the spectrum lines and the exchange interaction in course of collisions does not lead to the collapse of the entire spectrum into an exchange narrowed line. In such a situation, the shift of the spectrum lines and the collapse of the spectrum into a broad line can occur due to the spin coherence transfer induced by the dipole-dipole interaction.

We focus on the discussion of a relatively simple situation and consider the case of spin exchange between spin $S = 1/2$ particles. This choice can be supported, given that spin 1/2 particles, stable organic radicals with one unpaired electron, are usually used as spin probes. It is clear that from the point of view of the tasks, for which the spin probe method is most often used, it is very desirable that the exchange interaction be large enough. Indeed, it is best to use such spin probes so that spin exchange involving them can be considered as the equivalent spin exchange. In this situation, decoherence of spins and the spin coherence transfer due to the exchange interaction is given by the same parameter: the spin exchange rate constant K_{ex}. Of course, these arguments do not justify the choice for a detailed study of the case of equivalent exchange between particles with spins 1/2, if the purpose of the study is to identify the physical nature of the exchange integral in specific situations. For example, at the beginning of the development of studies of spin exchange, the important motif of these studies was to study the effect of ligands of paramagnetic complexes with $S \geq 1/2$, namely, the delocalization of unpaired electrons on the periphery of the complexes, on the exchange integral [1, 2]. The exchange integral characterizes the degree of the overlap of orbitals occupied by electrons, and in turn, this is important for understanding the electron transport reaction, the most important redox reactions (see discussion of this issue in [1, 2]).

Nowadays spin exchange attracts attention, first of all, in connection with the study of the molecular mobility and bimolecular collisions in complex environments, in particular in biological objects. For this purpose, nitroxide radicals proved to be good spin probes, since their participation, as a rule, implements the strong exchange situation which leads to equivalent spin exchange [1–3].

According to Eq. (5.15), the changes in the shape of the EPR spectrum in solutions with the change in the concentration of paramagnetic particles are determined by the spin decoherence rate $W_{sd} = KC + 1/T_{ddsd}$ and the spin coherence transfer rate $V_{sct} = K'_{ex}C - 1/T_{ddsct}$. When spin exchange can be treated as equivalent spin exchange we have $K = K'_{ex} = K_{ex}$. It is noteworthy that the contributions of the exchange and dipole-dipole interaction to the spin decoherence are added up, while their contributions to the electron spin coherence transfer are subtracted.

We emphasize that the spin coherence transfer rate V_{sct} is of fundamental importance for the transformations of the spectrum. The spin coherence transfer from the collision partners, i.e., the "reaction" spin coherence transfer, results in a mixed form of lines in the slow spin exchange region and gives the effect of the exchange narrowing of the spectrum in the fast spin exchange region. In the slow spin exchange region, the spectrum lines are broadened and shift as a function of what interaction makes the main contribution to the spin coherence transfer: exchange or dipole-dipole interactions. In the slow spin coherence transfer region, the resonance lines have a mixed form: they are a sum of absorption and dispersion contributions. In the fast spin coherence transfer region, all components of the spectrum merge into one homogeneously broadened line. In this case, the dipole-dipole interaction causes the concentration broadening ($1/T_{ddsd} + 1/T_{ddsct}$) ~ C. Note that the "rate" constant of a dipolar broadening of the spectrum in the exchange narrowing case is larger than that "rate" constant for the dipolar broadening of a single component spectrum in the slow spin exchange region, since in the last case the dipolar broadening of the k-th component of a spectrum is $1/T_{ddsd} + \varphi_k/T_{ddsct}$.

When the entire spectrum merges in a homogeneously broadened line in the case of nonequivalent spin exchange, the concentration broadening can contain the contribution linear on C caused by the exchange interaction as well as by the dipole-dipole interaction (see (5.30)).

The manifestation of spin exchange and dipole-dipole interactions of the spins in the shape of the EPR spectrum in the linear response case, in principle, make it possible to determine some combinations of the spin decoherence rate and the spin coherence transfer rate ("reaction" spin coherence transfer rate) caused by exchange and dipole-dipole interactions.

Our problem is to find the spin exchange rate between paramagnetic particles in solutions from the experimental shape of EPR spectra. At first glance, it may seem that this problem can be solved by numerically calculating the EPR spectrum using the general expressions (5.15) for different values of spin exchange parameters and other unknown parameters. In this way it is possible, in principle, to choose a set of unknown parameters that best simulates the spectrum observed in the experiment. In this way we can determine two combinations of unknown parameters $W_{sd} = KC + 1/T_{ddsd}$ and $V_{sct} = K'_{ex}C - 1/T_{ddsct}$ (see that (5.13 and 5.15)). When the equivalent spin exchange is realized then the spin exchange can be determined straightforwardly since under these conditions

$$W_{sd} \approx K_{ex}C + 5/T_{dd}; V_{sct} \approx K_{ex}C - 4/T_{dd}.$$

However, in nonequivalent spin exchange case, knowing of the two kinetic parameters W_{sd} and V_{sct} is not enough to determine three unknown kinetic parameters KC, $K'_{ex}C$ and $1/T_{dd}$. More information is needed to find the spin exchange rate.

In certain situations, even without the numerical simulation of the total spectrum, it is possible to obtain information about the spin exchange rate directly from the analysis of the specific features of the EPR spectrum shape. This is possible, e.g., when the experimental spectrum contains a number of resolved components. In general, the characteristic parameters of the spectrum are the position of maxima, the width of the components, the splitting between the components and the asymmetry of the spectrum components due to the dispersion contribution to the observed EPR signal in the presence of the spin coherence transfer. Often in spectroscopy, the position of the spectral line maximum gives directly the resonance frequency, and the line width is the true width of the observed resonance. But this is not the case if there is the spin coherence transfer between the spins (see below Eqs. (6.34 and 6.36)). But even in the presence of the spin coherence transfer, the positions of maxima and line widths observed in the experiment are giving information related to the true values of the resonance frequencies and widths. In such cases, the theory makes it possible to calculate these parameters using experimental EPR spectra.

In steady-state EPR spectroscopy, a derivative of the spectrum is usually recorded in the experiment. This derivative is called the "EPR spectrum". The characteristic properties of the EPR spectrum are given below.

6.1 The Ratio Between the Parameters of the Observed EPR Spectrum and the Physical Parameters of the Spin System. Simple Example

The aim of this section is to show that the analysis of specific features of the EPR spectrum shape can provide good supporting information on the contribution of the exchange and dipole-dipole interactions to the spin dynamics of paramagnetic particles in solutions.

The simplest situation is when the spins of the system have the same resonance frequency ω_0 and the same spin decoherence time T_2. In this case, the spectrum is a Lorentzian line

$$J_{EPR} = \text{const}\ (\gamma B_1 T_2/\pi)/\left(1 + (\omega - \omega_0)^2 {T_2}^2\right). \tag{6.1}$$

At half height, the width of (6.1) is $2/T_2$. The half-width of the resonance line (6.1) at half height is exactly equal to the spin decoherence rate

$$\Delta\omega_{1/2} = 1/T_2. \tag{6.2}$$

It was already noted that the derivative of (6.1) is recorded in experiment. For illustration, Fig. 6.1 shows the derivative of (6.1), which gives the shape observed in the experimental EPR spectrum. The zero position of the derivative of the spectrum gives the resonance frequency ω_0, i.e., in this case the resonance frequency is directly observable.

The positions of the maximum and minimum in Fig. 6.1 set the points of the greatest slope of the spectrum (6.1). The distance between the positions of the maximum and minimum of the derivative spectrum (Fig. (6.1)) is directly related to the width of the spectrum

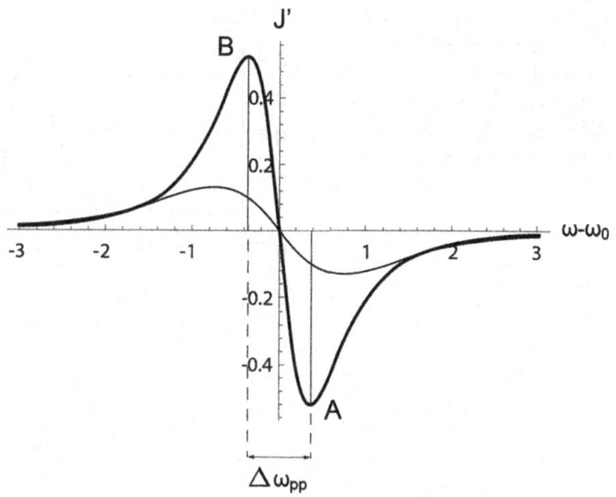

Fig. 6.1 Derivative of the Lorentzian line shape calculated for two values of the spin decoherence time $T_2 = 0.8$ (thin line) and 1.6 (thick line) in arbitrary units

$$\Delta\omega_{pp} = 2/\left(3^{1/2}T_2\right). \tag{6.3}$$

Thus, for the resonance Lorentzian line, the spin decoherence rate $1/T_2$ is easily found from the observed derivative of the spectrum (see (6.3)).

From the point of view of the problem of spin exchange, the Lorentzian form of lines in the EPR spectrum is expected only either at very slow spin exchange, when the contribution of spin exchange to the spin kinetics can be neglected, or at very fast spin exchange, when the exchange narrowing of the spectrum already occurs. Therefore, Eqs. (6.2 and 6.3) can be used to determine the spin exchange rate only in cases of very small or very high spin concentrations.

In the intermediate region of the spin exchange rate, the resonance lines have a mixed form: they are a sum of the symmetric Lorentzian absorption line and the asymmetric Lorentzian dispersion line with the same resonance frequency. Let us write the mixed form of the line in the form

$$j(x_0 - x) = \tau\left[1/\left(1 + (x_0 - x)^2\tau^2\right) + p(x_0 - x)\tau/\left(1 + (x_0 - x)^2\tau^2\right)\right], \tag{6.4}$$

where x_0 is the resonance frequency of the spins, x is the microwave field frequency and τ is the decoherence time of the spins, which determines the homogeneous width $\Delta\Omega$ of the line under consideration, $1/\tau = \Delta\Omega$. In Eq. (6.4), the parameter p specifies the fraction of the dispersion contribution to the observed EPR signal. As shown above, the p value in the slow spin exchange region increases proportionally to the spin coherence transfer rate (see, e.g., (5.32 and 5.45)).

Figure 6.2 shows the derivatives of the mixed line j(x) (6.4) for the two values of the dispersion parameter p. In Fig. 6.2, points A and B indicate the extrema of the derivative.

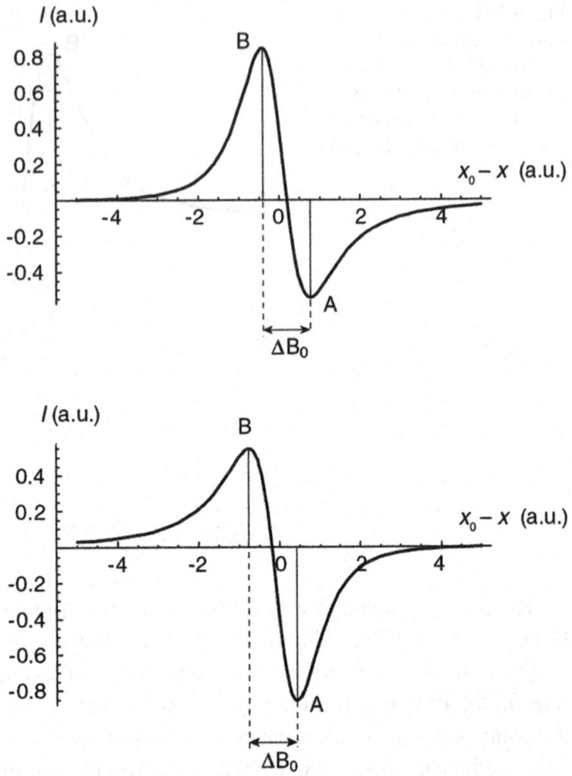

Fig. 6.2 Derivative of the spectrum line that has a mixed shape for p = 0.4 (upper curve) and p = −0.4 (lower curve). (Adapted with permission from Ref. [4])

It has already been noted above that the distance between extremes (see ΔB_0 in Fig. 6.2) derived in the case of a symmetric Lorentzian line is well characterized by the resonance width (see (6.3), where $\Delta\omega_{pp} = \gamma\ \Delta B_0$). For a symmetric line, the points of the maximum slope of the Lorentzian distribution on the left and right are at the same height. Therefore, the distance between the extremes of the derivative is a good characteristic of the resonance width (see (6.3)). In the situation (6.4) the resonance line shape is not symmetric, so the points of extremes of the derivative (see Fig. 6.2) are at different heights of the curve (6.4). Despite this, usually the distance between the two field values is called the width of the spectrum, ΔB_0 (see Fig. 6.2). For the resonance line (6.4), the distance between the points of the greatest slope (the distance between the points of the maximum and minimum of the derivative spectrum) is expressed as a product of the true half-width of the resonance $1/\tau$ and the function f(p), which depends only on the fraction of the dispersion contribution p and is symmetric with respect to p = 0 [4]

$$\Delta\omega_{pp} = (1/\tau)f(p), \tag{6.5}$$

$$f(p) = (2/p)3^{1/2}\left(1 + p^2\right)^{1/3} Im\left(\left(1 + ip\right)^{1/3}\right), f(-p) = f(p).$$

For $p < 1$

$$f(p) \approx 2\left(1 + (4/27)p^2\right)/3^{1/2} = 1.15 + 0.17p^2. \tag{6.6}$$

Thus, under the conditions of relatively slow spin exchange, the "width" of the resonance line, measured from the distance between the positions of the maximum and minimum of the derivative of this line, is determined by the product of a true line width and f(p) (see. (5.30 and 6.5)), e.g., in the case of the nonequivalent spin exchange

$$\Delta\omega_{k,\,pp} = \left(1/T_{2,k} + \left(KC - \varphi_k K'_{ex}C\right) + 1/T_{ddsd} + \varphi_k\left(1/T_{ddsct}\right)\right)f(p). \tag{6.7}$$

An important issue from this equation is that a new combination of unknown parameters is obtained from experimental EPR spectrum.

Another independent equation for unknown parameters can be obtained using positions of maxima of the spectrum or zeros of the derivative of the spectrum.

A maximum of j(x) (6.4) gives zero for its derivative. If $p = 0$, then the maximum (6.4) is observed at $x_0 = x$, as shown in Fig. 6.1. At $p \neq 0$, the signal maximum is no longer observed at the true resonance frequency, the EPR signal maximum (6.4) must be observed at [4].

$$x_0 = x + \left(\left(1 + p^2\right)^{1/2} - 1\right)/(p\tau). \tag{6.8}$$

In the slow spin exchange region $|p| < 0.5$ a good approximation is $x_0 \approx x + p/(2\tau)$ [4]. Thus, due to the dispersion contribution, the maximum of the line (6.4) is shifted by $\delta = p/(2\tau)$. Depending on the sign of p, the observed line is shifted to the right (when $p > 0$) or to the left (when $p < 0$).

In the case of slow spin exchange, this frequency shift can be described in more detail, if we use the width of the resonance lines calculated within the perturbation theory (see (5.30)) and the p value (see (5.32)). Given that for the k-th collective resonance in (6.4) it is necessary to substitute for $1/\tau$ the quantity, e.g., for the nonequivalent spin exchange case

$$\Delta\Omega_k = 1/T_{2,k} + 1/T_{ddsd} + \varphi_k\left(1/T_{ddsct}\right) + KC - \varphi_k K'_{ex}C(\text{see } (5.30)),$$

and instead of the parameter p

$$p_k = \left(K'_{ex}C - \left(1/T_{ddsct}\right)\right)\sum_{n \neq k} 2\varphi_n/(\omega_k - \omega_n)(\text{see Eq.}(5.32)),$$

we obtain

$$\delta_k \approx (1/2)\left(1/T_{2k} + 1/T_{ddsd} + \varphi_k(1/T_{ddsct}) + KC - \varphi_k K'_{ex}C\right)\left((K'_{ex}C - (1/T_{ddsct}))\right.$$
$$\left.\sum_{n\neq k} 2\varphi_n/(\omega_k - \omega_n)\right) = (1/2)\left(1/T_{2,k} + 1/T_{2ddsd} + \varphi_k(1/T_{2ddsct}) + KC - \varphi_k K'_{ex}C\right) p_k. \tag{6.9}$$

The experimental shift of the point at which the derivative of the spectrum passes through zero (the maximum of the resonance line) contains two contributions

$$\delta_k = \delta_{kcm} + \delta_{kdc}. \tag{6.10}$$

Both shifts are caused by the spin coherence transfer. One contribution is related to the change in the resonance frequency of the collective mode of spin coherence motion, and this shift is given by (see also (5.31))

$$\delta_{kcm} \approx (-\varphi_k/2)(K'_{ex}C - (1/T_{2ddsct}))^2 \sum_{n\neq k} 2\varphi_n/(\omega_k - \omega_n). \tag{6.11}$$

Formally, this shift can be written as

$$\delta_{kcm} \approx -(\varphi_k/2)\left(K'_{ex}C - (1/T_{2ddsct})\right)p_k.$$

This form may be useful when finding the spin coherence transfer rate from EPR data. This shift arises as the effect of the second-order spin exchange contribution to the resonance frequencies. It gives the real resonance frequency shift. Interestingly, formally the shift δ_{kcm} is given by the product of the spin coherence transfer rate $K'_{ex}C - (1/T2_{ddsct})$ and p_k-fraction of the dispersion contribution to the observed spectral line. Note that the resonance frequency shift (6.11) converges resonance lines independent of the sign of $V_{sct} = K'_{ex}C - (1/T_{2ddsct})$, i.e., independent of the relative contributions of the exchange and dipole-dipole interactions to the spin coherence transfer.

This component of the shift increases in proportion to the square of the spin concentration. It should be borne in mind that in the case of nonequivalent spin exchange, an additional shift in the frequency of collective modes associated with interference during the collision of the exchange interaction between two particles with spin-dependent interactions in each particle may occur. This issue was discussed in detail in the Chap. 2 (see (2.8.2)), and it was shown that this additional frequency shift is proportional to the spin concentration.

Another component of the shift (6.10) is directly related to the dispersion contribution to the observed signal. This shift arises as a result of the distortion of the resonance shape due to the dispersion contribution and it is given by the expression

$$\delta_{kdc} \approx (1/2)(\,1/T_{2,k} + 1/T_{ddsd} + KC)(K'_{ex}C - (1/T_{ddsct})\sum_{n \neq k} 2\varphi_n/(\omega_k - \omega_n)$$
$$= (1/2)(\,1/T_2 + 1/T_{2ddsd} + KC)p_k \tag{6.12}$$

This component of the shift of the position of the derivative zero is linear and quadratic in the spin concentration, and it is expressed by the product of p_k and the sum of the rates of all decoherence processes of spins, $1/T_{2,k}$ + $1/T_{2ddsd}$ + KC.

Thus, under the conditions of slow spin exchange, the dispersion contribution changes the shape of EPR lines: the resonance line becomes asymmetric and, as a result, the position of the maximum of the detected experimentally EPR line demonstrates a linear and quadratic in the spin concentration shift (6.9, 6.10, 6.11 and 6.12). In contrast to the resonance frequency shift given by (6.11) the shift of the position of line (6.12) depends on a sign of V_{sct}. As a result, this shift has opposite signs depending on what spin-spin interaction, exchange or dipole-dipole, provides the dominant contribution to the spin coherence rate: if the contribution of the exchange interaction dominates, this mechanism shifts the spectrum lines to the center of gravity of the spectrum, and if the dipole-dipole interaction dominates, this mechanism pushes the spectrum lines apart.

With the help of the above relations, the spin exchange rate can be evaluated from the experimental data on the width of the lines defined as the distance between the positions of the extrema of the EPR spectrum derivative and the shift of the maxima of resonance lines, which is determined by the zero position of the spectrum derivative.

Note that there may be spin-independent mechanisms of spectral diffusion induced, e.g., by a random modulation of anisotropic hyperfine interaction of unpaired electron with magnetic nuclei by rotational diffusion of paramagnetic particles. In this case, even at very low concentrations, the shape of the resonance lines may be a mixture of absorption and dispersion contributions. In this case, even at very low spin concentrations, the zero position of the spectrum derivative and the distance between its minimum and maximum points do not directly give the true resonance frequency and resonance width.

The experimental spectrum derivative makes it possible to evaluate directly the dispersion contribution to the signal. It was shown in [4] that the ratio of the minimum value of the derivative spectrum (A in Fig. 6.2) to the maximum value (B in Fig. 6.2), r, depends only on p. (To simplify the notation, A and B denote the values of the derivative at extreme points A and B). The analytical expression for r = A/B is quite complex, so in Fig. 6.3 we present only a plot of this function [4].

It is remarkable that the plot in Fig. 6.3 establishes a one-to-one correspondence between the two values. This plot can be considered as the dependence of the unknown p value on the parameter r, which can be found from the experimental EPR spectrum derivative. In the interval of $-$ 0.2 <p <0.2 a good approximation is

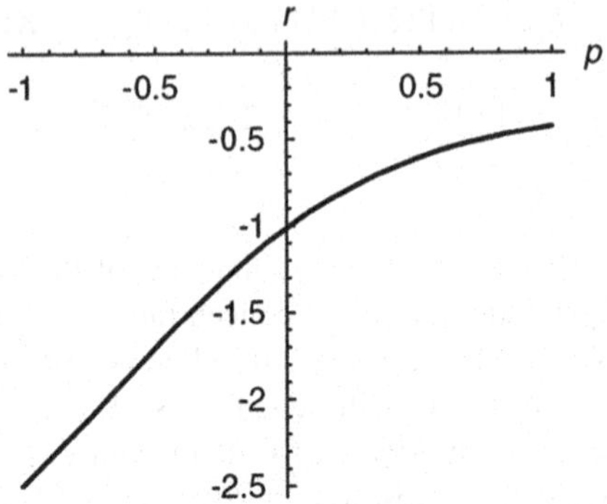

Fig. 6.3 Dependence of r = A/B on the parameter p - fraction of dispersion contribution to the resonance line. (Adapted with permission from Ref. [4])

$$p = \left(3^{1/2}/2\right)(1 + r) = 0.87\ (1 + r). \tag{6.13}$$

Using (6.13) or Fig. 6.3 the dispersion contribution (see the p-containing term in (6.4)) to the observed asymmetric line can be found directly from the experimental EPR spectrum derivative. Knowledge of the dispersion contribution p makes it easy to find the spin coherence transfer rate (see, e.g., (5.45)).

Thus, the asymmetry of the spectrum components (i.e., the fraction of the dispersion contribution) is the most direct method of determining the spin coherence transfer rate. To do this, we only need to know the initial splitting of lines in the EPR spectrum and the dispersion contribution p. Based on the theoretical considerations [1, 2], Bales et al. [5–7] developed an algorithm for determining the spin exchange rate, which is based on the phenomenological decomposition of the experimental spectrum into the sum of contributions of symmetric absorption lines and asymmetric dispersion lines.

Thus, experimental and theoretical studies of spin exchange led to the revision of the method for obtaining information concerning the spin exchange rate from the analysis of EPR spectra. The modern paradigm as the most important step involves finding the dispersion fraction in the shape of spectral lines.

Knowledge of the p value and the usage of Eq. (5.32)

$$p_k = \left(K'_{ex}C - 4/T_{dd}\right)\Big)\sum_{n \neq k} 2\varphi_n/(\omega_k - \omega_n)$$

make it possible to find a new equation for the spin coherence transfer rate and in this manner to determine with good accuracy the spin exchange rate from the analysis of EPR spectra of, e.g., nitroxide radical solutions (see, e.g., [5–12]).

If the components of the EPR spectrum remain resolved in a certain range of the spin concentration variation, the spin decoherence rate and spin coherence transfer rate due to exchange and dipole-dipole interactions can be found from the measurement of the concentration dependence of the distance between the positions of the maximum and minimum of the derivative spectrum over the field using Eqs. (6.5 and

6.7). From the concentration change of the position of the zero value of the spectrum derivative it is possible, in principle, to estimate the spin coherence transfer rate using (6.9, 6.10, 6.11 and 6.12).

But finding the spin coherence transfer rate this way in practice is not straightforward. The experimental shift of the maximum position consists of several contributions (see (6.9, 6.10, 6.11 and 6.12)). It is impossible to find the spin exchange rate when using only the experimental shift of the position of the derivative spectrum zero. This shift is giving one more "independent" equation which contains the unknown spin coherence transfer rate. In the early paradigm of spin exchange, it was assumed that the shift of the position of the derivative spectrum zero gives directly the shift of resonance frequencies. Indeed, in the early paradigm, the dispersion contribution to the observed signal and the contribution of dipole-dipole interaction to the spin coherence transfer are not taken into account. As a result, a very simple (but not correct!) formula for finding the spin exchange rate from the shift of the position of the derivative spectrum zero is used (see Eq. (6.11))

$$\delta_{\mathrm{kcm}} \approx (-\varphi_{\mathrm{k}}/2)((\mathrm{K}_{\mathrm{ex}}\mathrm{C})^2 \sum_{n \neq k} 2\varphi_n/(\omega_k - \omega_n));$$
$$\delta_{\mathrm{kdc}} = 0.$$

The current level of understanding of the problem shows that the shift of the position of the derivative spectrum zero can be applied, when determining the spin exchange rate, only in combination with other data, e.g., line broadening (see Eqs. (6.9, 6.10, 6.11 and 6.12)). Moreover, in the nonequivalent spin exchange case, experimental data given by the steady-state EPR spectra in the linear response region may not suffice to obtain all unknown spin exchange rates. Data obtained by other methods of EPR spectroscopy may be needed (see Chap. 7).

In the fast spin exchange region, when the exchange narrowing of the spectra is manifested, information about the spin exchange is contained in the Lorentzian line width, e.g., in the case of nonequivalent spin exchange for ^{14}N nitoxide radicals (cf. (5.17)

$$1/\mathrm{T}_{2\mathrm{eff}} \equiv 1/\mathrm{T}_{\mathrm{ddsd}} + 1/\mathrm{T}_{\mathrm{ddsct}} + (\mathrm{KC} - \mathrm{K}'_{\mathrm{ex}}\mathrm{C}) + 2a_{\mathrm{N}}{}^2/(3(\mathrm{K}'_{\mathrm{ex}}\mathrm{C} - 1/\mathrm{T}_{\mathrm{ddsct}}))$$
$$+ 1/\mathrm{T}_2.$$

Thus, the methods for estimating the spin exchange rate from the analysis of the characteristic features of the derivative spectrum, which is recorded in steady-state EPR spectroscopy in the linear response region, are developed. But these methods are applicable only in the region of intermediate spin concentrations, when the proton hyperfine structure has already undergone exchange narrowing, and nitrogen spectrum components are still resolved, or in the very fast spin exchange case, when the entire spectrum becomes exchange narrowed. The relationship between the experimental characteristics of the spectrum and the spin exchange rate in the analytical form is obtained only in these situations (see (6.5, 6.6, 6.7, 6.8, 6.9,

6.10, 6.11 and 6.12)) for slow spin exchange and (5.17) for fast spin exchange, respectively) [4, 11, 12].

The spin decoherence (spin dephasing) and the spin coherence transfer rates obtained by using Eqs. (6.5, 6.6, 6.7, 6.8, 6.9, 6.10, 6.11 and 6.12) are the estimates of these rates. It should be emphasized that Eqs. (6.5, 6.7, 6.9, 6.10 and 6.11) are obtained disregarding any overlap of the lines with different resonance frequencies. This overlap distorts the shape of the observed spectrum. Indeed, due to the overlap of different lines, the experimental positions of the lines can be shifted and their width can increase. Therefore, the parameters obtained from the analysis of the characteristic properties of the EPR spectrum using Eqs. (6.5, 6.7 and 6.9) should be considered only as reasonable estimates of these parameters.

A very detailed description of the current state of the technique for determining spin exchange rate constants in bimolecular collisions is given in [11, 12].

6.2 Potential of the Linear Response EPR Spectra for Determining the Spin Decoherence and the Spin Coherence Transfer Rates

6.2.1 General Speculations

In the previous sections, the main aspects of the modern paradigm of spin exchange and spin exchange manifestations in steady-state EPR spectra under linear response conditions were presented. It is a difficult task to determine the spin decoherence rate and the spin coherence transfer rate from the analysis of the experimental concentration dependence of the EPR spectrum shape. In the general case, only several combinations of the contribution of the exchange interaction in bimolecular collisions of paramagnetic particles and the dipole-dipole interaction between them to the spin decoherence and spin coherence transfer rate can be determined directly from the analysis of the EPR spectrum. In order to obtain information about bimolecular collisions of paramagnetic particles we are interested in, it is necessary to separate the contributions of these two interactions to the change in the spin state of paramagnetic particles.

The ways of determining the rates of the spin decoherence and the spin coherence transfer between spins for nitroxide radicals will be discussed in some detail here. Here we consider mainly what information can be directly obtained from the linear response steady-state EPR spectrum. To separate the contributions of the exchange and dipole-dipole interactions in the general case of nonequivalent exchange, it may be necessary to use other EPR methods as well. These EPR methods are discussed in Chap. 7.

How theoretical results presented in the previous sections can be adapted to study spin exchange between specific paramagnetic particles will be shown by the example of the analysis of the EPR spectrum of solutions of nitroxide radicals. Nitroxide free radicals are one of the favorite spin probes used to study spin exchange. With this in mind, we consider solutions of nitroxide radicals. These considerations can be properly projected onto other paramagnetic particles.

Several algorithms for determining the spin-exchange rate using EPR spectra are developed. The common algorithm (see, e.g., [1–3]) is based on the following widespread paradigm. The components of EPR spectra demonstrate concentration-dependent broadening and shifts of EPR lines at the low spin probe concentration. At high concentrations, all components of the EPR spectrum merge into a narrow single line at the center of gravity of the spectrum.

However, this widespread paradigm ignores some specific manifestations of spin exchange in the shape of the EPR spectrum. In the presence of spin exchange, the shape of the EPR spectra components can be the sum of the absorption and dispersion terms [1, 2, 4, 5]. For example, two external hyperfine components of the EPR spectrum of the nitroxide ^{14}N radical have the asymmetric shape in the slow spin coherence transfer case. The common algorithm does not properly handle the dipole-dipole interaction between paramagnetic particles: it ignores the spin coherence transfer caused by this interaction. According to the generally accepted theory, the dipole-dipole interaction in non-viscous solutions generates only spin decoherence [13, Chap. 8, Eqs. (VIII.79, VIII.89)]. However, in [14], Eqs.(4.20 and 4.75) I have shown (see also [15]) that the dipole-dipole interaction also causes the spin coherence transfer along with the exchange interaction. It should be noted that in the nonviscous solutions, the contributions of exchange and dipole-dipole interactions to the spin decoherence rate are added, but their contribution to the spin coherence transfer rate are subtracted in the experimental EPR conditions [4, 14, 15], see Eq. (4.6).

Another problem arises due to the hyperfine structure of the EPR spectra of the paramagnetic particles. For example, in the case of nitroxide radicals, EPR spectra have the hyperfine structure induced by the nitrogen nucleus and protons. There are three (in the case of ^{14}N) well-resolved nitrogen hyperfine components that are related to the sub-ensembles of nitroxide radicals with different projections of the nitrogen nuclear spins ($m = +1, 0, -1$ for ^{14}N). Each nitrogen component has a resolved/unresolved hyperfine structure induced by the interaction of an unpaired electron with protons. This proton-induced hyperfine structure and overlap of nitrogen components pose additional challenges for the determination of bimolecular spin exchange rates.

The EPR manifestations of spin exchange between nitroxide radicals were comprehensively investigated experimentally. The relevant results are reviewed in [1–3]. Bales et al. (see, e.g., [5, 6, 8]) developed a new approach for determining the spin exchange rate using EPR data based on the theoretical prediction [1, 2], see also [4, 16], that due to the coherence transfer the components of EPR spectra become asymmetric and can be presented as a sum of absorption and dispersion contributions. The theory also predicted the linear dependence of the fraction of the dispersion contribution on the spin coherence transfer rate in the slow spin exchange case. These theoretical predictions were confirmed in experiments [5–12] and they

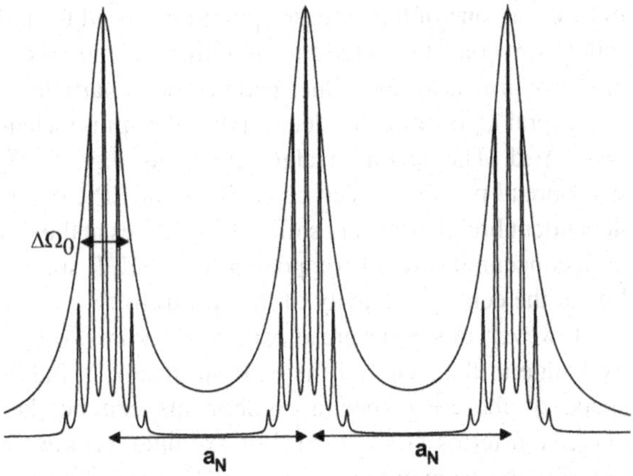

Fig. 6.4 Schematic presentation of the EPR spectrum for a model ^{14}N nitroxide radical with six equivalent protons. (Adapted with permission from Ref. [11])

demonstrated that the determination of the dispersion contribution to the shape of EPR spectra is the best method to determine the spin coherence transfer rate.

In real systems, the nitrogen hyperfine components of the EPR spectra of nitroxide radicals have the additional hyperfine structure due to the interaction of the unpaired electron with protons (see Fig. 6.4).

Nitrogen nuclei and protons lead to the hyperfine splitting of the EPR lines in different scales: the splitting by the nitrogen nucleus is much larger than the splitting due to protons. Therefore it is justified to introduce two characteristic spin concentrations: C_0 and $C*$. They correspond to the characteristic spin exchange rates, $K'_{ex}C$ (K'_{ex} is the spin coherence transfer rate constant) determined by the conditions $K'_{ex}C_0 = \Delta\,\Omega_0$ and $K'_{ex}C* = a_N$, respectively (see Fig. 6.4). When $C < C_0$, the nitrogen components manifest the inhomogeneous broadening induced by the hyperfine interaction with protons, since spin exchange inside the nitrogen components (spin exchange between proton hyperfine components) is relatively slow compared with the total width of the proton hyperfine splitting in this spin concentration range. Therefore the spin exchange effect on the EPR spectrum at $C < C_0$ has to be analyzed carefully. In the spin concentration range of $C_0 < C < C*$ the spin exchange rate is fast enough to merge each nitrogen component into the exchange narrowed homogeneous line. Qualitatively, the manifestations of the spin exchange in the EPR spectra of nitroxide radicals in the presence of the proton hyperfine structure are well-known (see, e.g., [1, 2] Chap. 3). For example, for the water solution of the stable free radical 2,2,6,6-tetramethyl-4-oxopiperidine-1-oxyl (TANONE) in the range of intermediate concentrations $(0.4 \div 1.7)\ 10^{-2}$ M/L ($C_0 < C < C*$) the line width depends on the spin concentration linearly, but the line width has the nonlinear part at very low concentrations $C < 0.3\ 10^{-2}$ M/L ($C < C_0$) (see Fig. 3.5 in [2] and the discussion therein). The manifestations of spin exchange in EPR spectra of nitroxide free radicals in the slow ($C < C_0$) and intermediate ($C_0 < C < C*$) spin exchange rate regions were studied in the solution of di-tert-butylnitroxide (DTBN) [17]. It was shown that spin exchange leads to the broadening and shifts of lines, and eventually to the collapse of the lines.

The theoretical results presented above make it possible to simulate the EPR spectra of paramagnetic particles in dilute solutions and to analyze their transformations when the spin concentration changes (see Eqs. (5.13, 5.14 and 5.15)). For numerical simulations, it is necessary to estimate scales of possible values of the spin decoherence and spin coherence transfer kinetic parameters. The binary collision rate Z can be estimated using the Smolukhovski result: $Z = 4\pi r_0 D\,C$, where r_0 is the collision radius of two spin probes, D is their mutual diffusion coefficient. To estimate the spin exchange rate, we assume that the exchange interaction is switched on suddenly at the collision of two spin probes. For this sudden collision approximation the maximum equivalent spin exchange rate constant is expected to be $K_{ex} = 2\,\pi\,r_0\,D$. If the concentration C is measured in mM/L, then $K_{ex} = 2\,\pi\,r_0\,D\,10^{-6}\,N_{A,}\,s^{-1}$ L/mM. In the EPR experiments, it is convenient to use the unit gauss (G) of the magnetic field strength. Thus $K_{ex} = 2\,\pi\,r_0\,D\,10^{-6}\,N_A/(1.76\,10^7) \approx 2\,r_0\,D\,10^{11}$, Gs L/mM. For example, for realistic molecular kinetic parameters $r_0 = 0.7$ nm, $D = 10^{-5}$ cm^2/s, one obtains $K_{ex} = 0.15$ Gs L/mM. In the course of numerical simulations we will use K_{ex} around this value. The contribution of the dipole-dipole interaction to the spin decoherence and spin coherence transfer rates were calculated numerically using expressions for $J^{(n)}(\omega)$ in [13, Eqs. (VIII.109, VIII.114)], see also Eqs. (3.25 and 3.26). The results of these numerical calculations are presented in Fig. 6.5. The dashed line shows $K_{ex} = 2\,\pi\,r_0\,D$. Figure 6.5 shows that in nonviscous

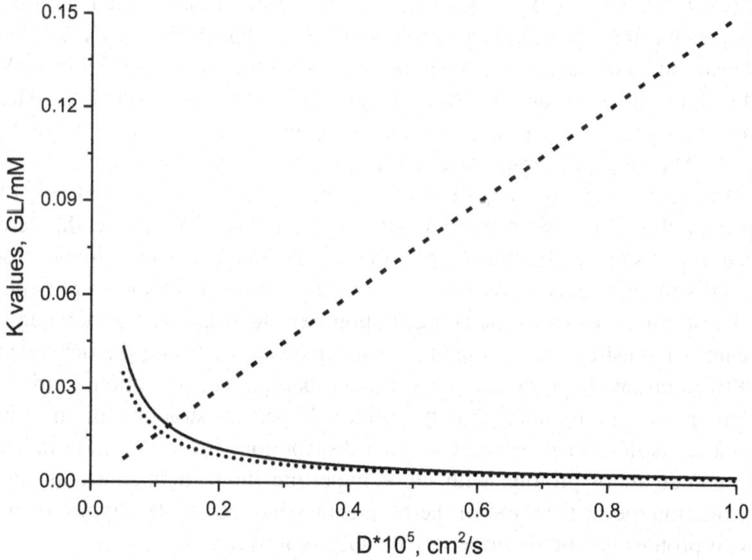

Fig. 6.5 Dependence of K_{dsd}, Gs L/mM (solid line) and K_{dsct}, Gs L/mM (dotted line), K_{ex}, G L/mM (dashed line) on the mutual diffusion coefficient of paramagnetic particles. Parameters used in these simulations: $a_N = 16$ Gs, $r_0 = 0.7$ nm, $\omega_0 = 3400$ Gs. (Adapted with permission from Ref. [11])

liquids, when D ~ 10^{-6}–10^{-5} cm^2/s, the contribution of the dipole-dipole interaction to the spin decoherence and spin coherence transfer is much less than the contribution of the exchange interaction (spin exchange). Note that K_{dsd} and K_{dsct} are almost equal when D ~ 10^{-6} –10^{-5} cm^2/s, $K_{dsd} \approx K_{dsct}$ (see Fig. 6.5).

According to the theory presented above (see Eqs.(5.13, 5.14 and 5.15)), the concentration dependence of the EPR spectrum shape is determined by two parameters W and V which characterize the additive contributions of the exchange and dipole-dipole interactions to the spin dephasing, W, and the spin coherence transfer, V. For example, in case of equivalent spin exchange

$$\begin{aligned} J_{EPR} &= -Im[M-\}; \\ M_- &= i\gamma B_1 M_0 G(\omega)/[1 + (K_{ex}C - 1/T_{ddsct})G(\omega)]; \\ W &= K_{ex}C + K_{dsd}C; \\ V &= K_{ex}C - K_{dsct}C. \end{aligned} \tag{6.14}$$

For the numerical simulation of the EPR spectrum, one has to calculate a function G(ω) (5.13)

$$G(\omega) = \sum_k \frac{\varphi_k}{\{-i(\omega_k + \delta_k - \omega) - 1/T_{2k} - (K_{ex}C + 1/T_{ddsd})\}}. \tag{6.15}$$

The form of this function is determined by the distribution of the resonance frequencies $\omega_k = \omega_0 + \Delta_k$. When this inhomogeneous distribution of resonance frequencies has the Gaussian form, the function G(ω) has the Voigt form if T_{2k} does not depend on k (see, e.g., [3]). When the spin coherence transfer is ignored, V = 0, all EPR lines in the G(ω) demonstrate the concentration broadening. The spin coherence transfer with the non zero rate V changes the observed spectrum given by (5.13). The coherence transfer leads to mixed resonance line shapes, changes the concentration broadening of the resonance lines, shifts the lines, and leads to the collapse of the EPR spectrum into one homogeneous line. If the dipole-dipole interaction gives the negligible contribution, the exchange narrowed line is observed at the fast spin coherence transfer case. If the dipolar contribution is not negligible, the collapsed line manifests the concentration-dependent dipolar broadening.

Figure 5.1 illustrates these qualitative statements concerning the dependence of the EPR spectrum shape on the paramagnetic particle concentration. During these calculations it was assumed that the proton hyperfine structure of the nitrogen components is described by the Gaussian distribution. If the nitroxide radical has 5–10 or even more protons with close hyperfine interaction constants, in good approximation the distribution of the frequencies induced by the hyperfine interaction with protons can be described by the Gaussian function,

$$f_m(\Delta) = (1/\mathrm{Sqrt}[2\pi\sigma])exp\left(-(\Delta - a_N m)^2/(2s)\right), \text{where } m = -1, 0, 1.$$

Here a_N m is the resonance frequency shift induced by the hyperfine interaction with the nitrogen nuclear spin and σ is the dispersion of the Δ distribution. Note that the width $\Delta\Omega_0$ of the Gaussian nitrogen component of the spectrum (see Fig. 6.4) can be characterized by the square root of dispersion, $\sigma^{1/2}$. Analogously, the numerical simulations of spectra can be performed for any hyperfine structure of the nitrogen components of the spectrum (see Fig. 6.4) and any spin concentration using the general solution (see Eqs. (5.13, 5.14 and 5.15)) . Comprehensive numerical simulations of the EPR spectrum of solutions of ^{14}N-containing nitroxide radicals, when the unpaired electron spin interacts with the magnetic nuclei of 12 equivalent hydrogen/deuterium atoms were presented in [11, 12].

Thus, Eqs. (5.13, 5.14 and 5.15) make it possible to simulate EPR spectra for any concentration of paramagnetic particles and for any hyperfine structure of the EPR spectra. Note that Eqs. (5.13, 5.14 and 5.15) include exchange and dipole-dipole interactions contributions to the spin decoherence and the spin coherence transfer. Thus, in principle, by varying the unknown parameters in Eqs. (5.13, 5.14 and 5.15) it is possible to find the best-fit parameters, which reproduce the experimental curves. This fitting approach can be considered as a universal (global) procedure for finding the spin exchange rate using the EPR spectrum. In principle, all parameters, W, V, $1/T_{2k}$, ω_k, and δ_k (see (5.13, 5.14 and 5.15)) may be found in this way. However, the implementation of this spectrum fitting is not straightforward since many variable parameters are involved. Under this condition, it is of real importance to pick up reasonable initial probe values of these parameters in the fitting procedure.

The following approach can be used as an option. First, the parameters $1/T_{2k}$ and ω_k are found by fitting experimental spectra detected at very low spin concentrations, when the exchange and dipole-dipole contributions can be neglected so that $W = 0$, $V = 0$, $\delta_k = 0$. Then at higher spin concentrations, fitting of experimental spectra can be done by varying unknown parameters W, V, δ_k. For nitroxide radicals, the situation is simplified, since the condition of equivalent spin exchange is often satisfied, and then $K'_{ex} = K_{ex}$. It is expected that the centers of the nitrogen components of the nitroxide radical EPR spectrum should have additional shifts of the resonance frequencies δ_{exk} which depend only on the projection of the nuclear spin of nitrogen since the protons create symmetric hyperfine structure of the nitrogen components of the EPR spectrum. Thus, for the ^{14}N and ^{15}N nitroxide radicals this shift has the same value, but different signs for the two extreme nitrogen components of the spectrum. For a ^{14}N nitroxide radical, the central nitrogen component of the spectrum has not this shift of resonance frequency.

To this end, it seems to be practically feasible to obtain W and V values from the fitting procedure.

The procedure for finding the spin exchange rate using the EPR spectrum fitting on the basis of general expressions (5.13, 5.14 and 5.15) is a consistent rigorous method of solving the problem. But it has some disadvantages. Different sets of fitting parameters can simulate the observed spectrum with comparable accuracy. To this end, it is desirable to obtain reasonable estimates for the spin dephasing rate W, spin coherence transfer rate V and resonance frequencies shifts δ. In fact, possible options for estimating these parameters directly from the experimental EPR spectra were already presented in the above discussion.

To specify more how general results can be applied to the nitroxide radical solutions it is helpful to take into account that there are three characteristic ranges of the nitroxide free radical concentrations:

I. $C < C_0$ ($V < 3\ \Delta\Omega_0$), slow spin coherence transfer;
II. $C_0 < C < C*$ ($3\ \Delta\Omega_0 < V < a_N$), intermediate spin coherence transfer rate, the proton hyperfine structure lines inside each nitrogen component merge into an exchange-narrowed homogeneous Lorentzian line;
III. $C > C*$, ($V > a_N$), exchange narrowing of the spectrum (see (5.17)).

The approximate solutions for the shape of the EPR spectrum relevant to these three spin concentration ranges were proposed in [11]. The case I was also very comprehensively studied in [12]. It is most difficult for the fitting procedure. But when the proton hyperfine structure of the nitrogen EPR spectrum components is well resolved, the interval $C < C_0$ can be divided into two sub-intervals

(Ia) $C < C_{00}$ ($K_{sct}C < a_p$), "super" slow spin coherence transfer (a_p is the proton hyperfine interaction constant) and
(Ib) $C_{00} < C < C_0$ ($a_p < K_{sct}C < \Delta\Omega_0$).

In the Ia case, the perturbation theory can be developed with respect to the small parameter $K_{sct}C/a_p < 1$. In the Ib case, the perturbation theory can be developed with respect to the small parameter $K_{sct}C/\Delta\Omega_0 < 1$. This was done in [11].

In the case III, the total spectrum is given by the Lorentzian line with the line width Eq. (5.17). In this case, we have the model situation given by Eqs. (6.1, 6.2 and 6.3).

6.2.2 *EPR Spectrum in the Case of Intermediate Spin Coherence Transfer Rates*

It was shown in Chap. 5 that for all cases the EPR spectrum is the sum of independent lines, each of which is the sum of the Lorentz absorption line and the Lorentz dispersion line. For arbitrary hyperfine structure of the nitroxide radical EPR spectrum we do not know analytically the frequencies, widths and oscillator strengths of all collective modes for the spin coherence of the system. However in some concrete cases we can get helpful approximate results. For example, in the intermediate case II

$$\Delta\Omega_0 < K_{ex}C < a_N \tag{6.16}$$

the spectrum of the ^{15}N- or ^{14}N-containing nitroxide radicals can be reduced to the model situation when only the hyperfine interaction with the nitrogen nuclear spin is taken into account, while the hyperfine structure of the EPR spectrum induced by the hyperfine interaction with all other magnetic nuclei can be considered phenomenologically by introducing the shorter transverse relaxation time (this reduction of the spectrum to model situation is presented in detail in [11]. In the spin concentration range (6.16), all proton/deuterium induced hyperfine structures of each nitrogen component are merged into a homogeneous Lorentzian line, which has, e.g., for ^{14}N nitroside radicals, the line width $1/T_2 + 3\sigma/|V_{sct}|$ (see [11], Eqs. (11)). Thus in the case (6.16) the EPR spectrum can be modeled as a spectrum of a system with two or three resonance lines for ^{15}N or ^{14}N nitroxide radicals, repectively. The rigorous theoretical analysis of manifestations of the exchange and dipole-dipole interactions in the shape of the EPR spectrum was presented for model systems with two or three resonance frequencies [4].

It was shown that in the intermediate spin exchange rate case II (the slow spin coherence transfer with respect to the splitting induced by the hyperfine interaction with nitrogen magnetic nucleus), the shape of EPR spectra of ^{14}N or ^{15}N nitroxides is presented as a sum of three or two independent components.

The outer components of the spectrum are asymmetric. These asymmetric components contain the absorption and dispersion contributions.

For the two frequencies model (^{15}N radical) and the three frequencies model (^{14}N radical), the resonance frequencies, the asymmetry parameters (p-contribution of the dispersion term), the shift of the maximum of the observed line due to its asymmetry, the line width determined as a distance between positions of minimum and maximum of the line in derivative EPR spectrum were obtained in [4] for arbitrary rate of the equivalent spin exchange. These results are rather simple in the case when $K_{ex}C < a_N$. For this situation, results are presented in Tables 6.1 and 6.2.

Here the following parameters are introduced: a is the nitrogen hyperfine interaction constant, T_2 is the transverse relaxation time of isolated radicals, C is the concentration of radicals (see (5.11)),

$$\begin{aligned}
V_1C &= (1/2)(K_{ex}C - 1/T_{ddsct});\\
W_1C &= K_{ex}C/2 + 1/T_{ddsd} + 1/(2T_{ddsct});\\
V_2C &= (1/3)(K_{ex}C - 1/T_{ddsct});\\
W_2C &= 2K_{ex}C/3 + 1/T_{ddsd} + 1/(3T_{ddsct})
\end{aligned}$$

Parameters δ_{kex} give the resonance frequencies shift due to the "interference" of the exchange interaction and the spin-dependent interactions in isolated paramagnetic particles in the course of their bimolecular collisions (see discussion of this effect in Sect. 2.8.2).

The theory developed in [4] was generalized with allowance for the superhyperfine structure of EPR spectra of nitroxide radicals in [11].

Table 6.1 Parameters of the EPR spectrum lines for the two resonance frequency system in the slow intermediate spin coherence transfer rate case when $K_{ex}C < a_N$

Line number	Resonance frequency	Asymmetry parameter, p	Shift of the maximum due to the dispersion contribution, $\Delta\delta_{dc}$ (see (6.12))	Line width
1	$\omega_0 - a/2 + V_1^2C^2/a + \delta_{1ex}C$	$-(K_{ex}C - 1/T_{ddsct})/a$	$-(V_1C/(aT_2) + V_1W_1C^2/a)$	$1.15(1/T_2 + W_1C)$
2	$\omega_0 + a/2 - V_1^2C^2/a - \delta_{1ex}C$	$+(K_{ex}C - 1/T_{ddsct})/a$	$+(V_1C/(aT_2) + V_1W_1C^2/a)$	$1.15(1/T_2 + W_1C)$

Adapted with permission from Ref. [16]

Table 6.2 Parameters of the EPR spectrum lines for the three resonance frequency system in the slow intermediate spin coherence transfer rate case when $K_{ex}C < a_N$

Line number	Resonance frequency	Asymmetry parameter, p	Shift of maximum due to the dispersion contribution, $\Delta\delta_{dc}$ (see (6.12))	Line width
1	$\omega_0 - a + \delta_{2ex}C + 3V_2^2C^2/(2a)$	$-(K_{ex}C - 1/T_{ddsct})/a$	$-(3V_2C/(2aT_2) + 3V_2W_2C^2/(2a))$	$1.15(1/T_2 + W_2C)$
2	ω_0	0	0	$1.15(1/T_2 + W_2C)$
3	$\omega_0 + a - \delta_{2ex}C - 3V_2^2C^2/(2a)$	$+(K_{ex}C - 1/T_{ddsct})/a$	$+(3V_2C/(2aT_2) + 3V_2W_2C^2/(2a))$	$1.15(1/T_2 + W_2C)$

Adapted with permission from Ref. [16]

The shape of the EPR spectrum of the nitroxide radical (Eqs. (5.13, 5.14 and 5.15)) can be simplified further for the concentration region of $C_0 < C < C*$ when the spin coherence transfer rate is less than the inhomogeneous broadening of the nitrogen components of the spectrum. In this case, all spins in the nitrogen component have the same resonance frequency and the additional shift of the outer components should be $\pm$ (1/6) (V^2/a_N) as it was already shown in [1, 2, 4]. Due to the spin exchange between spins inside nitrogen components of the spectrum, the additional homogeneous broadening of the nitrogen components appears, which is $3 < \Delta^2 > /|V|$ [11]. Thus, in the intermediate concentration region of $C_0 < C < C*$ Eqs. (5.13, 5.14 and 5.15) are reduced to the model situation considered in [4]. In this intermediate spin concentration case, the EPR spectrum, e.g. for ^{14}N nitroxide radicals, is a sum of three components given by the following equations (V and W are given by Eq. (6.14)) [11]

$$J_m \cong -(\omega_1/3)Re\{(1+im\,(V/a_N))G_m/(1+(1/3)V\,G_m)\}, m=0, \pm 1,$$
$$G_1 = 1/(-i(\omega_0 + a_N - \delta_{ex} - (1/6)\,(V^2/a_N) - \omega) - (W + 3 < \Delta^2 > /|V| + \Gamma_{01})), \tag{6.17}$$
$$G_0 = 1/(-i(\omega_0 - \omega) - (W + 3 < \Delta^2 > /|V| + \Gamma_{00})),$$
$$G_{-1} = 1/(-i(\omega_0 - a_N + \delta_{ex} + (1/6)\,(V^2/a_N) - \omega) - (W + 3 < \Delta^2 > /|V| + \Gamma_{0-1})).$$

In the second order of the perturbation theory for a small parameter $K_{ex}C/a_N < 1$, the collective independent lines J_m (6.17) for the equivalent spin exchange case under consideration can be presented as

$$J_m = (1/3)\omega_1 T_{2ef,m}(1 + m(V/a_N)\,(\omega_{res,m} - \omega)\,T_{2ef,m})/\left(1 + (\omega_{res,m} - \omega)^2 T_{2ef,m}{}^2\right), \tag{6.18}$$

where $\omega_{res,m}$ and $T_{2ef,m}$ are the frequencies and the characteristic life times of three resonances, respectively

$$\begin{aligned} \omega_{res,m} &= \omega_0 + m\,a_N - m\,\delta_{ex} - m(1/6)(V^2/a_N), m = 1,0,-1; \\ 1/T_{2ef,m} &= W - V/3 + 3\Delta^2 > /|V| + \Gamma_{0m} \\ &= (1/T_{ddsd} + 1/(3T_{ddsct}) + 3 < \Delta^2 > /|V| + 1/T_{2,m}). \end{aligned} \tag{6.19}$$

Outer components of the spectrum, J_1 and J_{-1}, contain the absorption and dispersion terms, while the central component J_0 contains only the absorption term.

Equations (6.17, 6.18 and 6.19) describe three components of the EPR spectrum of ^{14}N-containing nitroxide radicals for the intermediate concentration case (6.16). Analogous results for ^{15}N nitroxide radicals are presented in [4].

Note that the additional contribution $3 < \Delta^2 > /|V|$ to the line width appears in Eqs. (6.18 and 6.19), which describes the contribution of the spin coherence transfer inside the nitrogen component to the spin decoherence. Note that the spin coherence transfer rate inside the nitrogen component is V/3 and $< \Delta^2 >$ is the dispersion of the proton hyperfine interaction induced distribution of the resonance frequencies. If this distribution is a Gaussian, then $<\Delta^2> = \sigma$.

In the case Ib, the results analogous to Eqs. (6.17) were obtained in [12]. In the case of this very slow spin coherence transfer (very low spin concentration) the amplitude of the dispersion contribution to the nitrogen component of the spectrum does not depend on the proton hyperfine structure of the nitrogen EPR spectrum components. But the shape of the dispersion contribution as well as the shape of the absorption contribution of the nitrogen components of the spectrum depend on the proton hyperfine structure. In the absence of the proton hyperfine structure, the absorption and dispersion contributions are given by the real and imaginary parts of the Lorentzian form $(1/(1 + i\,z))$. In the case of the Gaussian distribution of the proton hyperfine structure the shape of the nitrogen components of the spectrum is described by Voigt form [3]. When the spin coherence transfer occurs the "nitrogen" component can be presented as a sum of Lorentzian lines of mixed shape. However, the total absorption and dispersion contributions to the nitrogen components are not described by simple functions like the real or imaginary parts single Lorentzian line or Voigt forms for several reasons: collective mode resonances have different widths; resonance frequencies of collective modes in general case are unknown, they are not to have Gaussian distribution; intensities of collective mode lines are different.

At high concentrations, when $V > a_N$, Eqs. (6.17, 6.18 and 6.19) are not valid. Under this condition it is necessary to use the general solution given by Eqs. (5.13, 5.14 and 5.15). But in this limiting case, the exchange narrowing effect operates and the total EPR spectrum of, e.g., ^{14}N the nitroxide free radical appears as the Lorentzian absorption curve with the concentration-dependent dipolar broadening $W - V = 1/T_{ddsd} + 1/T_{ddsct}$ (see Eqs.(6.14)):

$$J_{EPR} \cong \omega_1 M_{eq}\left(W - V + \Gamma + 2a_N^2/(3\,V)\right)/\left((\omega_0 - \omega)^2 + \left(W - V + \Gamma + 2a_N^2/(3\,V)\right)^2\right). \tag{6.20}$$

Here $\Gamma = (1/3)\,(\Gamma_{01} + \Gamma_{00} + \Gamma_{0-1})$ and a term $2\,a_N^2/(3|V|)$ gives the spin coherence transfer contribution to the line width [1, 2].

6.2.2.1 The EPR Spectrum Shape of ^{14}N Nitroxide Radicals in Slow Spin Exchange Region. Perturbation Theory

The most nontrivial result obtained in [11] and presented above in Eq. (6.18) is the contribution of dispersion to the shape of nitrogen components of the EPR spectrum. Using the perturbation theory for the small parameter $3\,V/a_N < 1$ it was shown in [11] that the contribution of the dispersion to the form of the extreme nitrogen components of the spectrum is practically independent of the transfer of spin coherence between proton hyperfine components within each nitrogen component of the spectrum. This section demonstrates how this result was obtained.

Consider the solution of the ^{14}N nitroxide radicals. The hyperfine interaction with nitrogen nuclei in nitroxides is much larger than the hyperfine interaction with protons/deuterons. In this case at the low radical concentration the expected EPR spectrum is presented schematically in Fig. 6.4. There are three bands which correspond to the projections of the nitrogen nuclear spins, m = +1, 0, −1. Each of these bands has an additional hyperfine structure due to the interaction of unpaired electrons with other (except for nitrogen) magnetic nuclei (protons, deuterons). The widths of these bands $\Delta\Omega$ (see Fig. 6.4) is less than the distances between centers of these bands, $\Delta\Omega < a_N$. In the wide range of the radical concentrations the rate of coherence transfer V can be much less than the nitrogen hyperfine splitting, $V < a_N(|K'_{ex}C - 1/T_{ddsct}| < a_N)$. In this case, the spin coherence transfer between spins which belong to different nitrogen EPR spectrum bands can be treated within the perturbation theory for non-degenerate resonance frequencies. Under the same condition the spin exchange inside nitrogen components of the EPR spectrum corresponds to the quasi-degenerate case.

The EPR spectrum shape is found from the solution of the system of Eq. (6.21) for the partial transverse magnetization M_k:

$$(-i(\omega_k + \delta_k - \omega) - \Gamma_k)M_k - W\,M_k + g_k V\,M = i\omega_1 g_k M_0, \tag{6.21}$$

where $W = KC + 1/T_{ddsd}$, $V = K'_{ex}C - 1/T_{ddsct}$.

We divide all sub-ensembles of nitroxide radicals into three groups, which correspond to three projections of the ^{14}N nuclear spins, m = 1, 0, −1, while a set of numbers {n} characterizes all hyperfine structure components in each nitrogen component of the EPR spectrum (see Fig. 6.4). Then in Eq. (6.21) $k \equiv \{m,n\}$, $M_k \equiv M_{mn} = \{M_{1n}, M_{0n}, M_{-1n}\}$, $g_k \equiv g_m g_n = (1/3)g_n$. In these notations, we re-write Eq. (6.21) as

$$\begin{aligned}
&(-i(\omega_{1n} + \delta_{1n} - \omega) - \Gamma_{01})M_{1n} - W\,M_{1n} + (1/3)g_n V\,(\Sigma M_{1n} + \boldsymbol{\Sigma M_{0n}} + \boldsymbol{\Sigma M_{-1n}})\\
&= i\omega_1 M_{1n,\,eq},\\
&(-i(\omega_{0n} + \delta_{0n} - \omega) - \Gamma_{00})M_{0n} - W\,M_{0n} + (1/3)g_n V\,(\Sigma M_{0n} + \boldsymbol{\Sigma M_{1n}} + \boldsymbol{\Sigma M_{-1n}})\\
&= i\omega_1 M_{0n,\,eq},\\
&(-i(\omega_{-1n} + \delta_{-1n} - \omega) - \Gamma_{0-1})M_{-1n} - W\,M_{-1n} + (1/3)g_n V\\
&(\Sigma M_{-1n} + \boldsymbol{\Sigma M_{1n}} + \boldsymbol{\Sigma M_{0n}}) = i\omega_1 M_{0-1n,\,eq}.
\end{aligned} \tag{6.22}$$

In these equations the terms marked in bold italic describe the spin exchange between spins with sufficiently large difference of their resonance frequencies. Their contributions to the solution of Eq. (6.22) can be found using the perturbation theory for the non-degenerate case. In the approximation linear over V the partial magnetizations M_{1n}, M_{0n}, M_{-1n} and the right-hand sides of Eq. (6.22) take the form

$$\begin{aligned}
M*_{mn}(\omega_{mn}) \cong{} & M_{mn}(\omega_{mn})\\
& + i(1/3)g_n V\Sigma M_{pq}(\omega_{pq})/(\omega_{mn} - \omega_{pq}) \quad (q, p \neq m),
\end{aligned} \tag{6.23}$$

The right-hand side of Eq. (6.22) is transformed analogously

$$\begin{aligned}
&M*_{mn,\,eq} = g_{mn}\Big(1 + iV\Sigma g_{pq}(\omega_{pq})/(\omega_{mn} - \omega_{pq})\Big)M_{eq} = M_{eq}(1/3)g_n(1+\\
&i(1/3)V\Sigma g_q(\omega_{pq})/(\omega_{mn} - \omega_{pq})\ )\ (q, p \neq m).
\end{aligned} \tag{6.24}$$

Due to the spin coherence transfer between different nitrogen components the resonance frequencies shift as

$$\omega*_{mn} = \omega_{mn} - g_{mn} V^2 \Sigma g_{pq}/(\omega_{mn} - \omega_{pq}); (q, p \neq m). \quad (6.25)$$

In Eqs. (6.23, 6.24 and 6.25) the denominators of $1/(\omega_{mn}-\omega_{pq})$ are the difference of the resonance frequencies of spins, which belong to different nitrogen components of the EPR spectrum. We assume that the difference $(\omega_{mn}-\omega_{pq})$ is mainly determined by the hyperfine interaction with nitrogen: $(\omega_{1n}-\omega_{0q})\cong a_N$, $(\omega_{0n}-\omega_{-1q})\cong a_N$, $(\omega_{1n}-\omega_{-1q})\cong 2a_N$. With this approximation Eqs. (6.23, 6.24 and 6.25) take the forms:

$$\begin{aligned}
&M*_{1n}(\omega_{mn}) \cong M_{1n}(\omega_{mn}) + i(1/3)g_n(V/a_N)\Sigma M_{0q}(\omega_{pq}) + i(1/3)g_n (V/2a_N)\Sigma M_{-1q}(\omega_{pq}),\\
&M*_{0n}(\omega_{mn}) \cong M_{0n}(\omega_{mn}) - i(1/3)g_n(V/a_N)\left(\Sigma M_{1q}(\omega_{pq}) + i(1/3)g_n(V/a_N)\Sigma M_{-1q}(\omega_{pq}\right),\\
&M*_{-1n}(\omega_{mn}) \cong M_{-1n}(\omega_{mn}) - i(1/3)g_n(V/2a_N)\left(\Sigma M_{1q}(\omega_{pq}) - i(1/3)g_n (V/a_N)\Sigma M_{0q}(\omega_{pq}\right),\\
&M_z*_{1n,\,eq} = g_n(1/3)\,(1 + i(1/2)(V/a_N))M_{eq},\\
&M_z*_{0n,\,eq} = g_n(1/3)M_{eq},
\end{aligned} \quad (6.26)$$

$$\begin{aligned}
&M_z*_{-1n,\,eq} = g_n(1/3)\,(1 - i(1/2)(V/a_N)\,)M_{eq},\\
&\omega*_{1n} = \omega_{1n} - (1/2)g_n(V^2/a_N),\\
&\omega*_{0n} = \omega_{0n},\\
&\omega*_{-1n} = \omega_{-1n} + (1/2)g_n(V^2/a_N).
\end{aligned}$$

Thus, for the small parameter $V/a_N < 1$ Eq. (6.22) take the form of the uncoupled equations for the transverse magnetizations of the nitrogen hyperfine components

$$\begin{aligned}
&(-i(\omega*_{1n} + \delta_{1n} - \omega) - W - \Gamma_1)M*_{1n} + (1/3)g_n V\, M*_1 = g_n i\omega_1(1/3)\\
&(1 + i(1/2)(V/a_N))M_{eq}, (-i(\omega*_{0n} + \delta_{0n} - \omega) - W - \Gamma_0)M*_{0n}\\
&+(1/3)g_n V\, M*_0 = g_n i\omega_1(1/3)M_{eq},\\
&(-i(\omega*_{-1n} + \delta_{-1n} - w) - W - \Gamma_{-1})M*_{-1n} + (1/3)g_n V\, M*_{-1}\\
&= g_n i\omega_1(1/3)\,(1 - i(1/2)(V/a_N)\,)M_{eq},
\end{aligned} \quad (6.27)$$

where $M*_1 = \Sigma M*_{1n}$, $M*_0 = \Sigma M*_{0n}$, $M*_{-1} = \Sigma M*_{-1n}$.

These equations give

$$M{*}_{1n} = (g_n/(-i(\omega{*}_{1n} + \delta_{1n} - \omega) - W - \Gamma_{01}))$$
$$(i\omega_1(1/3)\ (1 + i(1/2)(V/a_N))M_{eq} - (1/3)\ V\ M{*}_1);$$
$$M{*}_{0n} == (g_n/(-i(\omega{*}_{0n} + \delta_{0n} - \omega) - W - \Gamma_{00}))\ (i\omega_1(1/3)\ M_{eq} - (1/3)\ V\ M{*}_0);$$
$$M{*}_{-1n} = (g_n/(-i(w{*}_{-1n} + \delta_{-1n} - \omega) - W - \Gamma_{0-1}))$$
$$(i\omega_1(1/3)\ (1 - i(1/2)(V/a_N))M_{eq} - (1/3)\ V\ M{*}_{-1}).$$

The magnetizations $M{*}_1$, $M{*}_0$, $M{*}_{-1}$ can be found straightforwardly:

$$M{*}_m = C_m G_m/(1 + (1/3)VG_m), \tag{6.28}$$

where

$$G_m = \sum g_n/(-i(\omega{*}_{mn} + \delta_{mn} - \omega) - W - \Gamma_{0m}),$$
$$C_m = (i\omega_1(1/3)\ (1 + im(1/2)(V/a_N))M_{eq}, m = +1, 0, -1. \tag{6.29}$$

In the EPR experiments one observes the total tranverse magnetization $M = M_1 + M_0 + M_{-1}$. In the approximation linear over V/a_N (see Eqs. (6.23 and 6.26))

$$M_1 \cong M{*}_1 - i(1/3)\ (V/a_N)\ M{*}_0 - i(1/3)\ (V/2a_N)\ M{*}_{-1},$$
$$M_0 \cong M{*}_0 + i(1/3)\ (V/a_N)\ M{*}_1 - i(1/3)\ (V/a_N)\ M{*}_{-1},$$
$$M_{-1} \cong M{*}_{-1} + i(1/3)\ (V/2a_N)\ M{*}_1 + i(1/3)\ (V/a_N)\ M{*}_0.$$

Thus, the EPR spectrum is given as

$$J = -Im\{M\} = J_1 + J_0 + J_{-1},$$
$$J_m = -Im\{(1 + im(1/2)(V/a_N)M{*}_m\} \cong -(1/3)Im\{i\omega_1(1 + im(V/a_N))$$
$$G_m/(1 + (1/3)VG_m)\} = -(\omega_1/3)Re\{(1 + im(V/a_N))G_m/(1 + (1/3)VG_m)\}. \tag{6.30}$$

The main observation from this consideration is that in a reasonably good approximation the dispersion contribution to the outer components of the EPR spectrum does not depend essentially from the proton hyperfine splittings of nitrogen components and it is

$$p_m = mV/a_N, \text{if } V < a_N, m = \pm 1. \tag{6.31}$$

Exact results obtained for the simple 3 frequency model [16] presented in previous section (see Fig. 5.9) show that (6.31) is valid when $|V| < a_N/2$.

The last result (6.31) is of real importance. It opens a new way for determining the spin coherence transfer rate by measuring the contribution of the dispersion to the EPR spectrum.

6.3 Algorithms for Finding the Spin Decoherence and the Spin Coherence Transfer Rates from EPR Spectra of Nitroxide Radicals in Linear Response Case

Our task is to obtain the spin decoherence rates and the spin coherence transfer rates from experimental EPR spectra. This can be done by comparing numerically simulated spectra and experimental spectra. The simulations can be done using rigorous theoretical expressions for spectra (5.14 and 5.15) or approximate expressions obtained for low concentrations regions (6.17, 6.18, 6.19 and 6.30)). In practice, one uses this kind of simulations for testing and tuning parameters of a system studied, while probe parameters are estimated in some way (see below).

6.3.1 *Fitting Spectra*

In the linear response region, the kinetic equations for the spin exchange in the general case are (see (4.6))

$$\partial M_{k-}/\partial t = -i(\omega_k + \delta_k - \omega)M_{k-} - M_{k-}/T_{2k} - (KC + 1/T_{ddsd})M_{k-}$$
$$+ \varphi_k(\boldsymbol{K'_{ex}C} - \boldsymbol{1/T_{ddsct}})M_- - i\gamma B_1 M_{k0};$$
$$M_- \equiv M_x - iM_y \equiv \Sigma M_{n-}.$$

In these equations, the exchange interaction is responsible for the spin decoherence rate KC, the spin coherence transfer rate $\boldsymbol{K'_{ex}C}$ and the set of the resonance frequency shifts $\{\delta_k\}$. EPR spectrum calculated using these kinetic equations is given by Eqs. (5.13, 5.14 and 5.15). The shape of EPR spectrum in the linear response case is determined only by two combinations of the exchange and dipole-dipole contribution to the spin exchange rates (see (4.6))

$$v_1 = KC + 5/T_{dd} \text{ and } v_2 = K'_{ex}C - 4/T_{dd}.$$

In some cases the spin concentration C may be an unknown parameter which is to be determined by EPR spectroscopy.

Suppose that we know the spin concentration and by the best fitting procedure we find two parameters v_1 and v_2. To this end we have two linear equations with three unknown variables: KC, $K'_{ex}C$, $1/T_{dd}$. In this situation, there is no unique solution for these parameters except the equivalent spin exchange case when $K = K'_{ex} = K_{ex}$ (see, e.g., Fig. 2.5). Indeed, it is straightforward to get $K_{ex}C$ and dipole-dipole contributions separately in the case of the equivalent spin exchange when the same K_{ex} enters into v_1 and v_2 rates. We have

$$K_{ex}C = (4v_1 + 5v_2)/9,$$
$$1/T_{dd} = (v_1 - v_2)/9.$$

In the case of nonequivalent spin exchange when $K \neq K'_{ex}$ we need additional information to solve a problem of finding unknown parameters. There are developed several approaches.

According to the new paradigm of the spin exchange and its manifestation in the EPR spectrum, the first step is the determination of the fraction of the dispersion contribution to the experimental EPR spectrum. Then using (6.31) the spin coherence transfer rate can be obtained straightforwardly.

Two methods to subtract the dispersion contribution to the EPR spectrum are developed.

6.3.1.1 Fitting Method

It was shown theoretically [1, 2, 16] that at low and intermediate concentrations of paramagnetic particles the EPR lines correspond to excitation of collective modes and have the mixed shape, there are contributions of absorption and dispersion terms, and a dispersion contribution term is directly connected with the spin coherence transfer rate (6.31). Keeping this in mind, Bales and Peric suggested a fitting method. As the first step, they suggested to decompose experimental spectrum of nitroxide radicals into a sum of spectra of nitrogen components, while each of them is presented as a sum of absorption and dispersion terms [5–10]. Each nitrogen component of the EPR spectrum is modeled as a first-derivative Voigt absorption line plus a Lorentzian dispersion line. To describe the absorption term, the Voigt form was chosen since in the limit of the very low spin concentration, the proton hyperfine structure of nitrogen components can be approximated very well as the Gaussian distribution [3]. It was also shown (see [3]) that the Voigt form can be approximated as a sum of Lorentzian and Gaussian curves

$$Y = V_{pp}[\alpha A_L + (1 - \alpha)A_G] + V_{disp}D_L \tag{6.32}$$

where A_L and A_G are Lorentzian and Gaussian absorption spectra, D_L is Lorentzian dispersion curve, the fitting parameter $\alpha \rightarrow 0$ when $C \rightarrow 0$, as well as the amplitude of the dispersion term V_{disp} is zero when $C = 0$. Each spectral component has five adjustable parameters: the line width of the Voigt absorption spectrum ΔH_{pp}^{Voigt}, position of resonance field H_0, V_{pp}, V_{disp}, and α. The number of parameters is reduced because A_L, A_G, and D_L have the same ΔH_{pp}^{Voigt} and H_0. Note that according to [18] in Eq. (6.32) line widths of Lorentzian ($2/T_2$), dispersion σ of Gaussian and width of Voigt curve $\Delta\omega_{p-p}$ are connected as

$$\Delta\omega_{p-p} = \left(4\sigma + 1/\left(3T_2{}^2\right)\right)^{1/2} + 1/\left(3^{1/2}T_2\right). \tag{6.33}$$

From α the Voigt parameter $\chi = \Delta H_{pp}^G / \Delta H_{pp}^L$ is determined and using the Dobrayakov–Lebedev relation, ΔH_{pp}^G and ΔH_{pp}^L are found separately from the overall line width ΔH_{pp}^{Voigt}.

The fit outlined explicitly includes the overlap of lines. For partially resolved spectra, separate parameters for each line are easily determined to high precision both for simulated and experimental spectra. As lines merge, the initial guesses to input to the fitting routine get more difficult to estimate but still feasible. Note that a fraction of the dispersion contribution is $p = V_{disp}/V_{pp}$ (6.32).

From the five fitting parameters, the useful parameters, the spin exchange interaction and dipole-dipole interaction contributions to the rates of the spin decoherence and spin coherence transfer, can be found (see [19]).

This fitting method was successfully applied when studying the spin exchange in numerous solutions of nitroxide radicals (see, e.g., [5–12]). The comprehensive analysis performed led to several basic observations. The theoretical predictions concerning dispersion contribution to the EPR spectrum induced by the spin coherence transfer were confirmed. As a result, in modern science the measurement of the dispersion contribution became a key point in the study of spin exchange by using EPR spectra. Of a major importance is also the experimental confirmation of the theoretical prediction that the dipole-dipole interaction between spin probes in liquids, firstly, contributes to the spin coherence transfer and, secondly, its contribution to the coherence transfer is in anti-phase with respect to the coherence transfer induced by the exchange interaction, contributions of these interactions have opposite signs [7]. Remarkable results were obtained as a result of the thorough measurement of the shifts of resonance frequencies. In [7, 9], the additional shift of the resonance frequency due to the "interference" of the exchange interaction with the hyperfine interaction in the course of bimolecular collisions [20] was confirmed. Note that this shift can give unique information about finest details of bimolecular collisions, e.g., re-encounters of two paramagnetic particles in the course of their meeting in solution.

6.3.1.2 Two-Point Method

This method was developed in [4, 11, 12] and it addresses to the situations when nitrogen components of the nitroxide radical EPR spectrum are resolved. In the specific case of the nitroxide radicals, there is an interesting intermediate spin concentration case when, e.g., in case of ^{14}N nitroxide radicals $\Delta\Omega_0 < 3V < a_N$, where $\Delta\Omega_0$ is the width of the nitrogen components of the EPR spectrum (see Fig. 6.1). Under the condition $\Delta\Omega_0 < 3V$ each nitrogen component is merged to

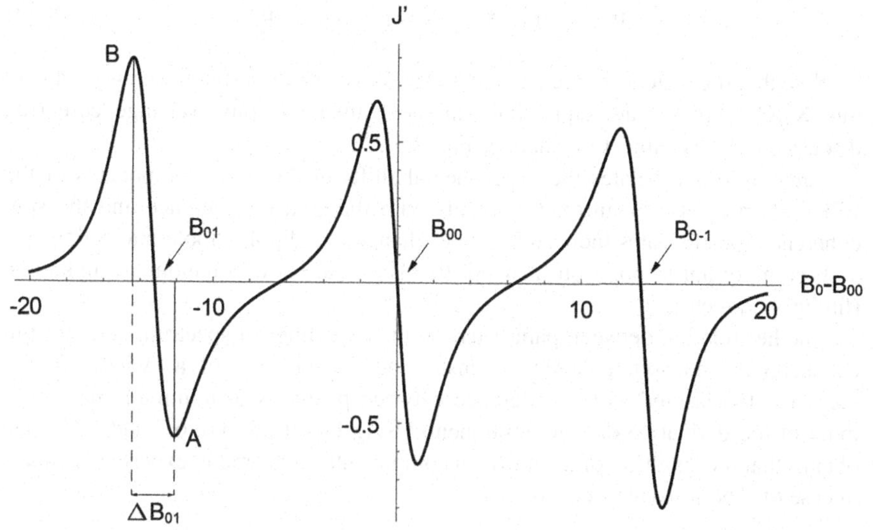

Fig. 6.6 Model derivative EPR spectrum of ^{14}N nitroxide radicals in (6.16) case calculated as a sum of three components given by Eq. (6.17)

the exchange narrowed homogeneous line as a result of the spin exchange *inside* the nitrogen components. The spin exchange between *different* nitrogen components broadens and shifts these homogeneous lines at condition $\Delta\Omega_0 < 3$ V. Thus under the conditions $\Delta\Omega_0 < | K'_{ex}C - 4/T_{dd} | < a_N$ one obtains the situation considered theoretically in ref. [4] and presented above (see Eqs. (6.17, 6.18 and 6.19) and Tables 6.1 and 6.2). For this situation, speculations presented above allow to suggest the algorithm [4] for finding the spin exchange rates by analyzing the concentration transformations of the nitroxide EPR spectrum.

The EPR spectrum of ^{14}N nitroxide radicals in the intermediate spin concentration range (6.16) is shown schematically in Fig. 6.6.

One can easily find from this spectrum the following parameters:

B_{0m} are the values of B_0 which give zero values of the derivative spectrum, they correspond to maximum values of the spectrum. Maxima correspond to points indicated in Fig. 6.6 as B_{01}, B_{00}, B_{0-1}. According to (6.9 and 6.2) the shifts of the positions B_m relative to their values at very low spin concentrations of (C→0) are (see Eq. (6.9), Table 6.2)

$$\delta B_m = (1/\gamma)\big(m b_{ex}C + (1/2)\big(1/T_{2,m} + 1/T_{ddsd} + \varphi_m/T_{ddsct} + \big(KC - \varphi_m K'_{ex}C\big)\big)p_m\big); m = \pm 1, 0, \varphi_m = 1/3 \text{ for } ^{14}\text{N case} \tag{6.34}$$

Note that $\delta B_0 = 0$;

Equation (6.34) leads to two different equations for m = 1 and m = −1 since for nitroxide radicals typically [1, 2] $T_{2,1} \neq T_{2,-1}$, while $\delta B_0 = 0$. The dispersion contribution parameter p for the relatively low spin concentration is (see (5.32 and 6.31))

$$p_m = m\left(K'_{ex}C - 1/T_{ddsct}\right)/a_N \equiv mp. \quad (6.35)$$

Note that the spin decoherence rate KC in (6.34), and the spin coherence transfer rate $K'_{ex}C$ in (6.35) are equal only in the equivalent spin exchange case (see discussion of this subject in Chap. 2, Fig. 2.5).

Equation (6.34) relates the experimental shifts of the outer components of the EPR spectrum of ^{14}N nitroxide radicals with the spin decoherence and the spin coherence transfer rates induced by the exchange and dipole-dipole interactions.

It is important to point out that Eq. (6.34) provide new equations for unknown (fitting) parameters.

Another relation between parameters of the experimental spectrum and the spin decoherence and spin coherence transfer rates comes out of a "width" of the component measured as the difference between positions of minimum and maximum of the derivative of that component. Using Eqs. (6.5, 6.6, 6.7 and 6.19) one obtains that the "widths" of the nitrogen components registered in experiment equals in case of ^{14}N nitroxide radicals (6.7)

$$\Delta B_0(m) = (1/\gamma)\left(1.15 + 0.17\, p_m{}^2\right)\left((K - 1/3K'_{ex})C + 1/T_{ddsd} + 1/(3T_{ddsct}) + 3 < \Delta^2 > /\left(K'_{ex}C - 1/T_{ddsct}\right) + 1/T_{2,m}\right). \quad (6.36)$$

Note that for the EPR spectrum of outer components $p_1 = -p_{-1}$, so that $p_1{}^2 = p_{-1}{}^2$, while $p_0 = 0$ for the central component. The widths of the outer components are not equal usually since they have different T_2 values (see [1, 2]).

One more experimental parameter arises from an asymmetry of the outer nitrogen components of the EPR spectrum (see Figs. 6.2 and 6.6). This asymmetry is characterized by the ratio of a minimal and maximum values of the derivative spectrum, $r_m = A_m/B_m$ (see Fig. 6.3). In the intermediate spin concentration region (6.16) this ratio is determined directly by the contribution p_m of the dispersion to the observed spectrum. It was shown [4, 16] that in the low spin coherence transfer rate case (6.13)

$$p_m = \left(3^{1/2}/2\right)(1 + r_m) = 0.87\,(1 + r_m). \quad (6.37)$$

From the positions of the maxima of the EPR spectra detected at low spin concentration one can find the hyperfine interaction constants with nitrogen nuclear spin and with protons and also the characteristic times $T_{2,m}$ of the spin decoherence of isolated paramagnetic particles. There are five more parameters KC, $K'_{ex}C$, $1/T_{ddsd}$, $1/T_{ddsct}$ and $\delta_{ex}C$ to be found. Of some help may be the fact that in ^{14}N nitroxide radicals one of the nitrogen components has considerably shorter T_2 compared with two other nitrogen components. So Eqs. (6.34 and 6.36) provide four new equations for unknown fitting parameters.

However, the problem can be simplified. Indeed, under typical conditions of the EPR experiments contributions of the dipole-dipole interaction to the spin dephasing and to the spin coherence transfer obey the equation (see (97))

$$1/T_{ddsct} \approx (4/5)\ (1/T_{ddsd}). \tag{6.38}$$

Thus four parameters, KC, $K'_{ex}C$, $1/T_{ddsd}$, and $\delta_{ex}C$ remain to be found. Using Eq. (6.37) the dispersion contribution, p, for the nitrogen component of the EPR spectrum is determined. In [12], this approach was called the two-point model.

Equation (6.35) gives for the spin coherence transfer rate

$$K'_{ex}C = p\ a_N + 1/T_{ddsct} = p\ a_N + (4/5)(1/T_{ddsd}). \tag{6.39}$$

This equation can be rewritten as

$$1/T_{ddsd} = (5/4)\ \left(K'_{ex}C - p\ a_N\right) \tag{6.40}$$

By inserting Eq. (6.40) into Eqs. (6.34 and 6.36) the problem is reduced to finding three unknown parameters KC, $K'_{ex}C$ and $b_{ex}C$. With the $K'_{ex}C$ found in that way one can find also $1/T_{ddsd}$ and $1/T_{ddsct}$ using Eq. (6.40).

In principle, $1/T_{ddsd}$ and $1/T_{ddsct}$ can be calculated theoretically (see Eqs. (3.25, 3.26 and 3.27)). Then $K'_{ex}C$ can be found directly from (6.39). The comparison of results obtained in different ways can give information about accuracy of a whole procedure. Otherwise, one can accept theoretical $1/T_{ddsd}$ and $1/T_{ddsct}$ values and insert these results into Eqs. (6.34, 6.35, 6.36 and 6.40). In this case, Eqs. (2.36 and 2.39) can serve as two independent sources for finding KC and $K'_{ex}C$. Ideally, they should give the same spin exchange rates. But they may differ since Eqs. (6.36 and 6.39) are based on several approximations. These approximate parameters can be further tuned by simulating the EPR spectrum (see (5.13, 5.14 and 5.15)) using those approximate parameters as probe ones.

The two-point model outlined was successfully tested in [11, 12]. This analysis of the EPR spectra of ^{14}N nitroxide radical solutions can provide reasonably good approximate parameters of a system. Note that spin exchange in [11, 12] was considered to be equivalent, therefore it was assumed that $K_{ex}C = K'_{ex}C$.

Thus, the EPR spectra of paramagnetic particles in dilute solutions make it possible to determine the spin decoherence rate and the quantum coherence transfer rate caused by the exchange and dipole-dipole interactions. The contributions of these interactions in different combinations can be found from the experimental spectra. By solving the system of algebraic equations for the spin dynamics rates obtained in this way at the expense of each of these interactions, the contribution of the exchange interaction and the dipole-dipole interactions can be found separately.

Additional information concerning the spin exchange can be obtained also by other EPR methods. They are shortly described in the next Chap. 7.

References

1. Zamaraev, K.I., Molin, Y.N., Salikhov, K.M.: Spin exchange, p. 317. Nauka, Sibirian branch, Novosibirsk (1977)
2. Molin, Y.N., Salikhov, K.M., Zamaraev, K.I.: Spin exchange. Principles and applications in chemistry and biology. Springer-Verlag, Berlin/Heidelberg/New York (1980)
3. Bales, B.L.: Inhomogeneously broadened spin-label spectra. In: Berliner, L.J., Reuben, J. (eds.) Biological magnetic resonance, vol. 8, pp. 77–130. Plenum Publishing Corporation, New York (1989)
4. Salikhov, K.M.: Contributions of exchange and dipole–dipole interactions to the shape of EPR spectra of free radicals in diluted solutions. Appl. Magn. Reson. **38**, 237–256 (2010)
5. Bales, B.L., Peric, M.: EPR line shifts and line shape changes due to spin exchange of nitroxide free radicals in liquids. J. Phys. Chem. B. **101**, 8707–8716 (1997)
6. Bales, B.L., Peric, M., Dragutan, I.: EPR line shifts and line shape changes due to spin exchange of nitroxide free radicals in liquids 2. Extension to high spin exchange frequencies and inhomogeneously broadened spectra. J. Phys. Chem. A. **106**, 4846–4854 (2002)
7. Bales, B.L., Peric, M., Dragutan, I.: EPR line shifts and line shape changes due to spin exchange of nitroxide free radicals in liquids 3. Extension to five hyperfine lines. Additional line shifts due to re-encounters. J. Phys. Chem. A. **107**, 9086–9098 (2003)
8. Bales, B.L., Meyer, M., Smith, S., Peric, M.: EPR line shifts and line shape changes due to spin exchange of nitroxide free radicals in liquids 4. Test of a method to measure re-encounter rates in liquids employing ^{15}N and ^{14}N nitroxide spin probes. J. Phys. Chem. A. **112**, 2177–2181 (2008)
9. Kurban, M.R., Peric, M., Bales, B.L.: Nitroxide spin exchange due to re-encounter collisions in a series of n-alkanes. J. Chem. Phys. **129**, 064501 (2008)
10. Bales, B.L., Meyer, M., Smith, S., Peric, M.: EPR line shifts and line shape changes due to spin exchange of nitroxide free radicals in liquids 6: separating line broadening due to spin exchange and dipolar interactions. J. Phys. Chem. A. **113**, 4930–4940 (2009)
11. Salikhov, K.M., Bakirov, M.M., Galeev, R.T.: Detailed analysis of manifestations of the spin coherence transfer in EPR spectra of ^{14}N nitroxide free radicals in non-viscous liquids. Appl. Magn. Reson. **47**, 1095–1122 (2016)
12. Bales, B.L., Bakirov, M.M., Galeev, R.T., Kirilyuk, I.A., Kokorin, A.I., Salikhov, K.M.: The current state of measuring bimolecular spin exchange rates by the EPR spectral manifestations of the exchange and dipole-dipole interactions in dilute solutions of nitroxide free radicals with proton hyperfine structure. Appl. Magn. Reson. **48**, 1399–1447 (2017)
13. Abragam, A.: Nuclear magnetism. Principles of nuclear magnetism. Oxford University Press, London (1961)
14. Salikhov, K.M., Semenov, A.G., Tsvetkov, Y.D.: Electron spin echo. Nauka, Sibirien branch, Novosibirsk (1976). in Russian
15. Galeev, R.T., Salikhov, K.M.: The theory of dipolar broadening of magnetic resonance lines in non-viscous liquids. Khim. Fizika. **15**, 48–64 (1996)
16. Salikhov, K.M.: Consistent paradigm of the spectra decomposition into independent resonance lines. Appl. Magn. Reson. **47**, 1207–1228 (2016)
17. Bales, B.L., Willett, D.: EPR investigation of the intermediate spin exchange regime. J. Chem. Phys. **80**, 2997–3004 (1984)
18. Salikhov, K.M.: The contribution from exchange interaction to the line shifts in ESR spectra of paramagnetic particles in solutions. J. Magn. Res. **63**, 271–279 (1985)
19. Lebedev, Y.S., Dobryakov, S.N.: Analysis of the EPR spectra of free radicals. Zhur. Structur. Khim. **8**, 838–853 (1967). (in Russian)
20. Halpern, H.J., Peric, M., Yu, C., Bales, B.L.: Rapid quantitation of parameters from inhomogeneously broadened EPR spectra. J. Magn. Reson. **103**, 13–22 (1993)

Chapter 7
Other Methods of Measuring the Spin Exchange Rates

Abstract The modern theory of the manifestation of the saturation effect in the EPR spectra is described in detail. The theory predicts the formation of the coupled states of the microwave field and the spin system (spin polaritons) under the conditions of the manifestation of the saturation effect. It is shown that in strong microwave fields the frequency of polaritons and the condition of exchange narrowing depend on the amplitude of the microwave field.

The potential of double electron-electron resonance, electron spin echo, and the saturation-recovery method are briefly described. The effect of spin exchange on the saturation factor, which affects the magnitude of the effect of the dynamic polarization of nuclei, is examined in more detail. All these methods can complement each other well when studying spin exchange.

Most of the data available in literature on spin exchange between paramagnetic particles have been obtained by analyzing the EPR spectra in the linear response when the saturation effect is negligible. Therefore above attention has been focused on the spin exchange manifestations in the shape of spectra when the saturation effect is not revealed. This method is attractive due to the simplicity and feasibility of the experiment. However the possibilities of this method might be limited. For example, in linear response region the EPR spectrum does not give any information about the spin-lattice relaxation, including information about contribution of the spin exchange to the kinetics of the spin-lattice relaxation. The EPR spectra detected in strong microwave field when so called saturation occurs allow to find the spin-lattice relaxation time and to find a spin-exchange rate constants. The analysis of the EPR spectrum shape is difficult in the case of poorly resolved multi component spectra. The electron spin echo methods allow to extract directly the irreversible spin decoherence kinetics, etc.

There are developed other EPR methods that complement the method of EPR spectra in the linear response region. These methods are shortly reviewed below.

K. M. Salikhov, *Fundamentals of Spin Exchange*,
https://doi.org/10.1007/978-3-030-26822-0_7

7.1 Continuous-Wave Saturation Method

7.1.1 Review of the Paramagnetic Resonance Saturation Theory

Usually, EPR spectra are recorded in relatively weak microwave fields, in which a linear response of the system to the action of the microwave field B_1 is observed. In this situation, the spin system is not "overheated" due to the absorption of microwave quanta, since spin-lattice relaxation manages to maintain the thermodynamic equilibrium of the spin system with the thermostat (lattice). In this linear response region, the removal of the energy absorbed by the spins to the lattice is not the limiting stage of the absorption process and, therefore, the EPR spectrum does not contain information about spin-lattice relaxation.

With an increase in the microwave field amplitude, the rate of "heating" of the spin system increases and it can become greater than the rate of energy removal from the spin system to the lattice, which is $1/T_1$ (T_1 is the spin-lattice relaxation time). In this situation, the energy removal of the spin system into the lattice becomes the limiting stage of the stationary absorption of the microwave energy. Absorption of the microwave quanta goes to saturation. The saturation effect is manifested in the shape of the EPR spectra. Therefore, from the analysis of the EPR spectra under saturation conditions, in principle, it is possible to determine all magnetic resonance parameters of the spin system, including the spin-lattice relaxation time. This problem has been discussed in a number of papers (see, e.g., [1–10]).

The saturation effect is described in most detail for a spectrum that consists of a single homogeneously broadened line. An example is the EPR spectrum under conditions of exchange narrowing of the spectrum (see, e.g., [11–14]). For a homogeneously broadened line, the shape of the observed absorption spectrum of magnetic resonance is found from the solution of the Bloch equations and is given by the formula ([5], Eq. (III.15)), which describes the linear response of the system and the saturation effect

$$J(\omega) = \omega_1 T_2 / \left(1 + (T_2(\omega - \omega_0))^2 + \omega_1{}^2 T_1 T_2\right) M_0. \tag{7.1}$$

In this well-known expression T_1 and T_2 are relaxation times of the longitudinal (spin-lattice) and transverse components of magnetization of the spins, respectively, ω_0 is the resonant frequency of spins, ω is the microwave field frequency, $\omega_1 = \gamma B_1$ is the Rabi frequency of spins in the microwave field B_1. Here M_0 is the equilibrium magnetization of the spin system. In this situation, for all B_1 values, the spectrum has a Lorentzian form with a half width at half height

$$\Delta\omega_{1/2} = (1/T_2)\left(1 + \omega_1{}^2 T_1 T_2\right)^{1/2}. \tag{7.2}$$

It shows that the observed spectrum broadens with the increase in B_1.

From the point of view of the saturation effect, the integral intensity of the line (7.1) is an important parameter. It is equal to

$$J_0 = \pi\omega_1/\left(1 + \omega_1{}^2T_1T_2\right)^{1/2}M_0. \tag{7.3}$$

The largest amplitude of the signal (7.1) is achieved at exact resonance, $\omega = \omega_0$, and with increasing B_1 it changes as

$$J(\omega_0 = \omega) \equiv J_{max} = \omega_1T_2/\left(1 + \omega_1{}^2T_1T_2\right)M_0. \tag{7.4}$$

At sufficiently small B_1, the linear response of the system is observed and the EPR signal increases with increasing B_1 as

$$J_{lin}(\omega) = \omega_1T_2/\left(1 + (T_2(\omega - \omega_0))^2\right)M_0, J_{lin}(\omega_0 = \omega) = \omega_1T_2M_0. \tag{7.5}$$

Two processes occur in the spin system. On the one hand, spins absorb the microwave field quanta. At resonance, the absorption rate is $\upsilon_{ab} = \omega_1{}^2T_2$ N, N is the number of spins. On the other hand, there is spin-lattice relaxation of spins with a rate $\upsilon_{sl} = N/T_1$. The ratio of the rates of these processes $q = \upsilon_{ab}/\upsilon_{sl} = \omega_1{}^2\,T_1\,T_2$ is the most important parameter. The absorption of the microwave field quanta reduces the difference between the populations of the ground and excited spin states, and therefore, the ability of the spin system to absorb new quanta decreases and the spin system can become saturated. This saturation effect of the spin system practically is not manifested if spin-lattice relaxation manages to restore quickly the equilibrium distribution of populations of spin states. When $q < 1$, the amplitude of the EPR signal increases linearly with the microwave field amplitude, since in this case the saturation effect is not pronounced. When $q > 1$, the saturation effect is manifested and the intensity (7.1) decreases with increasing B_1. The signal amplitude at exact resonance $\omega_0 = \omega$ with increasing ω_1 passes through a maximum at $q = 1$ (7.2). Note that in the case under consideration for $q \geq 1$, i.e., under saturation conditions, the signal amplitude at exact resonance is $q + 1$ times less than the value that would be observed in the absence of the saturation effect with the same ω_1 value (cf. 7.4 and 7.5). When the microwave field frequency deviates from the exact resonance, $\omega_0 \neq \omega$, the spectrum amplitude passes through a maximum at $\omega_1{}^2T_1T_2 = 1 + (\omega_0{-}\omega)^2T_2{}^2 > 1$, i.e., when the frequency is detuned from the resonance, the saturation effect occurs at larger amplitudes B_1 of the microwave field. For a given amplitude B_1, the center of the line is saturated more than its wings. This is the mechanism for broadening the resonance lines by the external microwave field B_1.

Thus, the intensity of the EPR signal with an increase in the microwave field amplitude behaves in a characteristic way, increases linearly at small B_1 and decreases as $1/B_1$ (7.1 and 7.4) at large B_1, and the integral intensity goes to the plateau $J_0 = (1/(T_1\,T_2)^{1/2})M_0$. The intensity of the spectrum and the integral intensity directly reflect the saturation effect. Therefore, the dependences of the integral intensity of the EPR spectrum (7.3) or the intensity of the EPR signal on B_1 (ω_1) at a fixed frequency of the microwave field (7.1 and 7.4) are called saturation curves. One can consider the saturation factor for resonant pumping of spins defined as (see 7.1 and 7.5)

$$(J_{lin}(\omega) - J(\omega))/J_{lin}(\omega) = \omega_1{}^2T_1T_2/\left(1 + \omega_1{}^2T_1T_2\right). \quad (7.6)$$

In the course of typical EPR experiments, the main contribution to the splitting of spin levels of isolated paramagnetic particles is made by the Zeeman interaction of spins with a constant external magnetic field. Its energy is $-M_zB_0$. Therefore, a good saturation parameter, which directly reflects the change in the energy of the spin system during the absorption of microwaves, is $(M_0–M_z)/M_0$. This parameter has been suggested in [10]. Here M_z is the magnetization along the constant magnetic field in the stationary saturation experiment. At the resonance pumping of spins ($\omega_0 = \omega$) for the considered model of a single homogeneously broadened line one has

$$(M_0 - M_z)/M_0 = \omega_1{}^2T_1T_2/\left(1 + \omega_1{}^2T_1T_2\right). \quad (7.6')$$

This is the same as (7.6).

The saturation provides several possibilities for determining the paramagnetic relaxation times T_1 and T_2. It is possible to measure the half-width of the spectrum at half the height at several values of the microwave field amplitude and present these data as (see (7.2))

$$\Delta\omega_{1/2}{}^2 = 1/T_2{}^2 + \omega_1{}^2(T_1/T_2) \quad (7.7)$$

From these data one can easily find T_2 and T_1, if the ω_1 values are known in the experiment. The maximum possible amplitude of the EPR signal (7.1) at resonance corresponds to $\omega_{1max}{}^2T_1T_2 = 1$. Here ω_{1max} denotes the value that gives the maximum EPR signal at the resonance pumping. It follows from this and from (7.7) that for the spectral line observed with $B_{1max}=\omega_{1max}/\gamma$ microwave field one has

$$\Delta\omega_{1/2} = \sqrt{2}/T_2, \quad (7.8)$$

Note that the relation (7.8) can be used as a criterion for determining whether the experimental spectrum is a homogeneously broadened Lorentzian line or not. The maximum amplitude of the EPR signal under saturation conditions is two times less than the value that can be expected if the linear response condition would be satisfied for all B_1, and it is (insert $\omega_{1max}{}^2T_1T_2 = 1$ into (7.4))

$$J_{max} = (1/2)(T_2/T_1)^{1/2}M_0. \tag{7.9}$$

Note that using the experimental data under saturation conditions and the Eqs. (7.8) and (7.9), it is possible to find the relaxation times T_2 and T_1, and it is not necessary to know the microwave field power. It will be seen below that in real systems it is necessary to know the amplitude B_1 of microwave field in order to determine the magnetic resonance characteristics from the analysis of spectra under saturation.

Useful information can also be obtained from the width of the saturation curve (7.3) at half height. The signal amplitude (7.3), e.g., in the case $T_1 = T_2$, is half of the maximum value $J_{max}/2$ at two values $\omega_1(-) = (2-\sqrt{3})/T_2$ and $\omega_1(+) = (2+\sqrt{3})/T_2$. Their difference is $\omega_1(+)-\omega_1(-) = 2\sqrt{3}/T_2$. Note that the ratio $\omega_1(+)/\omega_1(-)\approx 13.93$ (at $T_1 = T_2$).).

In the linear response region, the shape of the spectrum depends only on T_2 (see e.g., (7.3)), but does not depend on T_1. Therefore, the T_2 value is usually found from the spectrum analysis in the region of the linear response.

The results presented show that studying the saturation effect we can determine not only T_2 but also T_1. This possibility of determining T_1 was discussed in detail and implemented for nuclear spins in [1].

The well-known results on the analysis of the saturation effect for a single homogeneously broadened line show that the saturation effect can be analyzed using the dependences of different quantities on the microwave field amplitude: the resonance width, the signal amplitude and the integral intensity of the spectrum.

In real systems, the analysis of the saturation effect is more complex, since spectra usually have the inhomogeneous broadening, which is often much larger than the homogeneous resonance width. In EPR spectroscopy, inhomogeneous broadening can be produced, e.g., by the anisotropy of the g-tensor, the hyperfine interaction of unpaired electrons with magnetic nuclei, the splitting of spectra in the zero magnetic field due to the spin-spin interaction in particles with two or more unpaired electrons (the so-called splitting of EPR spectra in the zero magnetic field), and the spatial inhomogeneity of the constant magnetic field B_0.

For EPR spectroscopy, the effect of the spectrum saturation in the presence of inhomogeneous broadening was observed experimentally for the first time in the study of F-centers in KCl crystals [2]. The signal amplitude in the EPR experiment under saturation conditions does not decrease with the increase in B_1 as expected from Eq. (7.1) for a homogeneous line, and it goes almost to the plateau [2]. To explain the experimental results, a model called the T_1–T_2 model was proposed by Portis [2].

According to this model, the spectrum is presented as a set of homogeneously broadened lines of spin sub-ensembles with the same resonance frequencies ω_0 (Portis called them spin packets [2]). The shape of the spin packet spectrum is given by Eq. (7.1). It is assumed that there is no interaction between spins belonging to different spin packets and that there is no spin excitation transfer between spins and no spectral diffusion processes that could transfer spin coherence from one spin packet to another. The set of resonance frequencies of spin packets gives an inhomogeneous broadening of the spectrum. In the case of F-centers in KCl crystals, the inhomogeneous broadening of the EPR spectrum is caused by the hyperfine interaction of F-centers with magnetic nuclei. Let the inhomogeneous broadening be described by the function $h(\omega_0 - \omega_{00})$, where ω_{00} is the mean frequency of the inhomogeneously broadened spectrum. In this model, instead of Eq. (7.1), the shape of the observed absorption spectrum of magnetic resonance is given by the formula

$$J(\omega) = M_0\omega_1 T_2 \int_{-\infty}^{+\infty} h(\omega_0 - \omega_{00})/(1 + T_2(\omega - \omega_0))^2 + \omega_1{}^2 T_1 T_2)d\omega_0. \quad (7.10)$$

The case, when the inhomogeneous broadening $\Delta\Omega$ is very large compared with the line width of the spin packet 1/T2, i.e., the case $\Delta\Omega\ T_2 >> 1$, is considered in [2]. It is obtained from Eq. (7.10) that the saturation curve of the signal amplitude tends to the dependence of the form

$$J(\omega) \sim (\omega_1/\Delta\Omega)\left(1/\left(1 + \omega_1{}^2 T_1 T_2\right)^{1/2}\right). \quad (7.11)$$

Equation (7.11) shows that the signal amplitude under saturation conditions tends to the plateau, which was observed in the experiment [2]. Using Eq. (7.11), the value (T_1T_2) was determined in [2].

It is interesting to note that in this case at $q \geq 1$, i.e., under saturation conditions, the EPR spectrum amplitude (7.11) does not depend on the frequency of microwave field and is proportional to $\omega_1/(1 + \omega_1{}^2 T_1 T_2)^{1/2}$, but not to $\omega_1/(1 + \omega_1{}^2 T_1 T_2)$ valid for a single spin packet (see (7.1) at $\omega = \omega_0$). It appears that the amplitude of the spectrum measured at the resonance frequency of one of the spin packets within a wide loop of the inhomogeneous frequency distribution is saturated similarly to the integral intensity of a single spin packet (cf. Eqs. (7.3) and (7.6)). This is to be expected. Indeed, in the model with a broad distribution of resonance frequencies of spin packets at the specified microwave field frequency the resonant and the nonresonant spin packets contribute to the EPR signal at the same time and their contributions are summed. The accumulated contribution of all spin packets should coincide exactly with the integral intensity of the spectrum of a single spin packet if the statistical weights of the spin packets practically coincide. Actually, it was assumed in [2], since in Eq. (7.10) Portis took out from under the integral function $h(\omega_0 - \omega_{00})$ assuming that it varies slowly compared with the Lorentzian lines of spin packets.

If the inhomogeneous broadening is described by a Gaussian distribution with dispersion σ

$$h_{gaus}(\omega_0) = \left(1/(2\pi\sigma)^{1/2}\right) \exp\left(-(\omega_0 - \omega_{00})^2/(2\sigma)\right), \quad (7.12)$$

(7.10) can be rewritten as

$$J(\omega) = \left(\omega_1\pi/\left(1+\omega_1{}^2T_1T_2\right)^{1/2}\right) J_{voigt}(\omega)M_0. \tag{7.13}$$

Here $J_{voigt}(\omega)$ is a Voigt distribution function [15], which is a convolution of the Gaussian (7.12) and Lorentzian distributions normalized to one

$$\begin{aligned} f_{lorenz}(\omega) &= (T_2'/\pi)\left(1/\left(1+(T_2\prime(\omega-\omega_0))^2\right)\right), \\ T_2' &= T_2/\left(1+\omega_1{}^2T_1T_2\right)^{1/2}. \end{aligned} \tag{7.14}$$

For the Voigt form of the spectrum in the theory of the linear response Lebedev and Dobryakov [15] obtained a very useful ratio of the Voigt function width to the widths of the Gaussian and Lorentzian distributions. Using the renormalization of the Lorentzian distribution parameter T_2 by formula (7.14), their relationship can be generalized to the situation of the saturation effect.

Generalizing the result of [15] to arbitrary microwave fields, the width of the Voigt function, which is calculated as the distance between the values of the resonant frequency at the points of the maximum slope of the spectrum curve, can be written as

$$\begin{aligned} \Delta\omega_{p-p} &= \left(4\sigma + 1/\left(3T_2{}^{\prime 2}\right)\right)^{1/2} + 1/\left(3^{1/2}T_2'\right) = \left(4\sigma + \left(1+\omega_1{}^2T_1T_2\right)/\left(3T_2{}^2\right)\right)^{1/2} + \\ &\quad \left(1+\omega_1{}^2T_1T_2\right)^{1/2}/\left(3^{1/2}T_2\right). \end{aligned} \tag{7.15}$$

Under saturated conditions, when $q = \omega_1{}^2T_1T_2 >> 1, \sigma >> q/T_2{}^2$, the width (7.15) of Voigt functions is

$$\begin{aligned} \Delta\omega_{p-p} &\approx c_0 + c_1\omega_1 + c_2\omega_1{}^2, c_0 = 2\sigma^{1/2}, c_1 = \left(T_1/(3T_2)\right)^{1/2}, c_2 = T_1/\left(12T_2\sigma^{1/2}\right) \\ &= c_1{}^2/\left(4\sigma^{1/2}\right) \end{aligned} \tag{7.16}$$

These relations can be used to determine the magnetic resonance parameters σ and T_1/T_2 from the experimental data of the dependence of the spectrum width on the Rabi frequency. Of course, it is still necessary to know the microwave field amplitude (Rabi frequency ω_1) in the experiment.

A significant development of the T_1–T_2 model [2] was proposed in [3], in which the behavior of the spectrum shape at saturation for an arbitrary width of the Gaussian distribution of spin frequencies was studied. Spectra were calculated for different ratios between the inhomogeneous width and the width of the spin packet and for different the microwave field amplitude values. For illustration, Fig. 7.1 shows dependence on the Rabi frequency $\omega_1 = \gamma B_1$ of the EPR signal amplitude

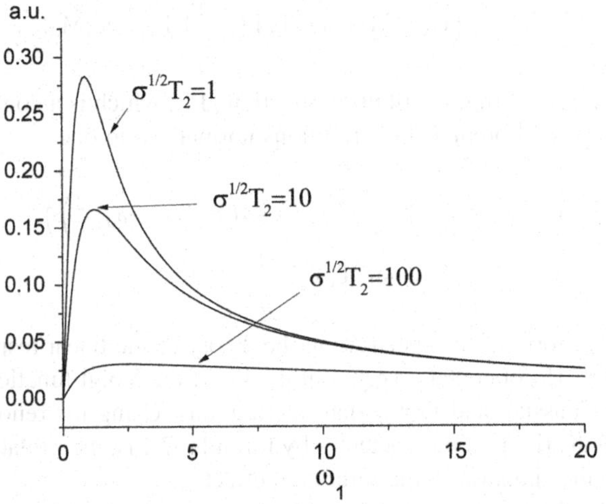

Fig. 7.1 Dependence of saturation curve of the absorption signal amplitude on ω_1 at the resonance pumping of spins in the center of the Gaussian frequency distribution. Saturation curves for three values of the ratio of inhomogeneous broadening to the width of the spin packet are given. This ratio can be characterized by the parameter $\sigma^{1/2}T_2$. It was assumed in these calculations that $T_1 = 3\ T_2$, $1/T_2 = 0.3$ Gs, ω_1 is given in Gs units. (Adapted with permission from Ref. [26])

calculated by (7.10), (7.12) and (7.13) when the resonance is observed in the center of the spectrum.

Figure 7.1 shows that as the inhomogeneous width of the spectrum increases, the signal decay under saturation conditions becomes more flat and with the increase in $\sigma^{1/2}T_2$, the signal dependence on the microwave field amplitude goes almost to the plateau. This dependence was observed in [2, 3]. Note that according to the curves in Fig. 7.1 for the T_1–T_2 model, the ω_{1max} value, at which the absorption signal reaches a maximum, slowly increases with the ratio of the inhomogeneous broadening width to the width of the spin packet (see also below Eqs. (7.30) and (7.31)).

Figure 7.1 shows that at $\sigma^{1/2}T_2 < 1$ the saturation curve of $M_y(\omega = \omega_0)$ passes through a maximum with increasing ω_1 and then slowly decreases, i.e., qualitatively behaves like a saturation curve for the homogeneous line (see Eq. (7.1) at $\omega = \omega_0$.). At $\sigma^{1/2}T_2 = 100$, saturation curve reaches a plateau when ω_1 increases, i.e., qualitatively behaves like a saturation curve for the spectrum with large inhomogeneous broadening (see (7.11)).

In [3], a method for determining the width of the spin packet from the saturation curves of the spectrum is proposed. For the T_1–T_2 model, the saturation curves of the spectrum, i.e., the dependence $J(\omega = \omega_{00})$ on ω_1, for different values of the ratio of the homogeneous width of the spin packet at half-height, $2/T_2$, to the width of the inhomogeneous Gaussian distribution of resonant frequencies, $2\ (\sigma)^{1/2}$ were calculated. For each of these curves, two ω_1 values were found, for which the amplitude of the observed value $J(\omega = \omega_{00})$ is half of its maximum value. The ratio of these two ω_1 values quite simply depends on the parameter $(1/(\sigma T_2^2))^{1/2}$. As a result, a calibration curve was found by Castner in [3], by which the ratio of the width of a homogeneous line of the spin packet and the inhomogeneous distribution of the resonance frequencies of the spin packets can be found from the analysis of the saturation curve for $J(\omega = \omega_{00})$. It can be noted that the calibration curve of Castner in the limit $(1/(\sigma T_2^2))^{1/2}$-->∞ coincides with the result which can be obtained from the analysis of the curve of saturation of a homogeneous line (7.3). In this case, as noted above, the signal amplitude (7.3) is half of its maximum value at $\omega_1(-) = (2-\sqrt{3})/(T_1T_2)^{1/2}$ and $\omega_1(+) = (2\sqrt{3})/(T_1T_2)^{1/2}$. The ratio of these two of ω_1 values is $(2+\sqrt{3})/(2-\sqrt{3}) = 13.93$ that coincides with the asymptotic value of the Castner calibration curve [3].

The results show that the saturation effect of EPR spectra is studied in detail for two situations: for a single homogeneously broadened line (spin packet) and for the T_1–T_2 model [2, 3] of an ensemble of independent spin packets with a Gaussian distribution of resonant frequencies of different spin packets. For these situations, algorithms for finding the magnetic resonance parameters T_1, T_2 are proposed.

In real systems, the assumption of independent spin packets, which underlies the T_1–T_2 model [2, 3], is often not fulfilled [1, 4]. Indeed, e.g., the spin-spin dipole-dipole interaction and exchange interaction between paramagnetic particles lead to the spin excitation transfer between spin packets with different resonance frequencies (see, e.g., [11–14, 16, 17]). In this case, the saturation effect depends not only on the absorption rate of microwave energy by the spin system and the spin-lattice relaxation rate of the spins but the important factor is the rate of transfer of energy of spin excitations and spin coherence transfer between different spin packets. In [4] it was proposed to call the redistribution of spin excitation between all spins (across the inhomogeneously broadened EPR spectrum) as cross-relaxation. Cross-relaxation can be induced, e.g., by spectral diffusion. It has already been noted above that spectral diffusion can be induced, e.g., by spin-spin interaction between electron spins of paramagnetic particles. Another important example of spectral diffusion is a random change in the local field of the hyperfine interaction of unpaired electrons with magnetic nuclei due to random mutual flip-flops of nuclear spins arising due to the dipole-dipole interaction between nuclei, i.e., due to nuclear spin diffusion [16–18].

The work [6] considers the situation of spectral diffusion, which is described by a random process without correlation: the resonance frequency of spins ω changes suddenly; the probability $p(\omega',\omega)$ of a sudden change in the value ω' by ω depends only on the frequency in the final state ω, and is equal to the stationary probability of distribution of the resonance frequencies of spins with frequency ω, i.e., $p(\omega',\omega) = \varphi(\omega)$. In [6], the distribution of resonant frequencies is assumed to be Gaussian Eq. (6.12). Sudden changes in frequency occur at the average rate $1/\tau$. In the case of F-centers [6], τ is the average lifetime of an electron in a state with a given resonance frequency of the electron spin. When electron jumps from one site to other one the electron enters the magnetic environment of the nuclei with the other projections of the nuclear spins. It is shown [6] that due to spectral diffusion inside the inhomogeneous contour, the EPR saturation curve qualitatively resembles the saturation curve of a homogeneous line (see Eq. (7.11)): the saturation curve of the inhomogeneously broadened spectrum does not give the plateau with an increase in the width of the distribution of the resonance frequencies of spin packets, as in the case without spectral diffusion (see Eq. (7.16)). It should be noted that the analysis of the saturation effect of EPR spectra taking into account spectral diffusion in [6] is carried out in the framework of the approximation, which does not take into account some fundamentally important manifestations of spectral diffusion. For example, the effect of resonance frequency shift, the mixed form of spin packet lines, and the effect of exchange narrowing at a high spectral diffusion rate are not taken into account [11–14, 19–21].

However, the model considered in [6] itself is of interest, since it can be applied in many cases both in EPR spectroscopy and in other systems. Often, the EPR spectrum or individual components of the spectrum of organic free radicals has an inhomogeneous broadening caused by the hyperfine interaction, and it can be approximated by the Gaussian distribution [22]. Nitroxide radicals are usually used as a spin probe. Due to the hyperfine interaction with nitrogen nuclei, the EPR spectrum has components that correspond to different values of the nitrogen nuclear spin projection. Each nitrogen component has a hyperfine structure due to the interaction with hydrogen ^{1}H atoms (or ^{2}H in deuterated radicals). Often the inhomogeneous broadening of each nitrogen component can be approximated by the Gaussian distribution of resonant frequencies [22]. The exchange interaction in the course of bimolecular collisions of radicals in dilute solutions induces spectral diffusion, which is described by a random uncorrelated process in the situation of equivalent spin exchange [11–14]. As shown in the previous sections, in general, the effect of the exchange and dipole-dipole interaction on the spin motion in solutions of paramagnetic particles is not reduced simply to the situation of spectral diffusion, which is described by one characteristic parameter of the spectral diffusion rate. In the general case, one has to divide the spin decoherence rate and spin coherence transfer (see kinetic equation (4.6)).

The saturation effect in radical solutions was considered in a series of papers [7, 8, 23]. In these works the coherence transfer between spins due to the dipole-dipole interaction was excluded. The initial kinetic equations took into account the coherence transfer caused by the exchange interaction of radicals in collisions. But the final formulas for the shape of the EPR spectrum, both in the linear response approximation [24] and under saturation conditions, were obtained without taking into account the spin coherence transfer. In fact, the saturation effect in the presence of spin exchange in [7, 8, 23] was considered under the same assumptions as in [6].

The consistent theory of the saturation effect in EPR of paramagnetic particles in dilute solutions is developed in [25–27]. This theory is a logical continuation of the new paradigm of spin exchange and its manifestation in the EPR spectrum under linear response conditions [19–21, 28]. The concept of collective spin modes, which are formed due to the spin coherence transfer, was further developed.

7.1.2 Peculiar Features of the Spectrum Under Saturation Conditions When Equivalent Spin Exchange Operates. System with Two Frequencies

In the case of the solution of paramagnetic particles, the shapes of the linear response EPR spectra in the presence of the spin coherence transfer induced by the exchange and dipole-dipole interactions were comprehensively discussed in [19–21, 28]. It was demonstrated that the spectrum observed in the experiment is a sum of contributions of independent collective modes [13, 14, 20]. These collective modes give resonance lines with mixed shapes: they contain Lorentzian absorption and Lorentzian dispersion contributions. The collective mode description provided a completely new interpretation of the exchange narrowing of spectra when the spin coherence transfer rate exceeds a definite critical value determined by the inhomogeneous broadening of spectra [20, 25].

Below results obtained for a model system which makes it possible to elucidate the role of collective modes in the spectra under saturation conditions in the presence of spectral diffusion are presented [25]. To get a better insight into physics of what is going on, the simplest model of two frequencies is chosen. In Sect. 7.5, this model was already described and its EPR spectrum under the linear response regime was analyzed. For this model the spectrum under saturation conditions was found analytically and was analyzed in detail [25]. It was confirmed that under the saturation conditions the spectrum is also a sum of independent lines, which have mixed shapes. This result was expected. But there were obtained also unexpected results concerning resonance frequencies and the exchange narrowing effect of spectra.

7.1.2.1 Model and Expression for the EPR Spectrum Shape

The two-frequency situation may occur in different systems. As it was pointed out already in Sect. 7.5, one possibility is an ensemble of paramagnetic particles with one unpaired electron with spin $S = 1/2$ and one magnetic nucleus with spin $I = 1/2$. For example, this may be a solution of free radicals where unpaired electrons interact with a ^{1}H or ^{15}N atom. In the external magnetic field $\mathbf{B}_0 = \{0,0,B_0\}$, the electron spin has the Zeeman splitting of energy levels: $\omega_0 = \gamma B_0$ in frequency units (γ is the electron gyromagnetic ratio). The hyperfine interaction $H_{hfi}=\hbar a S_z I_z$ splits the electron spin energy levels further. As a result, there are two sub-ensembles of paramagnetic particles which give the EPR lines at frequency $\omega=\omega_0 + a/2$ or $\omega=\omega_0 - a/2$. We denote the magnetization vectors of electron spins in sub-ensembles with different EPR frequencies as $\mathbf{M}_1$ and $\mathbf{M}_2$: $\mathbf{M}_1 = \{M_{1x}, M_{1y}, M_{1z}\}$, $\mathbf{M}_2 = \{M_{2x}, M_{2y}, M_{2z}\}$. For the sake of simplicity, it is supposed that the paramagnetic particles have the same paramagnetic relaxation times T_1 and T_2. Suppode, the circularly polarized $\mathbf{B}_1$ field is switched on in order to detect the spectrum: $\mathbf{B}_1 = \{B_1 \cos(\omega t), B_1 \sin(\omega t), 0\}$. Next, it is assumed that the exchange interaction in the course of bimolecular collisions produces the equivalent spin exchange. The spin exchange rate induced by bimolecular collisions of the paramagnetic particles is denoted as $V=K_{ex}C$. Here C is the concentration of paramagnetic particles, K_{ex} is the spin exchange rate constant [11–14]. It is worth to note that the equivalent spin exchange situation can be adequately described in terms of spectral diffusion.

For this model, the dynamics of the magnetization obeys the modified Bloch equations (see (4.5) and [11–14])

$$\begin{aligned}
\partial M_{1x}/\partial t &= -W\,M_{1x} - (\omega_0 + a/2 - \omega)\,M_{1y} + (1/2)V(M_{1x} + M_{2x}),\\
\partial M_{1y}/\partial t &= (\omega_0 + a/2 - \omega)\,M_{1x} - W\,M_{1y} + \omega_1 M_{1z} + (1/2)V(M_{1y} + M_{2y}),\\
\partial M_{1z}/\partial t &= -\omega_1 M_{1y} - W_1 M_{1z} + (1/2)\,V\,(M_{1z} + M_{2z}) + (1/2)\,W_0 M_0,\\
\partial M_{2x}/\partial t &= -W\,M_{2x} - (\omega_0 - a/2 - \omega)\,M_{2y} + (1/2)V(M_{1x} + M_{2x}),\\
\partial M_{2y}/\partial t &= (\omega_0 - a/2 - \omega)\,M_{2x} - W\,M_{2y} + \omega_1 M_{2z} + (1/2)\,V\,(M_{1y} + M_{2y}),\\
\partial M_{2z}/\partial t &= -\omega_1 M_{2y} - W_1 M_{2z} + (1/2)V(M_{1z} + M_{2z}) + (1/2)\,W_0 M_0.
\end{aligned} \tag{7.17}$$

Here M_0 is the equilibrium magnetization value,

$$W_0 = 1/T_1;\; W_{00} = 1/T_2; V = K_{ex}C;\; W = V + W_{00};\;\; W_1 = V + W_0. \tag{7.18}$$

Note that in Eqs. (7.17) the contribution of the dipole-dipole interaction is neglected. This is done to obtain as compact analytical expressions as possible. The main aim of studying this simple model situation is a demonstration of some specific features of the saturation effect when the spin coherence transfer operates. In the general case, to take into account the dipole-dipole interaction and the possibility of nonequivalent spin exchange it is necessary to use Eqs. (4.5) and (4.6) instead of (7.17).

The shape of the EPR spectrum is given by the steady-state $M_y = M_{1y} + M_{2y}$ value obtained as a steady-state solution of the system of Eqs. (7.17):

$$M_y = -\Big(M_0 W_0\, \omega_1 \Big((V + W_0)\Big((a/2)^2 (V + W_{00}) + W_{00}\Big((\omega - \omega_0)^2 + (V + W_{00})^2\Big)\Big) + W_{00}(V + W_{00})\omega_1{}^2\Big)\Big)/\text{Denominator};$$

$$M_z = -M_0 W_0 \Big((a/2)^4 W_1 + (a/2)^2\big((-V + W)\omega_1{}^2 - 2((V - W)W + (\omega - \omega_0)^2)W_1\big) + \Big((V - W)^2 + (\omega - \omega_0)^2\Big)\Big(W\omega_1{}^2 + \Big(W^2 + (\omega - \omega_0)^2\Big)W_1\Big)\Big)/\text{Denomenator};$$

$$\text{Denominator} = \Big(-(a/2)^4 W_0(V+W_0) - \Big(W_0\big((\omega-\omega_0)^2 + W_{00}{}^2\big) + W_{00}$$
$$\omega_1{}^2\Big)\Big((V+W_0)\big((\omega-\omega_0)^2 + (V+W_{00})^2\big) + (V+W_{00})\omega_1{}^2\Big) + (a/2)^2(2\,W_0$$
$$(V+W_0)\big((\omega-\omega_0)^2 - W_{00}(V+W_{00})\big) - (V^2 + 2\,W_0W_{00} + V\,(W_0+W_{00}))\omega_1{}^2)\Big). \quad (7.19)$$

This equation makes it possible to analyze the features of the spectral manifestations of the saturation effect as a function of the spin coherence transfer rate and the Rabi frequency ω_1. The behavior of the saturation parameters = $(M_0 - M_z)/M_0$ (see (7.5)) will be also analyzed later using M_z (7.19).

In two limiting cases, when either V or ω_1 is small enough, Eq. (7.19) is reduced to the known results. Indeed, Eq. (7.19) confirms that in the absence of the spectral diffusion, i.e., when $V = 0$, the spectrum is the sum of two independent Lorentzian lines with resonance frequencies $\omega_0 \pm a/2$. The shape of both resonance lines, their widths and amplitudes, for any values of the Rabi frequency ω_1 are given by Eq. (7.1). In this case, the width of individual lines increases when ω_1 increases as $\Delta\omega_{1/2} = (1/T_2)(1 + \omega_1{}^2T_1T_2)^{1/2}$ (see Eq. (7.2)).Under the saturation conditions, when $\omega_1{}^2T_1T_2 > 1$, the line width increases linearly with ω_1 as $\Delta\omega_{1/2} \approx \omega_1(T_1/T_2)^{1/2}$.

In the linear response limit, when $\omega_1{}^2T_1T_2 < 1$, but the spectral diffusion rate is not zero, Eq. (7.19) is reduced to the known result of the linear response theory (see, e.g., [11–14, 20]). In this case it is well known how spectra are transformed, when the spectral diffusion rate, V, increases. When V is small enough, i.e., $V < a$, Eq. (7.19) demonstrates the broadening of spectra lines, their resonance frequencies are shifting to the center of gravity of the spectrum. In contrast to the case $V = 0$, each line has a mixed shape: it is a sum of Lorentzian absorption and Lorentzian dispersion contributions [13, 14, 19–21]. When the spectral diffusion rate becomes equal to the splitting of two lines, $V = a$, two lines merge into a single Lorentzian line, which does not contain the dispersion term: the so-called exchange narrowing effect is manifested.

To illustrate the manifestations of the ω_1 and V values, two sets of spectra simulated using Eq. (7.19) are shown in Fig. 7.2: in the linear response case (ω_1 = 0.001 G, Fig. 7.2a) and under the saturation conditions (ω_1 = 2 G, Fig. 7.2b). In all cases the spectrum was calculated for several values of the spectral diffusion rate V.

Figure 7.2a shows the transformations of the two-frequency spectrum detected in the linear response case. As expected, when V grows, the two lines broaden, resonance frequencies shift to the center of the spectrum, and at $V = a$ = 10 Gs these two lines collapse practically into one homogeneous Lorentzian line.

Figures 7.2b shows that under the saturation conditions the spectral lines are broadened by the microwave (MW) field as it is expected. However, the MW field affects also the frequencies of resonances detected under saturation conditions. This is a non-expected observation. One of the consequences of the changing of the frequencies of resonances is clearly revealed in Fig. 7.2b. Indeed, at V = 10 Gs in the linear response limit the spectrum merges into the Lorentzian line (Fig. 7.2a). In the

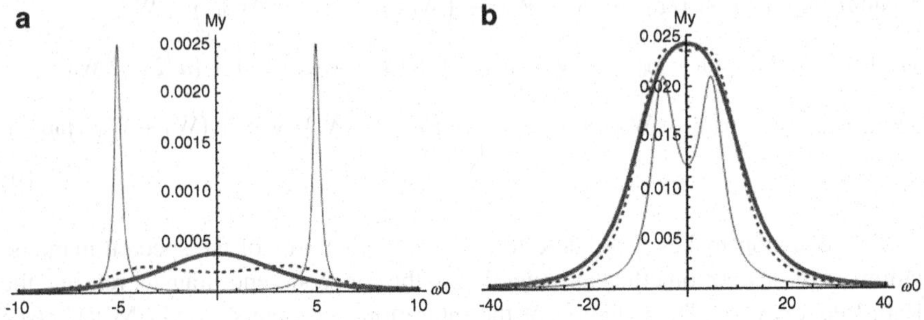

Fig. 7.2 Spectra calculated at different values of the B_1 field: (**a**) $\omega_1 = 0.001$ Gs, (**b**) $\omega_1 = 2$ Gs. The curves in Fig. 7.2a, b correspond to different spectral diffusion rates V = 0.2 Gs (thin solid line), 5 Gs dotted lines), 10 Gs (thick lines). Other parameters are $a = 10$ Gs; $W_0 = 0.05$ Gs; $W_{00} = 0.1$ Gs. Usually, frequencies in the EPR experiments are given in magnetic induction units. To transform the frequency and spectral diffusion rate given in gauss, one has to multiply them by $1.76\ 10^7$ and get them in angular frequency units 1/s (in fact, rad/s). Note that for Fig. 7.2a, b are used different frequency scales. (Adapted with permission from Ref. [25])

large B_1 field at V = 10 Gs the spectrum is not a single Lorentzian but is a sum of two lines with equal line widths but different resonance frequencies (see corresponding thick curve in Fig. 7.2b). In Figure 7.2b, the collapse of two lines into the exchange narrowed single line occurs at V around 20 Gs and 50 Gs for $\omega_1 = 1$ Gs and $\omega_1 = 5$ Gs, respectively. This means that under the combined action of the circularly polarized MW field and spectral diffusion (i.e., spin coherence transfer and spin excitation energy transfer) resonance frequencies change compared to the values in a case of the linear response. These new frequencies are presented below.

7.1.2.2 B_1 Dependence of the Resonance Frequencies of the Two-Frequency Model System in Presence of Equivalent Spin Exchange

The shape of the continuous-wave spectrum Eq. (7.19) is found by the steady-state solutions of Eqs. (7.17). Under the steady-state conditions, Eqs. (7.17) present the system of linear equations for components of the magnetization vector

$$\begin{aligned} \mathbf{M} &= \{M_{1x}, M_{1y}, M_{1z}, M_{2x}, M_{2y}, M_{2z}\} \\ \boldsymbol{L}\mathbf{M} &= -W_0\mathbf{M}_0, \\ \mathbf{M}_0 &= \{0, 0, M_0/2, 0, 0, M_0/2\}. \end{aligned} \tag{7.20}$$

The evolution operator $\boldsymbol{L}$ is given by Eqs. (7.17). In the presence of spectral diffusion, the operator $\boldsymbol{L}$ contains terms which couple $\mathbf{M}_1$ and $\mathbf{M}_2$ magnetizations of two sub-ensembles of spins considered in the model system. Due to this coupling there are formed collective modes of the evolution of all components of the magnetization vectors. Thus there are linear combinations of magnetization components, which evolve independently (compare with Eq. (7.17)). The collective modes approach in the case of the linear response was developed in Refs. [13, 14, 20]. In principle, all collective modes can give the resonance effect. Then the spectrum observed in the experiment is a sum of responses of the collective modes to the external alternating magnetic field B_1. Note that these collective modes can be excited differently by the external field B_1 [13, 14, 20]. Note that this approach was shown to be very instructive when studying manifestations of spectral diffusion in the EPR spectra in the linear response case, in particular, for the interpretation of the exchange narrowing effect [19–21, 29].

The resonance frequencies of the system under study are found as the eigenvalues of the evolution operator ***L***. One can find the eigenvalues of the evolution operator ***L*** using the fact that the determinant of ***L*** is the product of all of its eigenvalues. The determinant of matrix ***L*** can be straightforwardly factorized for ω–ω_0 containing terms (see Denominator (7.19)). One can see that Denominator (7.19) has the biquadratic form for the ω–ω_0 variable. Solving the secular equation.

$$\begin{aligned}&\Big(-(a/2)^4 W_0(V+W_0) - \Big(W_0\Big((\omega-\omega_0)^2 + {W_{00}}^2\Big) + W_{00}{\omega_1}^2\Big)\Big((V+W_0)\Big((\omega-\omega_0)^2 + (V+W_{00})^2\Big)+\\&(V+W_{00}){\omega_1}^2) + (a/2)^2\Big(2\,W_0(V+W_0)\Big((\omega-\omega_0)^2 - W_{00}(V+W_{00})\Big)-\\&(V^2 + 2\,W_0W_{00} + V(W_0+W_{00})){\omega_1}^2)) = 0.\end{aligned} \tag{7.21}$$

one finds that there are four eigenvalues which can give the resonance effect:

$$\lambda = \left\{\omega - \omega_0 \pm (1/2)^{1/2}\Big(R_1 \pm (R_2)^{1/2}\Big)^{1/2}\right\}. \tag{7.22}$$

The following notations are introduced here

$$\begin{aligned}R_1 &= (-1/(W_0(V+W_0))\,(W_0(V+W_0)\,(-(a^2/2) + V^2 + 2\,V\,W_{00} + 2\,{W_{00}}^2)\\&+(2\,W_0W_{00} + V(W_0+W_{00})){\omega_1}^2),\\R_2 &= (-a^2W_0(V+W_0)(V+2\,W_{00})\,(W_0(V+W_0)\,(V+2\,W_{00}) + (V+2\,W_0){\omega_1}^2)+\\&(V\,W_0(V+W_0)(V+2\,W_{00}) + V\,(W_0-W_{00}){\omega_1}^2)^2)/\Big({W_0}^2(V+W_0)^2\Big).\end{aligned} \tag{7.23}$$

There are only four eigenvalues containing ω–ω_0, since two of six equations, namely, for M_{z1} and M_{z2} (see Eq. (7.17)), do not contain the ω–ω_0 terms.

The eigenvalues (7.22) are complex numbers. For example, in the case $V = 0$ one has

$$R_1 = 2\left(a^2/4 - W_{00}{}^2\left(1 + \omega_1{}^2 T_1 T_2\right)\right), R_2 = -4a^2 W_{00}{}^2\left(1 + \omega_1{}^2 T_1 T_2\right),$$

so that the resonance frequencies are

$$\omega = \omega_0 \pm a/2 \pm i(1/T_2)\left(1 + \omega_1{}^2 T_1 T_2\right)^{1/2}. \tag{7.24}$$

The last result is an expected one. Indeed, in the case $V = 0$ two subensembles of spins happen to be uncoupled, so that under this condition we have to observe two resonance lines of isolated sub-ensembles.

The real parts of the eigenvalues (7.22) give the resonance frequencies of the collective modes of spins, while their imaginary parts give the widths of these resonances [20]. Since the eigenvalues (7.22) are the solutions of the biquadratic equation, there are always two pairs of complex conjugated values. In the model under study the eigenvalues can be further specified. It follows from Eqs. (7.23) that

$$\begin{aligned} R_1{}^2 - R_2 = {} & (1/(4\,W_0(V + W_0)))\Big(W_0(V + W_0)\ \left(a^2 + 4\,W_{00}(V + W_{00})\right)^2 + \\ & 4\left(a^2 + 4\,W_{00}(V + W_{00})\right)\ \left(V^2 + 2\,W_0 W_{00} + V\,(W_0 + W_{00})\right)\omega_1{}^2 \\ & +16\,W_{00}(V + W_{00})\omega_1{}^4\Big), \end{aligned} \tag{7.25}$$

so that $R_1{}^2 – R_2 > 0$ for any values of the magnetic resonance parameters.

When $R_2 = 0$, then the eigenvalues become (see (7.22))

$$\omega = \omega_0 \pm (1/2)^{1/2}(R_1)^{1/2}. \tag{7.26}$$

Note that at $R_2 = 0$ one has $R_1 < 0$:

$$\begin{aligned} R_1 = {} & -\Big(W_0{}^2(V + W_0)^2(V + 2\,W_{00})^4 + 2\,W_0(V + W_0)\ (V + 2\,W_0)\ (V + 2\,W_{00})^3\omega_1{}^2 \\ & +(4\,W_0 W_{00} + V(W_0 + W_{00}))\left(2\,V^2 + 4\,W_0 W_{00} + 3\,V(W_0 + W_{00})\right)\omega_1{}^4\Big)/(2\,W_0(V + W_0) \\ & (V + \ 2\,W_{00})\ \left(W_0(V + W_0)\ (V + 2\,W_{00}) + (V + 2\,W_0)\omega_1{}^2\right)) < 0. \end{aligned}$$

As a result, at $R_2 = 0$ the resonance frequencies are double degenerate

$$\omega = \omega_0 \pm i(1/2)^{1/2}(|R_1|)^{1/2}. \tag{7.27}$$

According to Eq. (7.27), the width of resonances are

$$\Delta\omega_{1/2} = (1/2)^{1/2}(|R_1|)^{1/2}. \tag{7.28}$$

Under the condition $R_2 = 0$ two spectral lines merge to a single line. It is expected that at this collapse the spin exchange rate is much larger than $W_0 = 1/T_1$ and

$W_{00} = 1/T_2$. Under $VT_1 \gg 1$ and $VT_2 \gg 1$ conditions the line width (7.28) can be approximated to (compare with the opposite limit, $V = 0$, case (7.2 and 7.24)).

$$\Delta\omega_{1/2} \approx V\left(1 + \omega_1{}^2(T_1 + T_2)/\left(T_2V^2\right)\right)^{1/2}. \tag{7.29}$$

Note, that (7.29) gives a width of resonances exactly at the condition of spectrum collapse, i.e. under condition $R_2 = 0$, when $V = V_c$ (see below).

The spectrum collapse appears when $R_2 = 0$. This gives the condition for the spectrum collapse

$$|\, a\, | = \left[\left(V\, W_0(V + W_0)\ (V + 2\, W_{00}) + V\ (W_0 - W_{00})\omega_1{}^2\right)^2/(W_0(V + W_0)\right.$$
$$(V + 2\, W_{00})\ \left(W_0(V + W_0)\ (V + 2\, W_{00}) + (V + 2\, W_0)\omega_1{}^2\right))]^{1/2}. \tag{7.30}$$

By solving this equation, it is possible to find the critical value of the spin exchange rate V_c, which provides the collapse of the spectrum into a single homogeneously broadened line. This value can be determined graphically as the intersection of two lines determined by the left- and right-hand side terms in Eq. (7.30).

Intersections of the solid curves with dashed line in Fig. 7.3 give critical values of the spectral diffusion rate, V_c, which lead to the spectrum collapse for a given set of parameters: a, ω_1, T_1, T_2. Figure 7.3 demonstrates that under saturation conditions $\omega_1{}^2T_1T_2 \geq 1$ the MW power affects this critical value V_c. In Fig. 7.3, the hyperfine interaction constant, a = 10 Gauss, is significantly greater than the MW field intensities chosen during the calculations. In this case, the critical rate V_c increases with increasing MW field intensity, ω_1 (see Fig. 7.3a). There is only one intersection point (one V_c value).

The situation is much more complicated when $a \leq \omega_1$ (see curves near the origin in Fig. 7.3a). For illustration purposes, see Fig. 7.3b where the curves in Fig. 7.3a

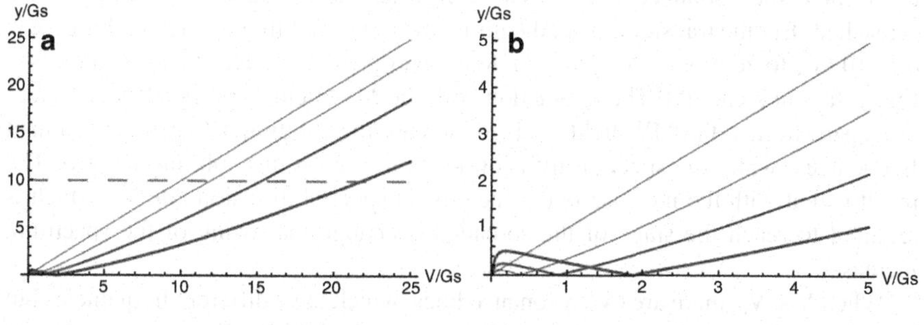

Fig. 7.3 Graphical determination of the V_c value which leads to the collapse of the spectrum into a homogeneously broadened line. Dashed line is $y = |a|$, which was chosen as $y = 10$ Gs during these calculations. Other curves correspond to the right-hand side of (18) for different intensities of the MW field $\omega_1 = 0.01, 0.5, 1, 2$ Gs. The greater the ω_1, the thicker the curve, so that the thickness of the curve increases from $\omega_1 = 0.01$ Gs to $\omega_1 = 2$ Gs. Parameters W_0, W_{00} are the same as in Fig. 7.1. (Adapted with permission from Ref. [25])

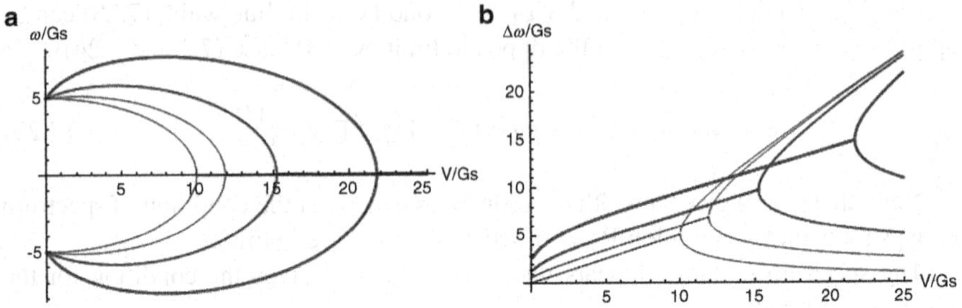

Fig. 7.4 Dependence of the resonance frequencies (**a**) and widths of resonances (**b**) on the spin exchange rate. The case $|a| > \omega_1$. Curves presented for $\omega_1 = 0.01, 0.5, 1, 2$ Gs; the greater the ω_1, the thicker the curve. Other parameters are: $a = 10$ Gs, $W_0 = 0.05$ Gs, $W_{00} = 0.1$ Gs. (Adapted with permission from Ref. [25])

near the origin are shown on a larger scale. It is seen that the horizontal line $y = |a|$ can cross the curves three times in Fig. 7.3b if the splitting a of two lines is small enough. We will show below how these features of curves in Fig. 7.3b are manifested in the features of the spectrum shape.

To better visualize the dependence V_c (ω_1), Fig. 7.4 presents the dependence of resonance frequencies, ω, and widths, $\Delta\omega$, calculated as the real and the imaginary parts of (7.22), respectively, for the case $a > \omega_1$.

Thin lines in Fig. 7.4a, b show the change of resonance frequencies and their widths with the increase in the spectral diffusion rate in the case of the linear response of the system, when the saturation effect can be neglected. It is known that when the spin exchange is slow enough, i.e., $V < a$, the splitting between frequencies in this case decreases monotonically with the increase in the V value and it becomes zero at $V = a$. At the same time, the width of the resonances is the same and it increases with the increase in the V value (see Fig. 7.4b).

When the saturation effect appears, the overall picture resembles the situation of the linear response, but there are also fundamental differences. With the increase in the V value, the resonance frequencies merge when the critical spectral diffusion rate is reached. But the transition from the initial frequency splitting, which in this case is $a = 10$ Gs, to zero is no longer monotonic, and passes through the maximum (see Fig. 7.4a, thick curves). The interaction with the MW field "pushes" these frequencies apart. In fact, the MW field tends to increase the splitting between the resonant frequencies, while the spectral diffusion tends to reduce this splitting to zero. The result is that with the increase in ω_1, the critical spectral diffusion rate V_c, which is required to reach the stage of the so-called exchange narrowing of the spectrum, increases.

When $V < V_c$, there are two resonance lines, which have different frequencies but their line widths are the same. When $V \geq V_c$, in principle, there are two resonances with the same frequency but their line widths are different: at $V > V_c$, one resonance demonstrates the exchange narrowing effect, its line width becomes less when the rate V increases (see Fig. 7.4b), and the width of another resonance increases when

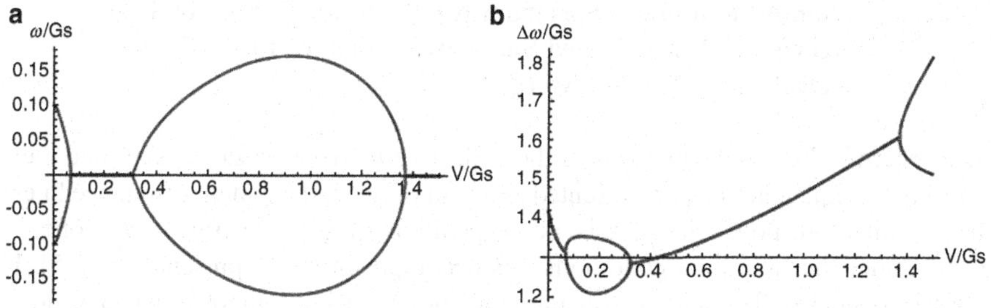

Fig. 7.5 Dependences of the resonance frequencies (on the left) and widths (on the right) on the spectral diffusion rate for a set of magnetic resonance parameters: $a = 0.2$ Gs; $\omega_1 = 1$ Gs, $W_0 = 0.05$ Gs, $W_{00} = 0.1$ Gs. (Adapted with permission from Ref. [25])

the V value increases. In experiment practically only one of two resonances, namely the narrow resonance, is detected since the integral intensity of the broad resonance tends to zero when $V \gg V_c$ [20]. One line corresponds to the in-phase motion of transverse components of magnetization of spins in both sub-ensembles. This line is narrow and accumulates almost the total intensity of the spectrum. Another line corresponds to the out-of-phase motion of transverse components of magnetization of spins and is almost not excited by the circularly polarized alternating magnetic field. It is the so-called dark state. These features of lines under the exchange narrowing conditions were studied comprehensively in [20].

According to Fig. 7.3b, in the opposite case when $|a|<\omega_1$, the dependence of the frequencies, ω, and widths, Δω, of collective modes on the spin exchange rate V is rather complicated. Their typical dependences are shown in Fig. 7.5.

Figure 7.5 (panel on the left) shows that in the intervals of the spectral diffusion rates V = {0.07, 0.3 Gs} and V > 1.37 Gs two resonance frequencies are degenerate. In these degenerate frequency cases, there are two resonance modes which have different line widths (see line width curves in the corresponding intervals in the right-hand panel). The integral intensities of two lines are not the same as well.

In the intervals of V = {0, 0.07 G} and V = {0.3, 1.37 G} resonance frequencies are equally shifted with respect to the center of gravity of the spectrum. The splitting of these two frequencies depends on the V and ω_1 values. In contrast to the degenerate case, in the non-generate situation the widths and integral intensities of two resonances coincide.

Note that as the spin exchange rate increases, the resonance width changes in a rather complicated way. This dependence is different for the cases of degenerate and non-degenerate resonance frequencies. For the case $a > \omega_1$ (see Fig. 7.4) and the case $a < \omega_1$ (see Fig. 7.5 in the interval V {0.3, 1.37 G}) both resonances have the same width, and this width increases with the increase in the V value. This result could be expected, since it is well known as the effect of line broadening caused by spectral diffusion, the rate of which is still insufficient for the collapse of the spectrum [11, 12]. However, Fig. 7.5 shows that at $a < \omega_1$, in the interval V {0, 0.07 G}, the resonance width decreases with the increase in the spin exchange rate.

7.1.2.3 Decomposition of the Spectrum Eq. (7.19) into a Sum of Two Independent Lines. Mixed Shape of Resonance Lines When Spectral Lines Are Not Merged

The spectrum Eq. (7.19) in terms of the collective modes (see Eq. (7.20)) and their collective frequencies can be presented as a sum of two independent resonance lines. In the linear response limit, this decomposition of the spectrum was done in [20]. Under the saturation conditions this decomposition was presented in [25]. If the eigenvalues of the evolution operator are known, this decomposition can be done straightforwardly as follows. In the V intervals when the resonance frequencies are not degenerate, i.e., the absolute values of real and imaginary parts of all four eigenvalues ω (7.22) are the same (see Figs. 7.3 and 7.4), the denominator of M_y Eq. (7.19) can be presented as (see Eq. (7.22))

$$\begin{aligned}\text{Denominator} = W_0(V+W_0)\left((\omega-\omega_0-s)^2+b^2\right)\\ \times\left((\omega-\omega_0+s)^2+b^2\right),\end{aligned} \tag{7.31}$$

$$s = Re(\Omega), b = Im(\Omega), \Omega = (1/2)^{1/2}\left(R_1 + (R_2)^{1/2}\right)^{1/2}.$$

Using Eq. (7.31), the EPR spectrum shape, M_y Eq. (7.19), can be presented as

$$\begin{aligned}My = (c_1 + c_2(\omega-\omega_0+s))/\left((\omega-\omega_0+s)^2+b^2\right)\\ +(c_1 - c_2(\omega-\omega_0-s))/\left((w-w_0-s)^2+b^2\right);\end{aligned} \tag{7.32}$$

Where

$$\begin{aligned}c_1 = M_0\Big(W_0(V+W_0)\,\big((a/2)^2(V+W_{00})+W_{00}\left(b^2+s^2+(V+W_{00})^2\right)\big)\omega_1\\ +W_0W_{00}(V+W_{00})\omega_1{}^3\Big)/\left(4W_0(V+W_0)\left(b^2+s^2\right)\right),\\ c_2 = M_0\Big(W_0\omega_1\Big((a/2)^2(V+W_0)\,(V+W_{00})+W_{00}(-b^2(V+W_0)\\ +(V+W_0)\,(-s+V+W_{00})\,(s+V+W_{00})\\ +(V+W_{00})w_1{}^2)\Big)\Big)/\left(4\,W_0(V+W_0)s\,\left(b^2+s^2\right)\right).\end{aligned} \tag{7.33}$$

Similar to the linear response case, the spectrum Eq. (7.32) is a sum of two lines, which have dispersion contributions with opposite signs. Equations (7.33) show that the dispersion contribution increases when $V \to V_c$, since under this condition $s \to 0$ and two resonance frequencies become degenerate. Denominator of c_2 (7.33) contains "s" as a multiplier, so that $c_2 \to \infty$ if $s \to 0$. But at this point the dispersion

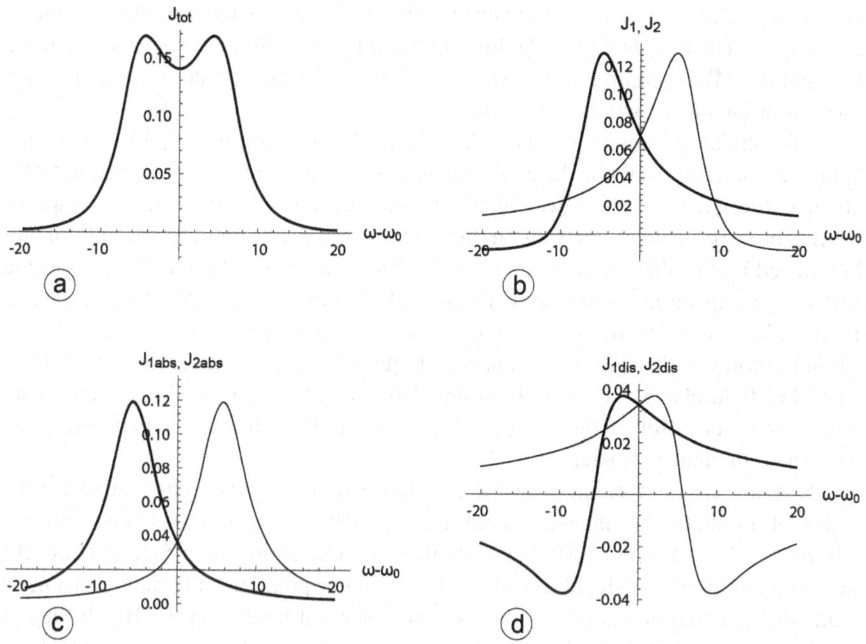

Fig. 7.6 Decomposition of the total spectrum into contributions of two independent collective modes which have resonances described by mixed shapes. The curve Fig. 7.6a presents the total spectrum, Fig. 7.6b shows the decomposition of the spectrum into two collective resonance lines, Fig. 7.6c presents the absorption parts of these lines, while Fig. 7.6d presents their dispersion parts calculated using Eq. (7.32). Parameters used in these simulations are: $a = 10$ Gs, $V = 3$ Gs, $\omega_1 = 1$ Gs, $W_0 = 0.05$ Gs, $W_{00} = 0.1$ Gs. (Adapted with permission from Ref. [25])

contributions of two resonance lines exactly cancel each other. In the region of the exchange narrowing there is no dispersion contribution.

Figure 7.6 illustrates the mixed shape of the spectrum lines with the resonance frequencies $\omega = \omega_0 + s$ and $\omega = \omega_0 - s$ when $V < V_c$.

The mixed form of collective mode resonances in the two-frequency model was discussed in detail in [20] in the case of the linear response of the system. Figure 7.6 illustrates that in the situation, when the saturation effect is manifested, the collective resonances are manifested in the spectrum as independent lines with a mixed shape qualitatively similar to the situation in the case of the linear responses. However a under saturation conditions the spectrum collapse appears at larger values of the spin coherence transfer rate compared to linear response case.

7.1.2.4 More Comments Concerning the Saturation Effect

In the absence of the spin coherence transfer, the MW field affects the shape of the spectrum exclusively via the broadening of the resonances existing in the system.

Note that the homogeneous broadening of the lines is associated with the lifetime of a spin state. The broadening of the lines induced by the MW field appears due to the fact that the MW pumping in the center of the line gives the greater saturation effect than pumping on the wings of the line.

Spin coherence transfer results in the transfer of the quantum coherence between spin sub-ensembles with different resonance frequencies. This fundamentally changes the nature of the effect of the alternating magnetic field on the shape of the magnetic resonance spectrum. In this situation the spectrum lines are additionally broadened by the alternating magnetic field. But spin coherence transfer leads to the fact that the quantum coherences of spin sub-ensembles with different resonance frequencies are coupled. As a result, the spin magnetization vectors of two spin sub-ensembles with different resonance frequencies perform a collective motion. The MW field changes collective modes. However, the MW field does not create collective modes itself in the absence of some kind of the interaction between spins, e.g., spin coherence transfer.

There are cases of changes in the resonance frequency of spins caused by the action of an alternating magnetic field, e.g., the Bloch-Siegert shift [5] or manifestations of "dressed" spins [30]. However there should be no Bloch-Siegert shift [29] in the model under consideration, since the circularly polarized alternating magnetic field is considered here, and not the linearly polarized field as in the Bloch-Siegert theory. The Bloch-Siegert shift should also occur in the absence of any interaction between the spins. In the considered problem, no shift of lines due to the alternating magnetic field is found in the absence of spin coherence transfer, i.e., when the spins move independently of each other.

The changes of resonance frequencies predicted in [25] and caused by the circularly polarized alternating MW field cannot be ascribed to the formation of "dressed" spins, since the "dressed" atom theories refer to linearly polarized alternating fields. The concept of "dressed" spins refers to the participation of multi photon absorption acts in strong alternating fields. Note that in contrast to the circularly polarized alternating field, the linear alternating magnetic field allows 3,5,..2n + 1 multi photon absorption even by a two-level system.

In strong alternating magnetic fields, when the saturation effect is manifested, the interaction of spins with the alternating magnetic field broadens the spectral lines and, accordingly, reduces the amplitude of the observed signal. This is the expected effect of the alternating magnetic field, since it is well known result in the case of non-interacting spins (see Eq. (7.2)). However, in the presence of spin coherence transfer, under the saturation conditions of the spectrum new effects appear associated with the interaction of spins with the alternating magnetic field: the alternating magnetic field changes collective modes and changes the resonance frequencies. The MW field pushes the resonance frequencies apart. The resonance frequencies associated with these new states depend on the splitting of the original resonance frequencies, the spin coherence transfer rate and the amplitude of the alternating magnetic field.

On the basis of qualitative considerations, based on the knowledge about the broadening of the lines of the spectrum by the alternating magnetic field, it seems to be impossible to predict the mentioned above peculiar manifestations of the alternating magnetic field in the form of the spectra under saturation conditions. To justify the unexpected effects which are predicted theoretically (see, e.g., Figs. 7.3 and 7.4), new experiments are needed to demonstrate experimentally that under the saturation conditions the exchange narrowing effect, i.e., the collapse of a spectrum into one homogeneously broadened Lorentzian line, occurs at higher spin coherence transfer rate compared to the linear response case.

The essence of the saturation effect consists in a phenomenon that there is a limit of microwave field power which can be pumped into the given spin system. At the low power an energy absorbed by the spin system increases proportionally to the density of MW field photons, i.e., proportionally to $\omega_1{}^2$. At the high power the steady-state (stationary) power absorption stops to depend on the power: the spin system is "saturated" and its ability to absorb power under stationary conditions is limited.

A parameter, which directly describes the saturation effect, is [10]

$$\mathrm{sat} = (\mathrm{M}_0 - \mathrm{M}_z)/\mathrm{M}_0. \tag{7.34}$$

Solving Eqs. (7.17) under stationary conditions one finds

$$\mathrm{M_z} = \mathrm{NomZ/Denomenator}, \tag{7.35}$$

where Denomenator is given by Eq. (7.18) while the nominator is given by

$$\begin{aligned}\mathrm{NomZ} == &-\mathrm{M_0W_0}\Big((a/2)^4\mathrm{W_1} + (a/2)^2\big((-\mathrm{V}+\mathrm{W})\omega_1{}^2 - 2((\mathrm{V}-\mathrm{W})\mathrm{W} \\ &+(\omega-\omega_0)^2)\mathrm{W_1}\big) + \left((\mathrm{V}-\mathrm{W})^2 + (\omega-\omega_0)^2\right)\left(\mathrm{W}\omega_1{}^2 + \left(\mathrm{W}^2 + (\omega-\omega_0)^2\right)\mathrm{W_1}\right)\Big).\end{aligned} \tag{7.36}$$

It is worth to note that at the exchange narrowing conditions, when the spin coherence transfer rate V is much higher than splitting of two resonance lines, which is *a* in the model system under consideration, a saturation parameter tends to (compare with (7.6))

$$\mathrm{sat} \rightarrow \omega_1{}^2\mathrm{T_1T_2}/\left(1 + \omega_1{}^2\mathrm{T_1T_2}\right) \tag{7.37}$$

if the spectrum is pumped at the frequency which coincides with a center of gravity of the distribution of resonance frequencies ($\omega_0 + a/2$ and $\omega_0 - a/2$ in the model under consideration). The result (7.37) is expected since it refers to a situation of the fast spin coherence transfer when the spectrum merges into the exchange narrowed Lorentian line with resonance frequency equal to the average frequency.

7.1.3 Shape of the EPR Spectrum in Strong Microwave Fields for Paramagnetic Particles with Spin 1/2. Equivalent Spin Exchange Case

7.1.3.1 General Solution

Let us consider an ensemble of paramagnetic particles with spin ½ in solution. Suppose they have EPR frequencies described by the distribution $\varphi(\omega_k)$ and paramagnetic relaxation times T_{1k} and T_{2k}. In solutions of nitroxide radicals all spins have the same T_1 but widths of hyperfine components may differ. In the course of the further consideration is assumed that $T_{1k} = T_1$. This ensemble of paramagnetic particles can be divided into sub-ensembles of isochromatic ones. Suppose that the exchange interaction in a course of bimolecular collisions leads to the equivalent spin exchange with the rate constant $K_{ex}C$. Taking into account that in EPR experiments Zeeman frequencies of electron spins are high enough, the contribution of the dipole-dipole interaction between spins can be expressed with only two characteristic parameter T_{dd} and T_{1d} (see Eqs. (3.31) and (4.4). For this system, components of magnetization vectors $\{M_{kx}, M_{ky}, M_{kz}\}$ of sub-ensembles with resonance frequencies $\{\omega_k\}$, respectively, obey the modified Bloch kinetic equations (see (4.4))

$$\partial M_{kx}/\partial t = -(1/T_{2k} + W_{sd})\, M_{kx} - (\omega_k + \delta_k - \omega)\, M_{ky} + \varphi_k V_{sct} M_x,$$
$$\partial M_{ky}/\partial t = (\omega_k + \delta_k - \omega)\, M_{kx} - (1/T_{2k} + W_{sd})\, M_{ky} + \varphi_k V_{sct} M_y + \gamma B_1 M_{kz},$$
$$\partial M_{kz}/\partial t = -\gamma B_1 M_{ky} - W\, M_{kz} + \varphi_k V M_z + \varphi_k W_0 M_0.$$

Here T_{2k} is the paramagnetic relaxation time of x, y projections of magnetization of the k-th sub-ensemble of spins, $\omega_k = \omega_0 + a_k$ is a sum of the Zeeman frequency and the hyperfine or fine splittings, δ_k is the possible contribution to the resonance frequencies due to the interference of the exchange interaction with, e.g., hyperfine interaction of an unpaired electron with magnetic nuclei. As it was already several times indicated, $W_{sd} = K_{ex}C + 5/T_{dd}$ gives the dephasing rate due to spin exchange and dipole-dipole interactions (see Eq. (4.4)); $V_{sct} = K_{ex}C - 4/T_{dd}$ gives the spin coherence transfer due to the exchange and dipole-dipole interactions

$$W = 1/T_1 + K_{ex}C + 2/T_{dd},\ V = K_{ex}C + 2/T_{dd};V_{sct} = K_{ex}C - 4/T_{dd};$$
$$W_0 = 1/T_1;1/T_1 = 1/T_{10} + 1/T_{1d};M_{x,y,z} = \sum_n M_{nx,ny,nz},$$

M_0 is the equilibrium magnetization of the system.

Under steady-state conditions one has to solve Eqs. (7.38)

$$
\begin{aligned}
&-(1/T_{2k} + W_{sd})\, M_{kx} - (\omega_k + \delta_k - \omega)\, M_{ky} + \varphi_k V_{sct} M_x = 0,\\
&(\omega_k + \delta_k - \omega)\, M_{kx} - (1/T_{2k} + W_{sd})\, M_{ky} + \varphi_k V_{sct} M_y = -\gamma B_1 M_{kz},\\
&\quad - \gamma B_1 M_{ky} - W\, M_{kz} + \varphi_k V M_z = -\varphi_k W_0 M_0.
\end{aligned}
\tag{7.38}
$$

Solving these equations we obtain

$$
\begin{aligned}
&M_x = M_0 S_2 W_0 \omega_1/\text{Denom};\\
&M_y = -M_0 W_0 \omega_1 \left(S_2{}^2 V_{sct} W + S_4\left(-1 + S_4 V_{sct} W + S_1 V_{sct} w_1{}^2\right)\right)/\text{Denom};\\
&M_z = -(M_0 W_0/V)\left(1 + \left(-1 + \omega_1{}^2 V_{sct} S_1 (1 - S_4 V_{sct} W)\right.\right.\\
&\left.\left.-V_{sct} W\left(-2S_4 + \left(S_2{}^2 + S_4{}^2\right) V_{sct} W\right)\right)/\text{Denom}\right);
\end{aligned}
\tag{7.39}
$$

Here the following notations are introduced:

$$
\begin{aligned}
&\text{Denom} = (-VS_2{}^2 V_{sct}\omega_1{}^2 + (-1 + (S_3 + S_5)V)(-1 + S_4 V_{sct} W + S_1 V_{sct}\omega_1{}^2) -\\
&V_{sct}((-1 + (S_3 + S_5)V)W + S_4 V\omega_1{}^2)(S_2{}^2 V_{sct} W + S_4(-1 + S_4 V_{sct} W + S_1 V_{sct}\omega_1{}^2));\\
&S_1 = \text{sum}(p_k/\text{den});\\
&S_2 = \text{sum}(p_k(-\omega_k - \delta_k + \omega)/\text{den});\\
&S_3 = \text{sum}(p_k(-\omega_k - \delta_k + \omega)^2/\text{den});\\
&S_4 = \text{sum}(p_k(1/T_{2k} + W_{sd})/\text{den});\\
&S_5 = \text{sum}(p_k(1/T_{2k} + W_{sd})^2/\text{den});\\
&\text{den} = \omega_1{}^2(1/T_{2k} + W_{sd}) + W((1/T_{2k} + W_{sd})^2 + (\omega_k + \delta_k - \omega)^2).
\end{aligned}
\tag{7.40}
$$

When T_{2k} does not depend on k, then $S_3 + S_5$ and S_4 are proportional to S_1 and these equations are rather simplified. Eqs. (7.39) and (7.40) give steady-state values of the magnetization at any power of the microwave field. So they make it possible to analyze the manifestations of the saturation effect in the shape of the EPR spectrum. The $M_y(\omega_0)$ values (7.39) give the EPR spectrum, while $M_z(\omega_0)$ can be used, e.g., to find a shape of a "hole" burned out in the spectrum with the strong microwave field, etc.

Equations (7.39) make it possible to simulate EPR spectra in the strong microwave field for any concentration of paramagnetic particles and for any hyperfine structure of the EPR spectra. By varying the unknown parameters in Eqs. (7.39) one can find the best-fit parameters, which reproduce the experimental curves. This fitting approach might be considered as a universal procedure for finding the spin exchange rate using the EPR spectrum. This fitting method of finding the spin exchange rate using the EPR spectra detected in the weak microwave field was comprehensively presented in Sect. 6.6 (see, in particular, Sects. 6.2.2 and 6.2.3). Note that the fitting method for the case of spectra detected in the strong B_1 field is less elaborated than in the case of the weak B_1 field. As it was presented in Sect. 6.6

in the case of the weak B_1 field (linear response case) the theory provides several simple methods to reasonably estimate the spin coherence transfer rate and the rate of spin decoherence, e.g., by using the so called two-point method (see Sect. 6.6.). In the case of the strong B_1 field, proper approaches will be presented below.

Note an evident but interesting property of the kinetic Eqs. (7.38). When the microwave field is switched off the time dependence of the longitudinal projections of magnetization M_z obeys the equation

$$\begin{aligned} \partial M_{kz}/\partial t &= -W\, M_{kz} + \varphi_k V M_z + \varphi_k W_0 M_0. \\ \partial M_z/\partial t &= -W\, M_z + V M_z + W_0 M_0. \end{aligned} \tag{7.41}$$

The solution of the last equation is given by the equation well known in the inversion-recovery method for measuring the spin-lattice relaxation time T_1 [31]

$$M_z(t) = M_z(0)\exp(-t/T_1) + M_0(1 - \exp(-t/T_1)). \tag{7.42}$$

Here $M_z(0)$ describes the initial M_z value. By sudstituting $M_z(t)$ into the first equation in (7.41), the recovery of the z-component of magnetization for the k-th subensemble of spins is

$$\begin{aligned} M_{kz}(t) = {} & \varphi_k M_0(1 - \exp(-t/T_1)) + \\ & \exp(-t/T_1)(M_z(0)\varphi_k + (M_{kz}(0) - M_z(0)\varphi_k)\exp(-Vt)). \end{aligned} \tag{7.43}$$

The total recovery of the equilibrium longitudinal magnetization proceeds via two processes: redistribution of spin excitation among all spins of the system and spin-lattice relaxation of spins. The excitation of spins of selected sub-ensembles is redistributed among all sub-ensembles of spins with the rate V, so that z-component of any subensemble tends to the $M_{kz}(0) \rightarrow M_z(0)\ \varphi_k$ value.

General solutions (7.39) and (7.40) were analyzed comprehensively for several systems [25–27]. Results obtained when studying the case of two resonance frequencies [25] were presented in Sect. 7.1.2.3. The results obtained for the case of the Gaussian distribution of the resonance frequencies [26] are presented below.

7.1.3.2 Analysis of the Saturation Effect for Free Radicals Which Have Gaussian Distribution of the Resonance Frequencies

When the unpaired electron of a free radical interacts with more than 5–6 protons/deuterons the hyperfine structure can be reasonably well described as of the Gaussian shape (see, e.g., [22]). In this case, it is expected that all paramagnetic particles have the same paramagnetic relaxation time T_2, i.e., $T_{2k} = T_2$ in Eqs. (7.38), (7.39) and (7.40). In addition, we suppose that shifts δ_k (see (7.38)) are negligible. These circumstances simplify the analytical treatment. For the further simplification, we suppose that the viscosity of the solution is low enough so that the exchange interaction in the course of bimolecular collisions gives the dominant contribution and the dipole-dipole interaction contribution can be ignored. To this end, the system considered is described by parameters (see Eqs. (7.38), (7.39) and (7.40))

$$W_{sd} = V_{sct} = W = V = V_{ex}, V_{ex} = K_{ex}C. \tag{7.44}$$

For these parameters, the general solution (7.39) and (7.40) is reduced to

$$\begin{aligned}
M_x &= \frac{1}{DET} M_0 S_2 W_0 \omega_1, \\
M_y &= \frac{1}{DET} \left[M_0 W_0 \omega_1 \left\{-S_2^2 V(V+W_0) + S_1(V+W_{00})\left(1 - S_1 V\left[(V+W_0)(V+W_{00}) + {\omega_1}^2\right]\right)\right\}\right] \\
M_z &= \frac{M_0 W_0}{V}\left[-1 - \left(-1 + S_1 V\left\{(V+W_0)(V+W_{00}) + {\omega_1}^2\right\}\right.\right. \\
&\left.\left. + V(V+W_0)\left\{-{S_2}^2 V(V+W_0) + S_1(V+W_{00})\left(1 - S_1 V\left\{(V+W_0)(V+W_{00}) + {\omega_1}^2\right\}\right)\right\}\right)/DET\right]
\end{aligned} \tag{7.45}$$

Here we denote:

$$\begin{aligned}
DET &= \left[-{S_2}^2 V^2 {\omega_1}^2 + \left\{1 - V(S_3 + S_1(V+W_{00})^2)\right\}\left\{1 - S_1 V((V+W_0)(V+W_{00}) + {\omega_1}^2)\right\}\right] \\
&- V(-{S_2}^2 V(V+W_0) + S_1(V+W_{00})\{1 - S_1 V((V+W_0)(V+W_{00}) + {\omega_1}^2)\})(W_0 - V\{-1 \\
&+ S_3(V+W_0) + S_1(V+W_{00})((V+W_0)(V+W_{00}) + {\omega_1}^2)\}) \\
W_0 &= 1/T_1; W_{00} = 1/T_2; T_{1V} = 1/(V+W_0)); T_{2V} = 1/(V+W_{00}); T = T_{2V}/(1 + {\omega_1}^2 T_{1V} T_{2V})^{1/2}; \\
S_1(\omega) &= T_{1V} T^2 \sum_k \varphi_k / \det(k), \\
S_2(\omega) &= T_{1V} T^2 \sum_k \varphi_k (\omega - \omega_k)/\det(k); \\
S_3(\omega) &= T_{1V} T^2 \sum_k \varphi_k (\omega - \omega_k)^2/\det(k); (\text{note}, S_3 = T_{1V}(1 - S_1/(T_{1V} T^2)), \\
\det(k) &= 1 + (\omega - \omega_k)^2 T^2.
\end{aligned} \tag{7.46}$$

Using equations presented above, one can calculate the behavior of the magnetization under saturation conditions.

We consider the Gaussian distribution of the resonance frequencies. In this case, one has

$$\begin{aligned}
S_{n+1}(\omega) &= (T_{1V} T^2) J_n, T = \frac{T_2 V}{\sqrt{1 + {\omega_1}^2 T_{1V} T_{2V}}} \\
J_n &= 1/(2\pi\sigma)^{1/2} \int_{-\infty}^{\infty} dx(\omega - \omega_0 - x)^n \exp(-x^2/(2\sigma))/\left(1 + (\omega - \omega_0 - x)^2 T^2, n = 0, 1, 2.\right.
\end{aligned} \tag{7.47}$$

Equations (7.45), (7.46) and (7.47) make it possible to simulate the EPR spectra and to analyze the manifestations of the saturation effect taking into account the spin coherence transfer and the spin excitation energy transfer caused by the exchange interaction.

The saturation effect for the model with the Gaussian distribution of resonance frequencies was theoretically studied in [6]. However, as it was already mentioned, in [6] the problem was solved under the additional approximation: in Eq. (7.38) the spin coherence transfer (terms containing V_{sct}) were ignored. The spin coherence transfer is of principal value that is why this problem was reconsidered in [26].

Suppose there was obtained a set of EPR spectra $J_{EPR}(B_0)$ detected by using the microwave field of different power. This detected spectrum is proportional to the partial derivative $\partial M_y/\partial B_0$, where M_y is given by (7.39). The $M_y(B_0)$ (7.39) should be compared with $\int_{-\infty}^{B_0} J_{EPR}(B_0')dB'_0 = A(B_0)$. By double integrating the observed spectrum one finds the integral intensity of the spectrum: $\int_{-\infty}^{\infty} dB_0 \int_{-\infty}^{B_0} J_{EPR}(B_0')dB_0' = \int_{-\infty}^{\infty} dB_0 A(B_0) = I_0$. It should be compared with $\int_{-\infty}^{\infty} dB_0 M_y(B_0, \omega_1) = I_0(\omega_1)_{theory}$ where M_y is given by (7.39).

There are several parameters of the observed EPR spectrum which can be used to characterize the saturation effect and to find the best fitting parameters: times T_1 and T_2, the spin decoherence rate W_{sd} and the spin coherence transfer rate V_{sct}. For fitting one can use $J_{ERP}(B_0,\omega,\omega_1) \sim \partial M_y/\partial B_0$; $A(B_0,\omega,\omega_1) \sim M_y$; and/or $I_0(\omega_1)_{exp.} \sim I_0(\omega_1)_{theory}$.

For the fitting procedure, some other characteristic features of the curves $J_{ERP}(B_0,\omega,\omega_1)$, $A(B_0,\omega,\omega_1)$, or $I_0(\omega_1)$ can be also used: the positions of maxima of $A(B_0)$, i.e., zeros of $J_{EPR}(B_0)$, the widths ΔB_0 of components of the observed spectrum J_{EPR}, etc. Some examples of the analysis of the manifestations of the saturation effect in the EPR spectra will be presented below.

For illustration of the saturation effect, let us consider the EPR spectrum when the pumping frequency is in resonance with the central frequency of the Gaussian distribution, i.e., $\omega = \omega_0$ (see (7.47)). Then one has (see (7.46) and (7.47))

$$
\begin{aligned}
S_1 &= (T_{1V}T^2)J_0;\\
J_0 &= \left(\pi/(2\sigma T^2)\right)^{1/2}\exp\left[1/(2\sigma T^2)\right]\mathrm{Erfc}\left[1/(2\sigma T^2)^{1/2}\right];\\
S_2 &= 0;\\
S_3 &= T_{1V} - S_1/T^2.
\end{aligned}
\tag{7.48}
$$

J_0 depends only on one parameter $(\sigma T^2)^{1/2}$ which is the ratio of the width of resonance frequencies to the effective width 1/T of spectral lines (see (7.47)). The behavior of J_0 is shown in Fig. 7.7.

By substituting (7.48) into (7.45) one obtains the amplitude of the EPR spectrum (not derivative!) when the microwave field pumps spins in the center of the Gaussian distribution (J_0, see (7.47))

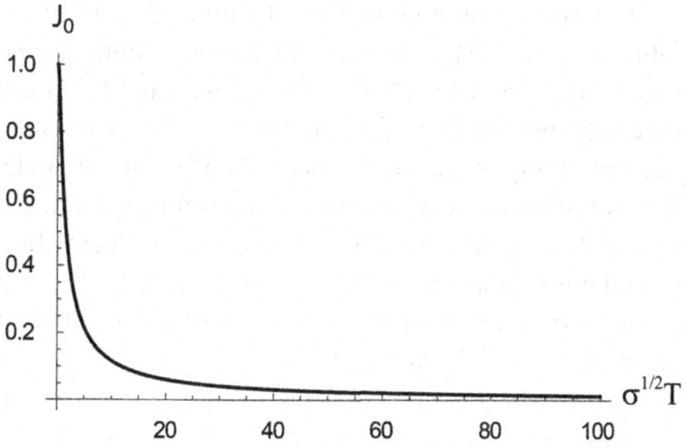

Fig. 7.7 Behavior of $J_0((\sigma T^2)^{1/2})$. At relatively small values of the parameter, $(2\sigma T^2)^{1/2} \leq 5$, J_0 decreases rapidly and then asymptotically goes to zero as $1/(\sigma^{1/2}T)$. (Adapted with permission from Ref. [26]).

$$M_y(\omega = \omega_0) = (M_0 J_0 T_{2V}\omega_1)/\left(1 + \omega_1{}^2 T_{1V}T_{2V} + V\,T_{2V}J_0\left(-1 + T_1 T_{1V}\omega_1{}^2\right)\right). \tag{7.49}$$

In fact, $M_y(\omega = \omega_0)$ (7.49) is valid not only for the Gaussian distribution of resonance frequencies, it is valid for *any symmetric distribution* of the resonance frequencies. Of course, J_0 (7.48) is valid only for the Gaussian distribution. For other cases, J_0 should be calculated for appropriate symmetric distributions.

In the case of the Gaussian distribution of resonance frequencies, some instructive results can be found.

For wide spectra with large dispersions σ when $\sigma T^2 >> 1$, up to the first approximation for a small parameter $1/(\sigma T^2)^{1/2}$ the saturation curve (7.49) is

$$M_y(\omega = \omega_0) \approx (\pi/(2\sigma))^{1/2}\omega_1\left(1 + \omega_1{}^2 T_{1V}T_{2V}\right)^{-1/2} M_0. \tag{7.50}$$

Note that when $V = 0$, Eq. (7.50) coincides with Eq. (7.11), the result obtained in [2, 3]. Result (7.50) is instructive. Within the limits of the applicability of Eq. (7.50) for wide distributions of resonance frequencies, when $\sigma^{1/2} > V$, $1/T_1$, $1/T2$, and ω_1, the saturation effect in the presence of spectral diffusion can be described by Portis T_1–T_2 model without taking into account the spectral diffusion ($V = 0$) by replacing the paramagnetic relaxation times T_1 and T_2 for effective times T_{1V} and T_{2V}, respectively. For example, the calibration curve for the dependence of the width of the saturation curve on the parameter $\sigma^{1/2}T_2$ suggested in [3] can be used as the dependence on $\sigma^{1/2}T_{2V}$. Similarly, it is possible to generalize the Lebedev-Dobryakov formula (7.15) and calculate the width of the spectrum using the formula.

$$\Delta\omega_{p-p} = \left(4\sigma + \left(1 + \omega_1{}^2 T_{1V}T_{2V}\right)/\left(3T_{2V}{}^2\right)\right)^{1/2} + \left(1 + \omega_1{}^2 T_{1V}T_{2V}\right)^{1/2}/\left(3^{1/2}T_{2V}\right). \tag{7.51}$$

However, it should be kept in mind that (7.50) gives the asymptotics of the saturation curve $M_y(\omega = \omega_0)$ for very large inhomogeneous broadening, but does not give an correct description of the saturation curve for real scales of inhomogeneous spectrum broadening. The analysis of the exact solution (7.49) shows that even without taking into account spectral diffusion, i.e. within the Portis T_1–T_2 model [2], the saturation curve $M_y(\omega = \omega_0)$ passes through a maximum and then decreases very slowly. Equation (7.51) does not predict this decay. In order to qualitatively correctly describe the behavior of the saturation curve (7.49), it is necessary to add at least the second approximation term of the expansion in a series in powers of the small parameter $1/(\sigma^{1/2}T)$ in (7.49):

$$M_y(\omega = \omega_0) \approx [(\pi/(2\sigma))^{1/2}\omega_1(1 + \omega_1^2 T_{1V}T_{2V})^{-1/2} - \omega_1/(\sigma T_{2V}) - \pi\omega_1 V(-1 + T_1T_{1V}\omega_1^2)/(2\sigma(1 + T_{1V}T_{2V}\omega_1^2))]M_0 \quad (7.52)$$

In contrast to (7.51), the shape of the saturation curve (7.52) depends on the spectral diffusion rate V explicitly. When analyzing the saturation effect, one of the interesting parameters is the amplitude of the B_{1max} field, at which the saturation curve reaches its maximum value. To find the theoretical value of this microwave field amplitude B_{1max} (Rabi frequency ω_{1max}), it is necessary to find the derivative of My (7.49) or its second-order approximation for the case of the wide distribution of resonance frequencies (7.52), equate it to zero and solve the resulting equation. It is important to pay attention to the fact that for the Gaussian distribution for the spectrum detected in the linear response regime, at $\sigma^{1/2} = V$, a nonhomogeneously broadened spectrum collapses into a homogeneously broadened Lorentzian line. Therefore, the considered approximation refers to a situation in which a collapse has not yet occurred. For this case, using approximate form (7.52) of the saturation curve and equating the derivative of (7.52) to zero, the following equation for determining ω_{1max} is obtained [26]

$$\sigma = (1/(2\pi T_{2V}^2(1 + T_{1V}T_{2V}\omega_1^2)))(2 - \pi T_{2V}V + 4\,T_{1V}T_{2V}\omega_1^2 + 3\pi T_1T_{1V}T_{2V}V\omega_1^2 + \pi T_{1V}T_{2V}^2V\omega_1^2 + 2T_{1V}^2T_{2V}^2\omega_1^4 + \pi T_1T_{1V}^2T_{2V}^2V\omega_1^4)^2. \quad (7.53)$$

This equation is easily solved in the absence of spectral diffusion. When $V = 0$, Eq. (7.53) gives (assuming $\sigma T_2^2 \gg 1$)

$$\sigma = \left(2\left(1 + T_1T_2\omega_1^2\right)^3\right)/\left(\pi T_2^2\right), \qquad \omega_{1\,max} \approx (1/(T_1T_2))^{1/2}\left(\pi\sigma T_2^2/2\right)^{1/6}. \quad (7.54)$$

This shows that ω_{1max} grows very slowly with the increase in the dispersion of the Gaussian frequency distribution. This is consistent with the curves in Fig. 7.1

From the analysis of Eq. (7.49), we also obtained an approximate formula for determining ω_{1max} for the case of relatively small inhomogeneous broadening, when $\sigma T_2^2 < 1$. In this situation

$$\omega_{1max} \approx \left(1 + \sigma T_2^2\right)/(T_1 T_2)^{1/2} \tag{7.55}$$

Equation (7.53) is reduced to a cubic equation for $T_{1V}\ T_{2V}\ \omega_1^2$, if the system parameters allow only the terms of higher degrees in this parameter to be left in the right-hand side of (7.53). Under this condition we obtain.

$$\omega_{1max} \approx (1/(T_{1V}T_{2V}))^{1/2}\left(2\pi\sigma T_{2V}^2/(2 + \pi V T_1)^2\right))^{1/6} \tag{7.56}$$

When $V = 0$, this estimation is reduced to (7.54).

It is interesting to note that at the sufficiently high spectral diffusion rate, when $VT_1 > 1$, $VT_2 > 1$,

$$\omega_{1max} \approx (2\sigma/\pi)^{1/6}(V/T_1)^{1/3}. \tag{7.57}$$

The last result can be interpreted following speculations suggested in Blombergen et al. work [4]. Consider a system which contains N spins and their resonance frequencies are distributed in the range of $\sigma^{1/2}$. Spins with resonance frequencies in an interval $\{(\omega - \omega_1/2), (\omega + \omega_1/2)\}$ will be excited by the microwave field with a frequency ω. The rate of absorbing of the microwave quanta by the spin system can be estimated as $\nu_{abs} = \omega_1^2 T_{2V}(\omega_1/\sigma^{1/2})N$. Suppose that $VT_1 > 1$, so that spin excitation is redistributed among all spins before spin-lattice relaxation will remove the energy of spin excitation to thermostat. The rate of the total excitation energy transfer to thermostat can be estimated as $\nu_{slr} = N/T_1$. At steady-state absorption, the rate of absorption of quanta should be equal to the rate of their removal to the thermostat. So from steady-state condition $\nu_{abs} = \nu_{slr}$ one obtains $\omega_{1max} \approx \sigma^{1/6}(1/T_{2V}T_1)^{1/3}$. By definition (7.46), $T_{2V} = T_2/(1 + VT_2)$ and $T_{2V} \rightarrow 1/V$, when $VT_2 > 1$. Thus these qualitative estimations give $\omega_{1max} \approx \sigma^{1/6}(V/T_1)^{1/3}$ which exceeds a value estimated with Eq. (7.57) only by a factor $(\pi/2)^{1/6} = 1.08$.

Based on the results of a consistent theory of saturation effect, in [26] were carried out numerical calculations of saturation curves $M_y(\omega = \omega_0)$ given by (7.49) for a number of parameters. From these curves, the corresponding values of the Rabi frequency ω_{1max} were found, which gives the maximum value of the saturation curve when pumped at the center of the spectrum. The approximate formula (7.56) gives an estimate of ω_{1max} with an accuracy of 50% when the condition

$V < \sigma^{1/2}$ is fulfilled. In a set of experiments performed with different B_1 one obtains the B_{1max} value. Then the value of γB_{1max} is equated to ω_{1max} given by (7.53), (7.54), (7.55), (7.56) and (7.57). The V and T_1 values obtained in that way can be used as trial values when modeling the saturation curve (7.49) or the derivative of the spectrum to obtain the values of V and T_1 which give the best fitting.

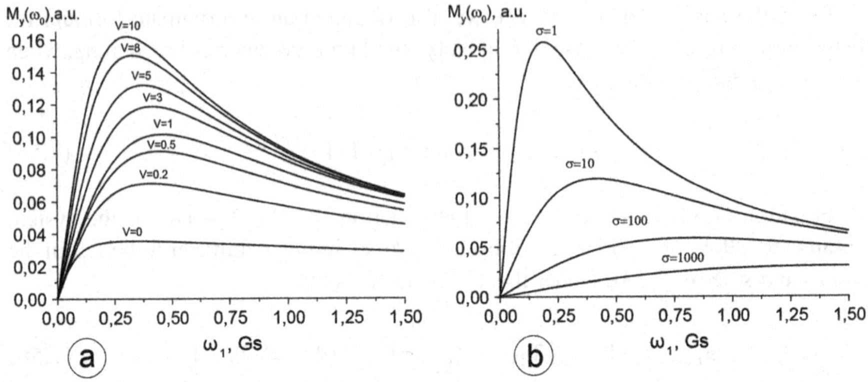

Fig. 7.8 Saturation curves of the amplitude of the EPR signal for different values of the spectral diffusion rate V with $\sigma = 10\ Gs^2$ (curves in Fig. 7.8a) and of the dispersion σ of the Gaussian frequency distribution when V is fixed and equals V = 3 Gs (curves in Fig. 7.8b). Other parameters used in these calculations: $1/T_2 = 0.1$ Gs; $1/T_1 = 0.1$ Gs. (Adapted with permission from Ref. [26])

For comparison, it is worth to remind that for the homogeneous Lorentzian line (7.1) the saturation curve determined as the ω_1 dependence of the amplitude of the M_y at $\omega = \omega_0$ has a form: $J(\omega) = \omega_1 T_2/(1 + \omega_1{}^2 T_1 T_2)\, M_0$, which leads to

$$\omega_{1max} = (1/T_1 T_2)^{1/2}.$$

Above was described one possible approach to estimate V and T_1 parameters by finding in experiment a value of B_{1max} which gives a maximum M_y (7.39). For this purpose one can use the saturation curve given by (7.49).

To get better insight into the problem, Fig. 7.8 shows the saturation curves (7.49) for a set of parameters: ω_1, V, σ.

A quick glance at Fig. 7.8 gives the impression that the given saturation curves are similar to the saturation curves obtained without taking into account spectral diffusion and given in Fig. 7.1. But a closer look shows that there are not only quantitative but also qualitative differences between the saturation curves in the situation V = 0 (see Fig. 7.1) and the situation V ≠ 0 (see Fig. 7.8).

The saturation curves of the EPR signal presented in Fig. 7.8b show that the microwave field amplitude, at which the signal maximum is reached, increases with increasing dispersion of the Gaussian distribution. This observation can be attributed to the fact that the fraction of resonant spins, which "absorb" the microwave field quanta, decreases when σ increases.

The effect of the spectral diffusion rate on the saturation effect may be due both to the influence of spectral diffusion on the stage of absorption of quanta by resonance spins due to the homogeneous broadening of spin levels, and on the stage of energy transfer of spins into the lattice due to the involvement of a larger amount of spins into the process of the transfer of energy of excited spins to the lattice. Therefore, the effect of spectral diffusion on the saturation effect turns out to be more complex. For

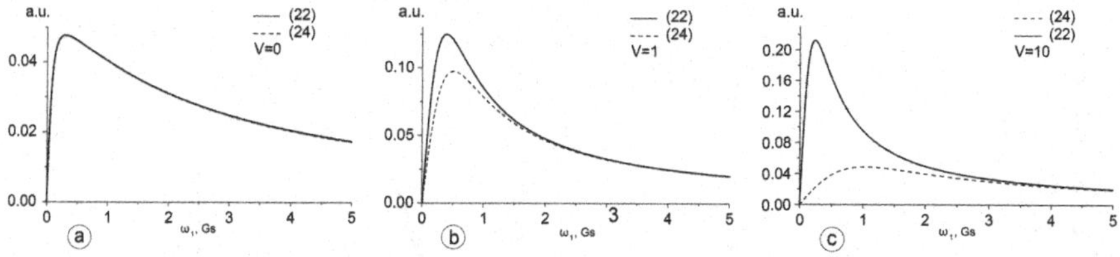

Fig. 7.9 Comparison of the saturation curves given by the Eq. (7.49) plotted with solid lines and by the approximate solution [6] plotted with dashed lines. There were used parameters: $\sigma = 5$ Gs^2, $1/T_1 = 1/T_2 = 0.1$ Gs. (Adapted with permission from Ref. [26])

small values of V with increasing V, the ω_{1max} increases as expected. For large values of V, when $V > \sigma^{1/2}$ the picture changes: in this range of V values, an increase in the rate of spectral diffusion causes a decrease in the microwave field amplitude (Rabi frequency), at which the saturation curve reaches its highest value (see Fig. 7.8a). Note that the approximate result (7.57) for the dependence of ω_{1max} on σ and V was derived for the slow spectral diffusion case when $V < \sigma^{1/2}$. When the spectral diffusion rate increases and $V > \sigma^{1/2}$ non-homogeneously broadened spectrum collapses into one homogeneously broadened line. To this end the saturation curve is given by (7.1) and at $\omega = \omega_0$ the saturation curve has the maximum at $\omega_{1max} = 1/(T_1\ T_2)^{1/2}$.

Thus, in the case of slow spectral diffusion, the spin excitation transfer from spins, which directly absorb microwave energy, to the remaining spins can be a "bottle- neck" of the whole process. The lower the transfer rate of the excitation V, the smaller the microwave field amplitude, which leads to the saturation effect. In the opposite case of fast spectral diffusion, the effect of exchange narrowing of the spectrum appears. This in turn leads to a decrease in ω_{1max} with an increase in V. In the last case the saturation curves behaves similar to ones presented in Fig. 7.1.

Changing of the dependence of ω_{1max} on the rate of spectral diffusion is related to the fact that in the region of slow exchange the spectrum retains some inhomogeneous broadening, while in the area of exchange narrowing (with fast spectral diffusion) the whole spectrum turns into a homogeneously broadened line. However, for the Gaussian contour, the transition from one kind of V dependence to another does not occur abruptly, at exactly the specified rate of spectral diffusion, $V = \sigma^{1/2}$. The center of the spectrum narrows faster than the wings; therefore, a complete collapse of the spectrum occurs in a certain interval of rates of spectral diffusion, the transition occurs smoothly. This is clearly seen in Fig. 7.8a: the maxima of the saturation curves form a "ridge", which represents an arc.

For the model with the Gaussian distribution of resonance frequencies and the spectral diffusion described by Eqs. (7.17), the saturation curve was calculated for the first time in [6]. However, in [6] an approximate solution was obtained. Instead of Eqs. (7.17), approximate Eqs. (7.75) were used. Fig. 7.9 shows the spectral saturation curves calculated by (7.49) and obtained by solving Eqs. (7.75).

The comparison of solid and dashed lines shows that the approximate saturation curves found in [6] deviate essentially from the exact saturation curves (7.49). This is

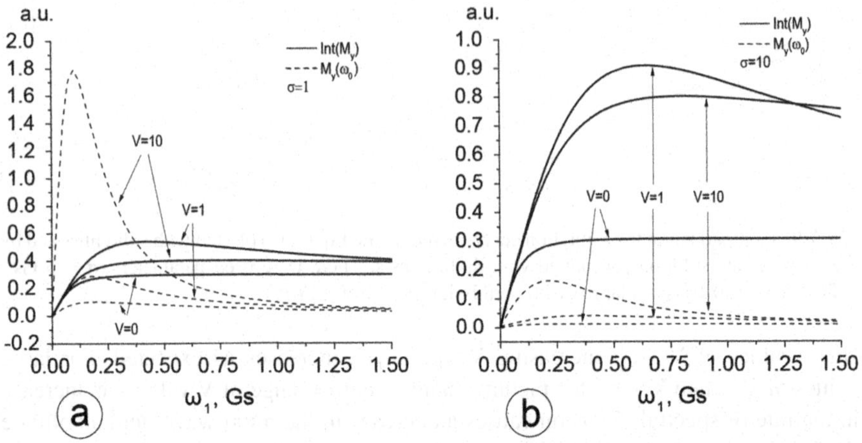

Fig. 7.10 Comparison of two functionals of the EPR spectrum which can serve as a characteristic saturation curves when analyzing the saturation effect of EPR using continuous wave EPR spectra: integral intensity of the spectrum Int(M_y) (see M_y (7.45)) and $M_y(\omega = \omega_0)$ (7.49). For these calculations were used $1/T_1 = 1/T_2 = 0.1$ Gs. (Adapted with permission from Ref. [26])

an expected result since in [6] the spin coherence transfer, which is responsible, e.g., for the exchange narrowing effect, was neglected.

The physics of the saturation effect is directly reflected by the integral intensity of the spectrum $\int_{-\infty}^{\infty} dB_0 M_y(B_0, \omega_1) = I_0(\omega_1)_{theory}$. Figure 7.10 (solid lines) shows the saturation curves for the integral intensity of the EPR spectrum calculated for the two values of the dispersion of the Gaussian distribution and three values of the rate of spectral diffusion by numerically integrating the expression for the spectrum M_y (7.45). For comparison, Fig. 7.10 (dashed lines) presents the saturation curves $M_y(\omega = \omega_0)$ (7.49) calculated for the same set of parameters as for solid lines. In both cases calculations are done following the general solution for M_y given by Eq. (7.45) but to characterize the saturation effect different functionals of the EPR spectrum are chosen: integral intensity of M_y and amplitude of M_y at pumping with the resonant microwave field, $\omega = \omega_0$ (7.49).

The comparison of curves in Fig. 7.10 does not make it possible to give preference to a particular saturation curve. Any of them can be used to compare the calculated saturation curve with the curve found from the experimental data.

The results of the theoretical studies of the manifestations of the saturation effect in the EPR spectra in a system of spins with a Gaussian distribution of resonance frequencies were presented above in the case where spectral diffusion can be described as a random process without correlation. This model can be implemented in many physical systems. A good example is the Doppler broadening of the optical transition frequency of atoms or molecules in gases. Random particle collisions cause spectral diffusion, which can be simulated by a random process without correlation. This model is directly related to the study of spin exchange caused by the exchange interaction between paramagnetic particles in bimolecular collisions.

The effect of equivalent spin exchange on the motion of the spin magnetization vector is fully described by the non-correlated spectral diffusion model.

From a pragmatic point of view, the saturation effect should be studied in order to determine the spin-lattice relaxation time, which cannot be measured by analyzing the shape of the spectra in a linear response situation. Of course, now there are pulse methods for measuring the spin-lattice relaxation time. But for many reasons, this does not mean that it is not necessary to study the saturation effect. Spectral diffusion creates problems also for pulse methods for the determination of the spin-lattice relaxation time.

The saturation effect of stationary spectra is a good resource for measuring the spin-lattice relaxation time, because not every laboratory has pulse spectrometers. The saturation effects should be studied in connection with possible applications of "enlightenment" of the medium in the microwave range. And, of course, the saturation effect is a very interesting physical phenomenon. It is a consequence of the strong interaction of the studied system with a strong external microwave field. As a result, we actually get a new "studied" system composed of the original system and the applied field.

What algorithm should be used to determine the spin-lattice relaxation time from the data on the saturation of the steady-state spectra? The most "simple" way is to model the spectrum numerically. For this purpose, there is an analytical form of the EPR spectrum depending on the magnetic resonance parameters, including the spin-lattice relaxation time (see (7.39)). If all the parameters except for the spin-lattice relaxation time are known, then it is quite feasible to choose the only one unknown parameter. It is more difficult if there are several unknown parameters. In this case, e.g., simple relations are obtained for the magnitude of the microwave field amplitude at which the saturation curve at the pumping in the center of the spectrum reaches its maximum value. They can be used to estimate an unknown parameter and use it as a trial value of the parameter in numerical simulation. There are formulated the conditions under which the saturation effect in a situation where there is spectral diffusion can be analyzed using a well-known theory based on the Portis T_1–T_2 model [2], if the parameters of the Portis model are changed in a certain way.

On the basis of numerous numerical experiments, the possible kinds of manifetation of the microwave field amplitude, of the spectral diffusion rate and of the inhomogeneous broadening on the saturation effect observed were demonstrated. These results can help to understand the experimental data on the saturation effect at a qualitative level and in the selection of parameters in the numerical simulation of spectra.

Thus, by analyzing the data on steady-state saturation one can determine the rate of energy redistribution in a spin system alongside with the spin coherence transfer rate and spin dephasing rate. The advantage of the method of steady-state saturation over the method based on measuring an exchange broadening or shifts of lines in the linear response region is that it allows the exchange rate to be measured at much lower concentrations of paramagnetic species. Indeed, in the region of a slow spin exchange, the transfer of excitation between the spins makes it possible to accelerate the process of relaxation of the states populations difference of those spins whose frequency is equal to or close to the frequency of the microwave field, since as a result of a mutual flip-flop of resonant and non-resonant spins, the excitation leaves the resonant spins and formally non-resonant spins enter as part of the lattice. There were suggested several approximate expressions for effective spin-lattice relaxation in the case of slow spin coherence transfer (see, e,g, [7, 8, 13, 14]). Interesting suggestion was presented by Freed [7, 8]: $T_{1ef} = T_1(1 + K_{ex}C\ T_1\varphi^*)/(1 + K_{ex}C\ T_1)$. For very slow spin exchange this expression is reduced to $1/T_{1ef} = 1/T_1 + K_{ex}C\ (1 - \varphi^*)$. When $K_{ex}C$ increases then the effective spin-lattice relaxation time tends to $T_{1ef} \to T_1\varphi^*$ (φ^* is statistical weight of resonant spins). Note that in the region of slow spin exchange, the effective transverse relaxation time is given by well known approximate expression: $1/T_{2ef} = 1/T_2 + K_{ex}C(1 - \varphi_k)$, the φ_k is the statistical weight of spins in the k-th component.

7.2 Electron-Electron Double Resonance (ELDOR)

The instruments and the different techniques of recording ELDOR spectra were developed by Hyde et al. [31]. A detailed theory of ELDOR has been developed for radicals with one magnetic nucleus having the spin I = 1/2 and I = 1 and successfully applied ELDOR for the first time to study spin exchange between nitroxide radicals in liquids and measured T_1 [31, 32].

There are several ways of conducting ELDOR experiments: steady-state and pulse variants. An ELDOR experiment consists of irradiating one part of the EPR spectrum with a microwave source and monitoring the resulting response at another portion of the spectrum with a second microwave source [31]. For steady-state ELDOR experiments there was designed a bimodal microwave cavity which may be used to irradiate a sample simultaneously at two different microwave frequencies [31]. In the pulse ELDOR technique the recovery of the signal intensity is recorded after the pumping is switched off.

Let us denote the amplitude and carrier frequency of the pump field by B_{1p} and ω_p, and the frequency at which the ELDOR effect is recorded through ω_{ob}. The saturating microwave field (pumping) takes the spins of the system to a non-equilibrium state, to a state with non-equilibrium polarization. On the one hand, the saturation effect is revealed, i.e., the difference between the populations of the states of the electron spins decreases and, therefore, the ability of the spins to absorb microwave field quanta decreases. This means that at the observation frequency ω_{ob} spins acquire non-equilibrium longitudinal polarization $M_z(\omega_{ob})$. On the other hand, the spins acquire transverse magnetization, passing into a quantum coherent state. When pumped at the frequency ω_p, due to the spin excitation transfer also the spins at the observation frequency ω_{ob} acquire a non-equilibrium polarization: the nonequilibrium longitudinal polarization $M_z(\omega_{ob})$ and quantum coherence (transverse magnetization $M_-(\omega_{ob}) = M_x(\omega_{ob}) - iM_y(\omega_{ob})$).

In the pulsed ELDOR experiment, the pump-generated transverse magnetization gives a free induction signal, FID. By measuring in the experiment the FID amplitude, the decay kinetics of the FID, one can find rates of the spin decoherence and the spin coherence transfer. If a microwave pulse is applied at the frequency ω_{ob}, then the signal of the primary spin echo can be observed. The decay kinetics of the envelope of the echo signals with an increase in the delay time of the pulse at the frequency ω_{ob} allows one to find the rates of the spin coherence transfer and spin decoherence. The last protocol of observing the saturation effect means that the echo is detected after only one microwave pulse on the observation frequency ω_{ob}: initial quantum coherent spin state is created by pumping field on spins with frequency ω_p and is transferred via the spin exchange process to spins with a frequency of observation ω_{ob}.

In fact, the presence of a free induction signal may make it difficult to use ELDOR to determine the relaxation times and transfer rates of spin excitation between spins. Thus, methods are requested to suppress the free-induction signal [32]. One possibility for dealing with the problem is separating the signal observed. The free-induction signal does not depend on a power of the observing microwave field while saturation-recovery signal is proportional to the amplitude B_{1ob} of the observing field. Then separating of the saturation-recovery signal can be done by modulating the B_{1p} at low frequency and using phase detection.

Non-equilibrium longitudinal spin polarization M_z (ω_{ob}) can be detected using different protocols. They are based on the fact that the EPR signal from spins with the frequency equal to $\omega_k = \omega_{ob}$ is proportional to M_z (ω_k). For monitoring of relaxation of the M_z (ω_k) to its equilibrium value $M_0\varphi_k$ one can use (see 7.43)

$$\begin{aligned} M_{kz}(t) = {} & M_0(1 - \exp(-t/T_1))\varphi_k + \\ & \exp(-t/T_1)\,(M_z(0)\varphi_k + (M_{kz}(0) - M_z(0)\varphi_k)\exp(-Vt)), \end{aligned} \tag{7.58}$$

$V = K_{ex}C + 2/T_{dd}$ is the spin energy transfer rate between spins (see (4.5) and (4.6)), $1/T_1 = 1/T_{10} + 1/T_{1d}$.

For example, the time resolved steady-state EPR spectrum intensity detected at $\omega_{ob} = \omega_k$ using a weak microwave field with amplitude B_{1ob} can be theoretically described as follows.

A long pumping pulse create a non-equilibrium steady state magnetization of each k-th sub-ensemble of spins. This steady state $\mathbf{M}_{k,stp}$ are given by Eqs. (7.38) and (7.39). When pumping MW field is switched off, further time development of z-components of magnetization vectors are given by Eq. (7.58) where initial values of $M_{kz}(0)$ will be equal to a steady state values of z-components $M_{kz,stp}$, $M_{z,stp}$ in a long pumping pulse.

Then, we have to remove possible contributions of the free induction signal when observing further the EPR spectrum in the observing MW field. One possible option is applying a short pulse of gradient of B_0 field.

To calculate the spectrum in a weak observing MW field we have to solve the equations (5.2) where the term $i\gamma\varphi_k M_0$ should be substituted by the term $i\gamma M_{kz}$. The value of M_{kz} equals

$$M_{kz} = M_0(1 - \exp(-t/T_1))\varphi_k + \exp(-t/T_1)\left(M_{z,stp}(0)\varphi_k + \left(M_{kz,stp}(0) - M_{z,stp}(0)\varphi_k\right)\exp(-Vt)\right).$$

In this case Eqs. (5.5) are non homogeneous linear algebraic equations for the transverse magnetizations M_{k-} which have nonzero independent variables on the right hand side of the equal sign. Solution of this system of linear equations will depend linearly on those independent variables. Thus, each line in EPR spectrum observed by the observing MW field as well as integral intensity of the spectrum will tend to the equilibrium intensity as a sum of three terms

$$J_k = J_{1k} + J_{2k}\exp(-t/T_1) + J_{3k}\exp(-Vt - t/T_1),$$
$$J = \sum_k J_k = J_1 + J_2\exp(-t/T_1) + J_3\exp(-Vt - t/T_1). \quad (7.59)$$

This approach to the ELDOR detecting the saturation effect was formulated and used in series of works (see, e.g., [31–34]). This method can be called as long-pulse saturation recovery experiment.

The important feature of (7.59) is that the time dependence of a signal observed is totally determined only by the redistribution of the energy inside the spin system (V) and further energy exchange with thermostat ($1/T_1$) (see (7.59)).

For any other EPR method the kinetics of recovery to the equilibrium state is described by (5.15) and (7.58). This kinetics makes it possible to determine the spin-lattice relaxation time and the excitation redistribution rate given by $V = K_{ex}C + 2/T_{dd}$ (see (4.5) and (4.6)).

Thus, direct measurements of the spin excitation transfer rate between spectrum components can be made by various two-frequency methods in a pulse and steady-state regimes. In the steady-state experiment one is recording the relative alteration of the monitor signal intensity in response to pumping at another frequency

$$R = 1 - I/I_0, \quad (7.60)$$

where I_0 and I are the amplitudes of the EPR signal without and with pumping, respectively.

In conclusion, it should be noted that though the methods based on spectroscopic manifestations of saturation effect in the strong microwave field and ELDOR methods are rather complicated, they can provide us with information which is inaccessible via simpler techniques based on the analysis of the EPR spectra shapes. The spin exchange transfer rates that can be measured by these more complicated methods are limited by the $1/T_1$ values, while for the methods based on the analysis of the shapes of spectra in the linear response case the limit is given by $1/T_2$. Since often $1/T_1 < 1/T_2$ the methods outlined above in this section permit one in principle to measure smaller values of $K_{ex}C$.

7.3 Saturation Recovery Methods

The kinetics of recovering equilibrium magnetization after a pump pulse produces non-equilibrium magnetization can be monitored using microwave pulses at the same frequency as the pump frequency. It is not necessary to use microwave fields with different frequencies as in ELDOR methods.

Consider a solution of nitroxide radicals. Suppose that the pumping microwave field is converting the electron spins of one of the nitrogen components of the EPR spectrum (denote it as the k-th component) to a state with equal probability for the projections of electron spins m = +1/2 and m = –1/2. Such a state can be created, e.g., by applying the $\pi/2$ pulse of a strong microwave field. The result is a state in which $M_{kz}(0) = 0$. But the $\pi/2$ pulse creates also a non-zero transverse magnetization of the spins. This transverse magnetization can be destroyed by giving a short pulse of the magnetic field gradient. In this situation, the recovery of equilibrium magnetization is described by the equation (see (7.43))

$$\begin{aligned} M_{kz}(t) = M_0(1-\exp(-t/T_1))\varphi_k + \\ \exp(-t/T_1)\, M_z(0)\varphi_k(1-\exp(-Vt)), \end{aligned} \tag{7.61}$$

where V is the rate of the excitation energy redistribution between spins. By measuring this magnetization at different times after switching off the pump pulse, we obtain a recovery curve of the equilibrium magnetization. Fitting curve (7.61) to the experimental curve, it is possible to find the unknown parameters: $M_z(0)$, T_1 and V. If the pump pulse does not rotate all spins in the observed k-component of spectrum and the $M_{kz}(0) \neq 0$, the fit to the experimental curve should be performed using the general solution (7.43) instead of (7.61). Note that an interesting case is when the pump pulse inverts the population of spins in the k-th component, i.e., $M_{kz}(0) = -M_0\,\varphi_k$. In this case, the recovery of the longitudinal magnetization of spins is described by the curve (see (7.43))

$$\begin{aligned} M_{kz}(t) = \varphi_k M_0(1-\exp(-t/T_1)) + \\ \exp(-t/T_1)\,(M_z(0)\varphi_k + (-M_0 - M_z(0))\varphi_k\exp(-Vt)). \end{aligned} \tag{7.62}$$

It can be seen that due to the spin excitation transfer at the rate of V, the population differences of spins are equalized and then this population difference at the rate of $1/T_1$ tends to an equilibrium value. If the pump pulse inverts the population of all spins, we obtain $M_{kz}(0) = -M_0\,\varphi_k$, $M_z(0) = -M_0$. In this case, both the total longitudinal magnetization and the longitudinal magnetization of the spins of the k-th component are described by the same dependence.

$$M_{kz}(t) = \varphi_k M_0(1-\exp(-t/T_1)) - \exp(-t/T_1)\,M_0\varphi_k. \tag{7.63}$$

This equation is the basis of the inversion-recovery method, which is widely used to measure the spin-lattice relaxation time. Longitudinal magnetization can be monitored at different times, e.g., by observing a primary echo signal with a fixed time interval between two microwave pulses that form an echo signal [17, 35–37]. In the work [37] was confirmed that the recovery of longitudinal magnetization is described by the sum of two exponentials at intermediate concentrations of spins while at high concentration of spins it is single exponential. These observations are in accordance with Eq. (7.62).

Long-pulse saturation recovery experiments using the equal frequencies for both, pumping and observing, microwave pulses was implemented in [37], the spin-lattice relaxation times and the spin exchange rates were found for two nitroxide radicals which are widely used in Overhauser dynamic nuclear polarization experiments: Carboxy proxyl (CP, 3-carboxy-2,2,5,5-tetramethyl-1-pirrolidinyloxy) in water and Tempol (TP, 4-hydroxy-2,2,6,6-tetramethylpiperidin-1-oxyl) in toluene. The spin exchange rates and T_1 for these radicals obtained in [37] by the inversion recovery method using pulse sequences π-T- $\pi/2$-τ-π-τ -echo signal, and π-T- $\pi/2$-free induction signal, and long-pulse saturation recovery of the CW EPR spectra were in good agreement.

The above reasoning illustrates that the kinetics of the recovery of equilibrium magnetization after the action of the pump pulse gives information about the rate of redistribution of spin excitation (cross-relaxation) caused by spin exchange and dipole-dipole interaction. In principle, there may be other mechanisms of cross-relaxation [4], but the contributions of spin exchange and dipole-dipole interaction can be distinguished, since they give a contribution to cross-relaxation, which depends on the concentration of spins.

7.4 Electron Spin Echo

The electron spin echo technique can also be used to study spin exchange between paramagnetic particles in solution [17, 36]. For example, the kinetics of the decay of the primary spin echo signal, which is formed by the sequence of two microwave pulses, gives directly the characteristic time of the spin decoherence. The primary spin echo is a direct method to study irreversible relaxation of the spin coherence. The kinetics of decay of the so-called stimulated spin echo, induced by three pulses, provides directly information about the relaxation of the longitudinal magnetization of spins or the relaxation of populations of the spin states with different projection to the external constant magnetic field direction.

The electron spin echo method is a useful complement to other methods of spin exchange study. For example, the analysis of exchange broadening, as well as of the saturation effects or ELDOR are rather time-consuming tasks for systems with unresolved or poorly resolved multicomponent EPR spectra. The electron spin echo method makes it possible to avoid the masking influence of the effects of inhomogeneous broadening and to reveal the processes of irreversible longitudinal and transverse relaxation of spins [17, 18, 36].

If the paramagnetic relaxation of spins in a liquid were described by the phenomenological Bloch equations (see [38]), the decay of the primary spin echo signal would directly give the transverse relaxation time [17, 36–39]

$$V(2\tau) = V(0)\exp(-2\tau/T_2). \tag{7.64}$$

Here τ is a time interval between two pulses forming the primary echo signal. In this case, the decay of the stimulated spin echo signal is described as

$$V(2\tau + T) = V(0)\exp(-2\tau/T_2)\exp(-T/T_1). \tag{7.65}$$

Here T is the time interval between the second and third pulses forming the stimulated spin echo signal.

In sections (5.4 and 5.5) it was shown that the spin coherence transfer caused by the exchange interaction in bimolecular collisions of paramagnetic particles and the dipole-dipole interaction of spins form collective modes of spin coherence motion [20]. Each such mode is defined by the characteristic frequency of the paramagnetic resonance and the "width" of the resonance, i.e., by the characteristic time $T*_{2k}$ of the paramagnetic relaxation of the given mode of motion. In fact, every mode of movement of the spins coherence can be described using equations of the type of the Bloch equations. Therefore, taking into account the spin exchange and the contributions of the dipole-dipole interaction, the spin echo signals of paramagnetic particles in solutions can generally be represented as

$$\begin{aligned} V(2\tau) &= V(0)\sum_k \varphi_k \exp(-2\tau/T*_{2k}); \\ V(2\tau + T) &= V(0)\sum_k \varphi_k \exp(-2\tau/T*_{2k})\, f(T). \end{aligned} \tag{7.66}$$

Function f(T) in (7.66) describes the relaxation of M_{kz} component of magnetization in the T interval. According to the discussion in Sect. 7.3, it is expected that at $T \ll T_1$ f(T) will decay as exp $(-VT-T/T_1)$, while at $T \geq T_1$ f(T) decays as $\exp(-T/T_1)$. Here $V=K_{ex}C + 2/T_{dd}$ (see (7.38)).

In the case of the slow spin coherence transfer, these characteristic times are (see (7.38))

$$1/T*_{2k} = 1/T_2 + K_{ex}C(1 - \varphi_k) + 5/T_{dd} + 4\varphi_k/T_{dd}, \tag{7.67}$$

where φ_k is the statistical weight of spins which contribute to the k-th EPR line in the spectrum.

In the case of the fast spin coherence transfer, the contribution of this process to the echo signal decay is given by [39]

$$V(2\tau) = V(0)\exp(-2\tau/T_{2eff});$$
$$V(2\tau + T) = V(0)\exp(-2\tau/T_{2eff})\exp(-T/T_1). \qquad (7.68)$$

Here T_{2eff} is given by (5.17)

$$1/T_{2eff} \equiv 9/T_{dd} + 2a_N^2/(3(K_{ex}C - 4/T_{dd})) + 1/T_2. \qquad (7.69)$$

As previously noted, the dipole-dipole interaction of spins in solution contributes to the spin-lattice relaxation of the total longitudinal magnetization of spins (see (3.28 and 3.31)). If this contribution is taken into account, Eqs. (7.64), (7.65), (7.66), (7.67), (7.68) and (7.69) should use $1/T_1 = 1/T_{10} + 1/T_{1d}$ (see (4.5)).

To illustrate the manifestation of spin exchange in the decay of spin echo signals, let us consider an example of a solution of paramagnetic particles whose EPR spectrum consists of two components with splitting *a* due to the hyperfine interaction of an unpaired electron with one magnetic nucleus with spin ½, e.g., ^{15}N or ^{1}H. Kinetic equations for magnetization of spins relevant to this model situation are given by Eqs. (7.17). Note that in the time intervals between microwave pulses, which form the echo signals, in (7.17) terms proportional to ω_1 (i.e., B_1) should be omitted. Echo signals for this model situation were studied in Refs. [40–42].

Kinetic equations for this model situation taking into account both the exchange and dipole-dipole interactions are (see (7.17))

$$\partial M_{1-}/\partial t = -i(\omega_0 + a/2 - \omega)M_{1-} - M_{1-}/T_2 - V_{sct}M_{1-} + V_{sct}M_{2-};$$
$$\partial M_{2-}/\partial t = -i(\omega_0 - a/2 - \omega)M_{2-} - M_{2-}/T_2 - V_{sct}M_{2-} + V_{sct}M_{1-}; \qquad (7.70)$$

$$\partial M_{1z}/\partial t = -W_1M_{1z} - V_{pt}M_{1z} + V_{pt}M_{2z} + (1/2)\,W_1M_0;$$
$$\partial M_{2z}/\partial t = -W_1M_{2z} - V_{pt}M_{2z} + V_{pt}M_{1z} + (1/2)\,W_1M_0.$$

Here are introduced the notations:

$$1/T_2 = 1/T_{20} + 9/T_{dd},\ W_0 = 1/T_{10}; W_1 = W_0 + 1/T_{1d},\ V_{pt} = (1/2)\,K_{ex}C + 1/T_{dd};$$
$$V_{sct} = (1/2)(K_{ex}C - 4/T_{dd}),$$

T_{10} and T_{20} are times of the longitudinal and transverse magnetizations relaxation of isolated spins, respectively.

Suppose that the microwave pulses have the frequency $\omega = \omega_0 + a/2$ and rotate only the magnetization vector $\mathbf{M}_1$ and do not rotate $\mathbf{M}_2$ vector. In equilibrium $\mathbf{M}_k = (0, 0, M_0/2)$, k = 1, 2. Thus after the first $(\pi/2)_x$-pulse $\mathbf{M}_1 = (0, M_0/2, 0)$ while $\mathbf{M}_2 = (0, 0, M_0/2)$. Then solving Eqs. (7.70) one finds $\mathbf{M}_k(\tau) = (M_{kx}(\tau), M_{ky}(\tau), M_{kz}(\tau))$. At the moment $t = \tau$ the second pulse, π-pulse, is turning the $\mathbf{M}_1$ vector around x-axis so that $\mathbf{M}_1(\tau + 0) = (M_{1x}(\tau), -M_{1y}(\tau), -M_{1z}(\tau))$, $\mathbf{M}_2(\tau + 0) = \mathbf{M}_2(\tau)$. Other way, the π_x-pulse converts M_{1-} to its complex conjugate value and inverts M_{1z} values. Solving Eqs. (7.70) with the initial values $\mathbf{M}_k(\tau + 0)$, one finds the spin echo signals.

The primary echo is created according to the pulse sequence protocol (θ_1–τ–θ_2–τ ($v(2\tau)$ echo)) and stimulated echo protocol is (θ_1–τ–θ_2–T–θ_3–τ ($v(2\tau + T)$ echo)), where θ_k are the angles through which spins are turned by the corresponding microwave pulses which form echo signals.

Kinetics of the spin echo signal depend on a pattern of excitations of spins by the microwave pulses. In the case when the microwave pulses rotate both magnetization vectors $\mathbf{M}_1$ and M_2 one has (see [17], pp. 212–222, [40, 41])

$$
\begin{aligned}
&v(2\tau) = v(0)\exp(-2\tau(1/T_2 + V_{sct}))\left[\left(\cosh(R\tau) + (V_{sct}/(R)\sinh(R\tau)\right)^2\right. \\
&\left.+(a^2/(4R^2))\sinh^2(R\tau)\right]; v(2\tau + T) = v(0)\exp(-2\tau(1/T_2 + V_{sct}))\,\exp(-W_1T) \\
&\left[(\cosh(R\tau) + (V_{sct}/(R))\sinh(R\tau))^2 + \exp(-V_{sct}T)(a^2/(4R^2))\sinh^2(R\tau)\right],
\end{aligned}
\tag{7.71}
$$

where $R^2 = V_{sct}{}^2 - a^2/4$.

According to (7.71), in the case of slow spin coherence transfer, when $2|V_{sct}| < |a|$,

$$
\begin{aligned}
&v(2\tau) = v(0)\exp(-2\tau(1/T_2 + V_{sct}))[1 + (V_{sct}/a)\sin(a\tau)]; \\
&v(2\tau + T) = v(0)\exp(-2\tau(1/T_2 + V_{sct}))\,\exp(-W_1T)\,[\cos^2(a\tau/2)+ \\
&+(V_{sct}/a)\sin(a\tau) + \exp(-V_{sct}T)\,\sin^2(a\tau/2)];
\end{aligned}
\tag{7.72}
$$

$1/T_2 + V_{sct} = 1/T_{20} + 9/T_{dd} + (1/2)(K_{ex}C - 4/T_{dd}) = 1/T_{20} + \mathbf{7/T_{dd}} + \mathbf{(1/2)K_{ex}C}$; $W_1 = 1/T_{10} + 1/T_{1d}$.

Thus, in the slow coherence transfer case the transverse relaxation time is reduced by the spin decoherence. This corresponds exactly to the homogeneous broadening of resonance lines in the case under consideration.

It is somewhat unexpected that the echo signal decay is modulated. It would be interesting to observe this effect of ESEEM in the experiment. This effect is of interest in itself, as it results from the spin coherence transfer between spins during random bimolecular collisions of spin probes. On the one hand, random collisions of particles. On the other hand, regular oscillations. It is remarkable that the modulation amplitude (depth) is V_{sct}/a. If it were possible to measure the modulation depth of the decay curve of the spin echo signal, it would be possible to determine directly the rate of spin coherence transfer V_{sct}. It is very interesting that the amplitude of the modulation of the spin echo signal decay can directly give a parameter p which determines a contribution of dispersion mode to the observed in the experiment line of the continuous wave (CW) EPR spectrum in linear response regime (cf. (5.45)).

According to (7.71), in the case of fast spin coherence transfer, when $2|V_{sct}| > |a|$,

$$
\begin{aligned}
&v(2\tau) = v(0)\exp\left(-2\tau\left(9/T_{dd} + a^2/(8|V_{sct}|)\right)\right); \\
&v(2\tau + T) = v(0)\exp\left(-2\tau\left(9/T_{dd} + a^2/(8|V_{sct}|\right)\,\exp(-W_1T).
\end{aligned}
\tag{7.73}
$$

The rate of the spin echo signal decay in the limit of fast coherence transfer is determined practically only by the dipole-dipole interaction contribution to the spin dephasing.

The study of the decay kinetics of electron spin echo signals at different concentrations and temperatures allows us to determine the rate of bimolecular spin exchange. In principle, this method has certain advantages in comparison with the method of CW EPR spectra shape analysis. The fact is that the phenomenon of spin echo allows in many cases to avoid masking effects of inhomogeneous broadening of EPR spectra. For example, the inhomogeneous broadening of the nitrogen components of the EPR spectrum of nitroxide radicals, caused by the hyperfine interaction of an unpaired electron with proton nuclei, significantly complicates the extraction of the contribution of spin exchange to the broadening of the nitrogen components of the EPR spectrum at low concentrations of radicals, when the spin exchange is relatively slow and there is no collapse of the nitrogen components of the spectrum in homogeneously broadened resonance lines.

However, spin echo techniques also face problems. The spin echo does not completely eliminate the effects of inhomogeneous resonance broadening if there is spectral diffusion and cross-relaxation. It has already been noted above that in the case of slow spin exchange, the decay of spin echo signals is described as the sum in which each term decreases with its characteristic time (see (7.66)). The observed spin echo signal depends very significantly on the pattern of the spin excitation by the microwave field pulses that form the signal (see, e.g., [42, 43]. Usually the total inhomogeneous width of EPR spectrum (measured in magnetic field induction units) is much larger than the amplitude B_1 of the fields used in the EPR experiments. This means that the experiment is accomplished by selective resonant frequency excitation of the spins.

The results of numerous works (see e.g., [35, 39, 44–46]) confirmed that spin echo techniques can be successfully applied to measure the spin exchange rate. The main result of these works is that the spin exchange rate found from the electron spin echo data coincides reasonably well with that found by analyzing the shape of the CW EPR spectrum.

7.5 Dynamic Nuclear Polarization

One of the remarkable phenomena in the magnetic resonance is the orientation of nuclear spins during microwave pumping of EPR transitions of paramagnetic particles (see, e.g., [47–49]). The essence of this phenomenon is that the pumping of EPR transitions creates a nonequilibrium magnetization of electron spins. Paramagnetic relaxation processes restore the equilibrium magnetization. The hyperfine interaction of unpaired electrons with magnetic nuclei often contributes to this relaxation [3, 5, 10]. Therefore, in the course of restoring the equilibrium magnetization of electron spins, in turn, the nuclear spins acquire a non-equilibrium magnetization (non-equilibrium spin polarization), which in magnitude can far exceed the equilibrium magnetization of nuclear spins. As a result of such dynamic polarization of nuclear spins (DNP), the intensity of nuclear magnetic resonance (NMR) signals increases and the sensitivity of the NMR method increases significantly. Therefore, the study of the DNP effect and the search for optimal conditions to increase this effect are of great interest, including the increase in the sensitivity of magnetic resonance imaging and production of targets with polarized nuclei.

The effect of the DNP is proportional to the EPR saturation parameter. In turn, the saturation effect depends on spin exchange (see, e.g., [50–54]).

In strong microwave fields, the magnetization of electron spins can change significantly. We have already noted that in inversion-recovery experiments a short microwave field pulse changes the direction of the magnetization vector **M** to opposite. Note that in liquids at equilibrium, $\mathbf{M}_0$ is directed along a constant magnetic field B_0, so the π pulse puts the system into a state with inverted populations of spin states, so that the magnetization becomes equal to $\mathbf{M} = \{0,0,-M_0\}$. A significant deviation of the magnetization from the equilibrium can also occur under the action of moderately strong microwave fields, if this field is applied long enough. In this situation, the saturation effect and the component of the magnetization vector on the directions of the external magnetic field may also change significantly. The saturation effect can be characterized by the saturation parameter (see (7.6) and (7.6′)) $s = (M_0 - M_z)/M_0$. For example, in the situation of the linear response of the system to the microwave field $M_z \approx M_0$ and we get almost zero saturation parameter. In this situation, the DNP effect is not expected.

A strong microwave pumping field tends to align the populations of the spin states with different projections on the direction of the B_0 field, i.e., seeks to produce a state with $M_z = 0$. In this situation, the saturation parameter $s = 1$.Thus, one of the optimal conditions for increasing the effect of the DNP is to implement the greatest

$$s = (M_0 - M_z)/M_0. \tag{7.74}$$

For illustration, the results of calculations of the saturation parameter for a model paramagnetic particle with one unpaired electron ($S = 1/2$), whose hyperfine interaction with a single magnetic nucleus with spin $I = 1/2$ splits the energy levels of electron spins, are presented below. This is, e.g., when the unpaired electrons interact with a ^1H or ^{15}N atom. The manifestations of the saturation effect in the shape of the EPR spectrum for this model system were discussed in detail in Sect. (7.1.2). The expression for the longitudinal magnetization M_z (7.19) for an arbitrary microwave field power was also derived there. Using (7.19), it is possible to calculate the saturation parameter. Some results are presented in Figs. (7.11 and 7.12).

Shown in Fig. 7.11, the curves demonstrate that at a given spin exchange rate and a fixed frequency of the microwave field, the saturation parameter first increases, and then tends almost to a plateau with increasing microwave field amplitude (the Rabi frequency). This could be expected on the basis of known results on the saturation effect in the absence of the spin exchange (see Eqs. (7.6) and (7.6′)).

The dependence of the saturation parameter on the spin exchange rate demonstrates a more complex behavior (see Fig. 7.12). When pumping at the resonant frequencies of isolated particles, the saturation parameter passes through a maximum

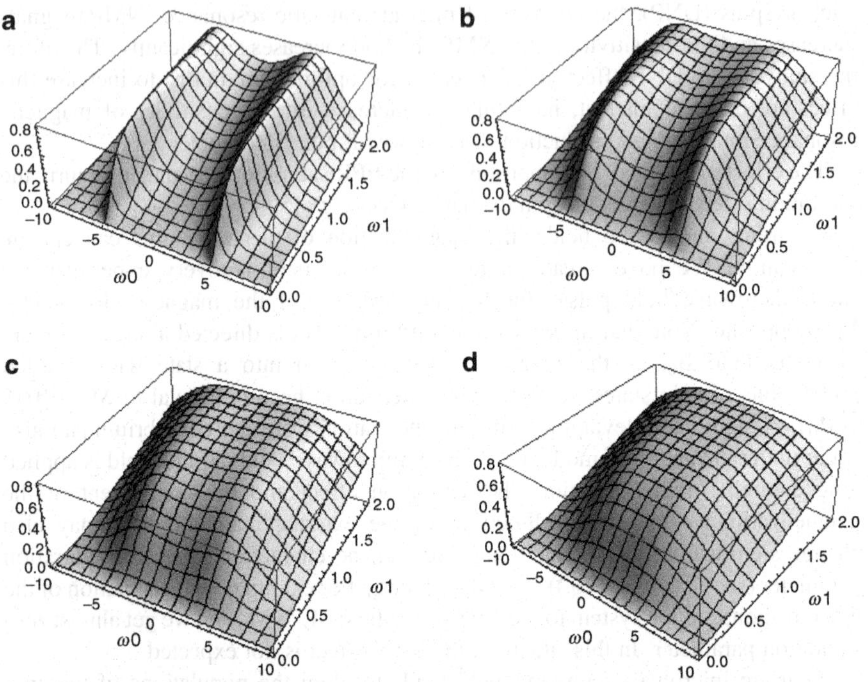

Fig. 7.11 Simulation of the saturation parameter *s* for the two-frequency model system. The curves show the dependence of s on the Zeeman frequency of the electron spins $\omega_0 = (g\beta/\hbar)B_0$ and Rabi frequency $\omega_1 = (g\beta/\hbar)B_1$. Other parameters used in these calculations are: the hyperfine interaction constant $a = 10$ Gs; the spin-lattice relaxation rate $1/T_1 = 0.05$ Gs; spin decoherence rate $1/T_2 = 0.1$ Gs; the spin exchange rate V is 0.5 Gs (**a**), 2 Gs (**b**), 5 Gs (**c**), 7 Gs (**d**). All frequencies and rates are given in the magnetic induction units. To transform them to 1/s units one has to multiply on $1.76\ 10^7$

with increasing spin exchange rate (see Fig. 7.12a, b). There are several mechanisms for interpreting the spin exchange effect on spin system saturation.

With the increase in the spin exchange rate, the effective decoherence (T_2) and spin-lattice relaxation (T_1) times become shorter. Therefore, the characteristic saturation factor $\omega_1{}^2\ T_1T_2$ decreases with the increase in the spin exchange rate V, and the saturation effect becomes less pronounced.

As the spin exchange rate increases, the resonance frequencies change [14, 25]; therefore, at a fixed pump frequency, with increasing V, the spins are not in resonance with the microwave field and because of this the saturation parameter decreases (cf. Eq. (7.1)). When pumping in the center of the spectrum, the saturation parameter increases with increasing V for the model system under consideration (see Fig. 7.12c, d). At low V at this pump frequency, there are no resonant spins and the saturation is negligible. With increasing V, exchange narrowing of the spectrum occurs, the spins are in resonance with the pump field, and the saturation parameter increases with increasing spin-exchange rate (see. Fig. 7.12c, d). Note that in reality, the shift of the resonant frequencies and the collapse of the spectrum into one homogeneously broadened line is due to the spin coherence tranfer caused by the exchange and dipole-dipole interactions.

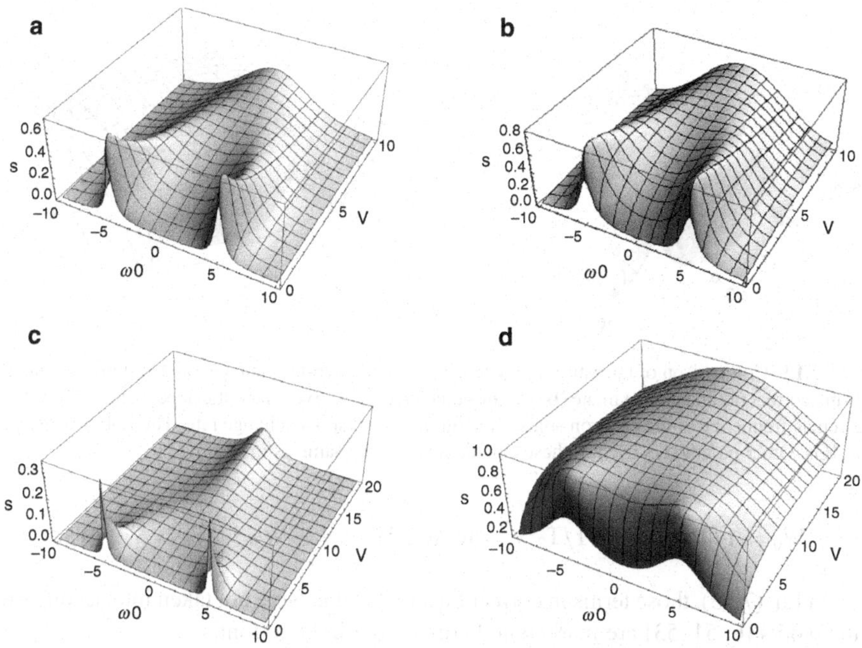

Fig. 7.12. Simulation of the saturation parameter *s* for the two-frequency model system. The curves show the dependence of s on the Zeeman frequency of the electron spins $\omega_0 = (g\beta/\hbar)B_0$ and spin exchange rate. Other parameters used in these calculations are the same as in Fig. 7.11: the hyperfine interaction constant $a = 10$ Gs; the spin-lattice relaxation rate $1/T_1 = 0.05$ Gs; the Rabi frequency is 0.3 Gs (**a**), 0.6 Gs (**b**), 0.1 Gs (**c**) and 2 Gs (**d**)

At a certain solution viscosity, the total contribution of these interactions to the spin coherence transfer rate can vanish. In this case, the saturation parameter will increase and tends to $s = 1$ with increasing microwave field amplitude B_1 (see Eq. (7.6)).

In the theory of EPR spectra under conditions of the linear response, the spin coherence transfer caused by the spin exchange has always been the focus of attention of researchers. However, when studying the saturation effect, the "reaction" effect during the spin coherence transfer was disregarded [6, 51–53]. This means that instead of Eqs. (7.17), the following equations were used

$$
\begin{aligned}
\partial M_{1x}/\partial t &= -W\,M_{1x} - (\omega_0 + a/2 - \omega)\,M_{1y} + (1/2)\,V\,(M_{1x} + \mathbf{M_{2x}}),\\
\partial M_{1y}/\partial t &= (\omega_0 + a/2 - \omega)\,M_{1x} - W\,M_{1y} + \omega_1 M_{1z} + (1/2)V(M_{1y} + \mathbf{M_{2y}}),\\
\partial M_{1z}/\partial t &= -\omega_1 M_{1y} - W_1 M_{1z} + (1/2)\,V\,(M_{1z} + M_{2z}) + (1/2)\,W_0 M_0,\\
\partial M_{2x}/\partial t &= -W\,M_{2x} - (\omega_0 - a/2 - \omega)\,M_{2y} + (1/2)\,V\,(\mathbf{M_{1x}} + M_{2x}),\\
\partial M_{2y}/\partial t &= (\omega_0 - a/2 - \omega)\,M_{2x} - W\,M_{2y} + \omega_1 M_{2z} + (1/2)\,V\,(\mathbf{M1y} + M_{2y}),\\
\partial M_{2z}/\partial t &= -\omega_1 M_{2y} - W_1 M_{2z} + (1/2)\,V(M_{1z} + M_{2z}) + (1/2)\,W_0 M_0.
\end{aligned}
\tag{7.75}
$$

Here M_0 is the equilibrium magnetization value,

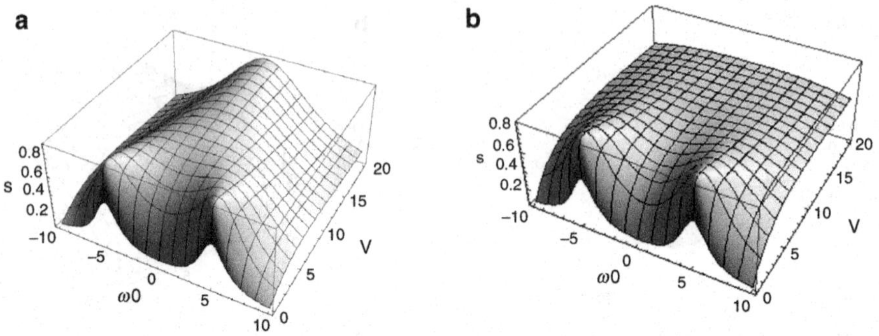

Fig. 7.13 Comparison of the saturation parameter *s* for the two-frequency model system calculated using correct (**a**) and approximate (**b**) kinetic equations. The curves show the dependence of s on the Zeeman frequency of the electron spins $\omega_0 = (g\beta/\hbar)B_0$ and spin exchange rate. The Rabi frequency is 1Gs. Other parameters used in these calculations are the same as in Fig. 7.11

$$W_0 = 1/T_1;\ W_{00} = 1/T_2; V = KexC;\ W = V + W_{00};\ \ W_1 = V + W_0.$$

In Eq. (7.75), those terms in correct Eqs. (7.17) that were not taken into account in the works [6, 51–53] are marked and crossed out in large font.

Figure 7.13 shows the dependences of the saturation parameter calculated using the kinetic Eqs. (7.17) and "approximate" Eqs. (7.75) for the same parameters. It can be seen that the saturation parameter can be significantly underestimated in the approximate calculations. Such an observation was also made in [26] in connection with the saturation effect in the situation when the resonance frequencies of the spins have a Gaussian distribution.

The enhancement of the NMR signal due to the DNP effect is proportional to the saturation parameter of the electron spins in the microwave field. Therefore, in principle, the DNP-detected saturation parameter of electron spins in a microwave field can also be used to determine the spin exchange rate [50]. This approach may become a working method for determining the spin exchange rate, but so far theoretically it has been developed less than methods based on the analysis of the shape of the EPR spectra.

References

1. Bloembergen, N., Purcell, E.M., Pound, R.V.: Relaxation effects in nuclear magnetic absorption. Phys. Rev. **73**, 679–712 (1948)
2. Portis, A.M.: Electronic structure of F centers: saturation of the electron spin resonance. Phys. Rev. **91**, 1071–1078 (1953)
3. Castner, T.G.: Saturation of the paramagnetic resonance of a V center. Phys. Rev. **115**, 1506–1515 (1959)
4. Bloembergen, N., Shapiro, S., Pershan, I.S., Artiiani, J.O.: Cross-relaxation in spin systems. Phys. Rev. **114**, 445–459 (1959)

5. Abragam, A.: The principles of nuclear magnetism. Clarendon Press, Oxford (1961)
6. Wolf, E.L.: Diffusion effects in the inhomogeneously broadened case: high-temperature saturation of the F-center electron spin resonance. Phys. Rev. **142**, 555–569 (1966)
7. Freed, J.H.: Theory of saturation and double-resonance effects in ESR spectra. J. Chem. Phys. **43**, 2312–2332 (1965)
8. Freed, J.H.: Theory of saturation and double resonance effects in electron spin resonance spectra 2. Exchange vs dipolar mechanisms. J. Phys. Chem. **71**, 38–51 (1967)
9. Bloembergen, N., Wang, S.: Relaxation effects in para- and ferromagnetic resonance. Phys. Rev. **93**, 72 (1954)
10. Abragam, A., Bleaney, B.: Electron paramagnetic resonance on transitions ions. Clarendon Press, Oxford (1970)
11. Kivelson, D.J.: Theory of ESR line widths of free radicals. J. Chem. Phys. **33**, 1094–1106 (1960)
12. Currin, J.D.: Theory of exchange relaxation of hyperfine structure in electron spin resonance. Phys. Rev. **126**, 1995–2001 (1962)
13. Zamaraev, K.I., Molin, Y.N., Salikhov, K.M.: Spin exchange, p. 317. Nauka, Sibirian branch, Novosibirsk (1977)
14. Molin, Y.N., Salikhov, K.M., Zamaraev, K.I., Exchange, S.: Principles and applications in chemistry and biology. Springer, Berlin/Heidelberg/NY (1980)
15. Lebedev, Y.S., Dobryakov, S.N.: Analysis of the EPR spectra of free radicals. Zhur. Structur. Khim. **8**, 838–853 (1967). in Russian
16. Milov, A.D., Salikhov, K.M., Tsvetkov, Y.D.: Spin diffusion and kinetics of spin-lattice relaxation of hydrogen atoms in glass-like matrices at 77K. Fizikatverdogotela. **14**, 2259–2264 (1972)
17. Salikhov, K.M., Semenov, A.G., Tsvetkov, Y.D.: Electron spin echo and its application, p. 342. Nauka, Sibirianbranch, Novosibirsk (1976)
18. Salikhov, K.M., Tsvetkov, Y.D.: Spin-spin interactions in solids as studied using electron spin echo method. In: Kevan, L., Schwartz, R. (eds.) Time-domain ESR spectroscopy. Wiley, New York (1979)
19. Salikhov, K.M., Bakirov, M.M., Galeev, R.T.: Detailed analysis of manifestations of the spin coherence transfer in EPR spectra of ^{14}N nitroxide free radicals in non-viscous liquids. Appl. Magn. Reson. **47**, 1095–1122 (2016)
20. Salikhov, K.M.: Consistent paradigm of the spectra decomposition into independent resonance lines. Appl. Magn. Reson. **47**, 1207–1228 (2016)
21. Bales, B.L., Bakirov, M.M., Galeev, R.T., Kirilyuk, I.A., Kokorin, A.I., Salikhov, K.M.: The current state of measuring bimolecular spin exchange rates by the EPR spectral manifestations of the exchange and dipole-dipole interactions in dilute solutions of nitroxide free radicals with proton hyperfine structure. Appl. Magn. Reson. **48**, 1399–1447 (2017)
22. Bales, B.L.: Berliner, L.J., Reuben, J. (eds.) Biological magnetic resonance, vol. 8, pp. 77–130. Plenum Publishing Corporation, New York (1989)
23. Eastman, M.P., Kooser, R.G., Das, M.P., Freed, J.H.: Studies of Heisenberg spin exchange in ESR spectra 1. Linewidth and saturation effects. J. Chem. Phys. **51**, 2690–2709 (1969)
24. Freed, J.H., Fraenkel, G.K.: Theory of linewidths in electron spin resonance spectra. J. Chem. Phys. **39**, 326–348 (1963)
25. Salikhov, K.M.: Peculiar features of the spectrum saturation effect when the spectral diffusion operates: system with two frequencies. Appl. Magn. Reson. **49**, 1417–1430 (2018)
26. Salikhov, K.M., Khairuzhdinov, I.T.: Theoretical investigation of the effect of saturation of the EPR spectrum taking into account spectral diffusion in a system with a Gaussian distribution of the resonance frequencies of spins. ZhETP. **155**, 806–823 (2019)
27. Bakirov, M.M., Salikhov, K.M., Peric, M., Schwartz, R.N., Bales, B.L., Simple, A.: Accurate method to determine the effective value of the magnetic induction of the microwave field from the continuous saturation of EPR spectra of Fremy's Salt solutions. Representative values of T_1. Appl. Magn. Reson. **50**, 919–942 (2019)
28. Salikhov, K.M.: Contributions of exchange and dipole–dipole interactions to the shape of EPR spectra of free radicals in diluted solutions. Appl. Magn. Reson. **38**, 237–256 (2010)

29. Bloch, F., Siegert, A.: Magnetic resonance for nonrotating fields. Phys. Rev. **57**, 522–527 (1940)
30. Dalibard, J., Cohen-Tannoudji, C.: Dressed-atom approach to atomic motion in laser light: the dipole force revisited. J. Opt. Soc. Am. B. **2**, 1707–1720 (1985)
31. Hyde, J.S., Chien, J.C.W., Freed, J.Y.: Electron-electron double resonance of free radicals in solution. J. Chem. Phys. **48**, 4211–4225 (1968)
32. Huisjen, M., Hyde, J.S.: Saturation recoverymeasurements of electron spin-lattice relaxation times of free radicals in solution. J. Chem. Phys. **60**, 1682 (1974)
33. Stunzhas, P.A., Bendersky, V.A., Blumenfeld, L.A., Sokolov, Y.A.: Double electron –electron resonance of triplet excitons I. Opt. Spectrosc. **XXVIII**, 278–283 (1970)
34. Stunzhas, P.A., Bendersky, V.A.: DEER for two frequencies spectrum considering spin exchange with sudden collisions model. Opt. Spectrosc. **30**, 1041–1046 (1971)
35. Braendle, R., Krueger, G.J., Mueller-Warmuth, W.: Impulsspektroskopischeuntersuchungen der Elektronenspinrelaxation in freienRadikalen. Z. Naturforsch. **25a**, 1 (1970)
36. Schweiger, A., Jeschke, G.: Principles of pulse electron paramagnetic resonance. Oxford university press, Oxford (2001)
37. Biller, J.R., McPeak, J.E., Eaton, S.S., Eaton, G.R.: Measurement of T_{1e}, T_{1N}, T_{1HE}, T_{2e} and T_{2H} by pulse EPR at X-band fornitroxides at concentrations relevant to solution DNP. Appl. Magn. Reson. **49**, 1235–1251 (2018)
38. Mims, W.B.: Phase memory in electron spin echoes, lattice relaxation effects in $CaWO_4$: Er, Ce, Mn. Phys. Rev. **168**, 370–389 (1968)
39. Milov, A.D., Salikhov, K.M., Tsvetkov, Y.D.: Electron spin echo studies of magnetic relaxation in liquids: solutions of 2, 4, 6-tri-t-butylphenoxyl. Chem.Phys. Lett. **8**, 523–526 (1971)
40. Zhidomirov, G.M., Salikhov, K.M.: On the theory of spectral diffusion in magnetically diluted solids. Zhur. Experim. Teoret. Fiz. **56**, 1933–1939 (1969)
41. Salikhov, K.M., Dzuba, S.A., Raitsimring, A.M.: The theory of electron spin echo signal decay resulting from dipole-dipole interactions between paramagnetic centers in solids. J. Magn. Reson. **42**, 255–276 (1981)
42. Salikhov, K.M., Khairuzhdinov, I.T., Zaripov, R.B.: Three pulse ELDOR theory revisited. Appl. Magn. Reson. **45**, 573–619 (2014)
43. Salikhov, K.M., Khairuzhdinov, I.T.: Four-pulse ELDOR theory of the spin ½ label pairs extended to overlapping EPR spectra and to overlapping pump and observer excitation bands. Appl. Magn. Reson. **46**(57–83), (2015)
44. Schwartz, R.N., Jones, L.L., Bowman, M.K.: Electron spin echo studies of nitroxide free radicals in liquids. J. Phys. Chem. **83**, 3429–3434 (1979)
45. Stillman, A.E., Schwartz, R.N.: Study of dynamical processes in liquids by electron spin echo spectroskopy. J. Phys. Chem. **85**, 3031–3040 (1981)
46. Collauto, A., Barbon, A., Brustolon, M.: First determination of the spin relaxation properties of a nitronyl nitroxide in solution by electron spin echoes at X band: a comparison with tempone. J. Magn. Res. **223**, 180–185 (2012)
47. Overhauser, A.: Psrsmagnetic relaxation in metals. Phys. Rev. **89**, 689–699 (1953)
48. Overhauser, A.: Polarization of nuclei in metals. Phys. Rev. **92**, 411 (1953)
49. Solomon, I.: Relaxation processes in a system of two spins. Phys.Rev. **99**, 559–565 (1955)
50. Bates, R.D., Drozdoski, W.S.: Use of nitroxide spin labels in studies of solvent-solute interactions. J. Chem. Phys. **67**, 4038 (1977)
51. Armstrong, B.D., Han, S.: A new model for Overhauser enhanced nuclear magnetic resonance using nitroxide radicals. J. Chem. Phys. **127**, 104508-(1-10) (2007)
52. Prandolini, M.J., Denysenko, V.P., Gafurov, M., Lyubenova, S., Endeward, B., Bennati, M., Prisner, T.E.: First DNP results from a liquid water-tempol sample at 400 Mhz and 260 GHz. Appl. Magn. Reson. **34**, 399–407 (2008)
53. Sezer, D., Gafurov, M., Prandolini, M.J., Denysenkov, V.P., Prisner, T.F.: Dynamic nuclear polarization of water by a nitroxide radical: rigorous treatment of the electron spin saturation and comparison with experiments at 9.2 Tesla. Phys. Chem. Chem. Phys. **11**, 6638–6653 (2009)
54. Tuerke, M.T., Bennati, M.: Comparison of Overhauser DNP at 0.34 and 3.4 T with Fremy's salt. Appl. Magn. Reson. **43**, 129–138 (2012)

Chapter 8
In Conclusion: Brief Summary of New vs Existing Paradigm of Spin Exchange

Abstract The differences between the existing and new paradigms in three projections are briefly formulated: In calculations of the efficiency of spin exchange in one elementary act; In the manifestations of the transfer of spin coherence in the form of EPR spectra; The protocol for determining the spin exchange rate from EPR spectroscopy data.

The new paradigm of spin exchange is fundamentally different from the existing paradigm because:

1. There is different description of the dynamics of the spins of the unpaired electrons of paramagnetic particles in solutions.
2. The description of the shape of EPR spectra differs.
3. The method of finding by EPR spectroscopy the rates of changes in the spin state of unpaired electrons of paramagnetic particles caused by their exchange interaction in bimolecular collisions is different.

Note that in the discussed system of paramagnetic particles in the solution there are two spin-spin interactions: exchange and dipole-dipole. Both interactions change the state of the spins of the particles. The goal is to obtain the contribution of exchange interaction in bimolecular collisions of particles to changes in the state of spins.

Unlike the exchange interaction, the dipole-dipole interaction is long-range, and therefore its contribution to spin dynamics cannot be described in terms of bimolecular collisions. Sometimes exchange and dipole-dipole interactions can cause virtually identical changes in the state of spins. For example, they can cause a mutual flip-flop of spins.

In this publication, terminology like spin decoherence, spin coherence transfer, and spin excitation energy transfer is used for both types of interactions. As a generic term spin exchange is used to describe any changes in the state of electron spins associated with the exchange interaction of particles in bimolecular collisions.

K. M. Salikhov, *Fundamentals of Spin Exchange*,
https://doi.org/10.1007/978-3-030-26822-0_8

8.1 The Dynamics of the Spins of the Unpaired Electrons of Paramagnetic Particles in Solutions

8.1.1 An Elementary Act of Spin Exchange in Bimolecular Collisions. Minor Interactions with Major Consequences

Existing paradigm assumes that the elementary act of changing the state of spins of unpaired electrons of particles in their bimolecular collisions is caused only by the Heisenberg interaction. This means that during the elementary act of "spin exchange" the total projection of the spins of colliding particles on any axis is preserved. This situation can be called equivalent spin exchange [1, 2].

New paradigm introduces non-equivalent spin exchange, along with an equivalent exchange. New paradigm also recognises at least three different types of changes in the state of spins, i.e. three types of elementary acts in which the spins of unpaired electrons participate:

1. decoherence of spins,
2. transfer of spin coherence between spins
3. transfer of spin excitation energy.

The existing paradigm operates with just equivalent spin exchange, the rate constant of all three processes can be expressed in terms of one common rate constant (see, for example. (2.56 and 4.5)).

In the case of nonequivalent spin exchange, each of these three processes is characterized by its effective radius, or its rate constant of the bimolecular process (see, for example, (4.4 and 4.6)).

Since new paradigm takes into account three spin processes it requires three rate constants of the bimolecular processes, respectively; instead of one rate constant in the existing paradigm. That allows the new paradigm to more adequately describe the spin exchange in real systems.

Let us imagine that the analysis of experimental data of EPR spectroscopy gives different values for the spin decoherence rate and the spin coherence transfer rate. Staying within the existing paradigm, this observation would be impossible to explain. The new paradigm allows consideration of the possibility that different elementary processes of spin exchange have different rates.

How can spin exchange become nonequivalent?

This case takes place if spin-dependent interactions of isolated paramagnetic particles have to be taken into account at the moment of particles collision, e.g., the hyperfine interaction of unpaired electrons with magnetic nuclei (HFI) or the energy levels splitting in a zero magnetic field for paramagnetic particles that have more than one unpaired electron (ZFS).

Typically, these interactions are several orders of magnitude smaller than the exchange interaction between the particles when they collide. Therefore, it may seem that neglecting, for example, the HFI and the ZFS at the time of collision of the particles is justified.

However this is not true for a generic case. The meeting of the two particles in solution (including re-encounters of two given particles) may last long enough to allow the effect of small interactions like the HFI and the ZFS during the meeting was comparable or even higher than Planck's constant. And then it is impossible to neglect these small interactions.

I call this the minor interactions with major consequences.

In this case, a situation of nonequivalent spin exchange can be realized (see, for example, Fig. 2.5).

If the spin collision partner has a strong spin-lattice interaction, then the spin exchange becomes nonequivalent as well (see Sect. 2.8.6).

In the case of nonequivalent spin exchange, the spin decoherence rate constant is greater than the spin coherence transfer rate constant. In this case, the rate constant of coherence transfer can vanish, and the rate constant of spin decoherence can become greater than the value that would have had the rate constant of decoherence, if at the same value of the exchange integral one would consider the spin exchange equivalent and neglect, for example, the HFI and the ZFS (see Fig. 2.5).

In the situation of nonequivalent exchange, there is a shift in the resonance frequency of spins (see Sect. 2.8.2, [3]).

To date, we have in the experiment convincing evidence of the manifestation of small spin-dependent interactions in the elementary act of spin exchange (see Sects. (2.8.1, 2.8.2 and 2.8.3), e.g. [4–6]).

Consistent accounting of interactions small compared to the exchange interactions of spins of colliding paramagnetic particles is an important component of the new paradigm of spin exchange.

8.1.2 Collective Modes of Motion of the Spins Due to Spin Exchange in the Bimolecular Collisions of Particles

The fundamental difference between the new paradigm of spin exchange and the existing paradigm is a multiparticle approach to the description of the motion of all spins in solution.

The new paradigm is based on the fact that the transfer of spin coherence and the transfer of spin excitation between spins during random bimolecular collisions of particles form collective modes of motion of coherence of all spins and collective modes of motion of longitudinal polarization of all spins in solution [7–10].

These collective modes are formed by the effect of transfer to a given spin of coherence or transfer of spin excitation from the collision partner.

This is something like the collision reaction in classical mechanics according to Newton's third law.

8.1.3 Contribution of Dipole-Dipole Interaction of Paramagnetic Particles to Spin Dynamics in Solutions

Along with the exchange interaction, the dipole-dipole interaction contributes to the paramagnetic relaxation of paramagnetic particles in solutions depending on the spin concentration. Therefore, the spin exchange paradigm should take into account the contribution of the dipole-dipole interaction to the paramagnetic relaxation of spins.

The new paradigm and the existing paradigm treat the transfer of spin coherence caused by dipole-dipole interaction in completely different ways.

The theory of paramagnetic spin relaxation caused by the dipole-dipole interaction of spins in a situation where the random translational diffusion of paramagnetic particles modulates the dipole-dipole interaction of their spins has been developed for a long time and is presented in many books (see, eg., [11], Chapter VIII). In this theory, the contribution of the dipole-dipole interaction to the transfer of spin coherence is neglected.

In the existing paradigm, it is believed that the dipole-dipole interaction of spins in solutions does not cause the transfer of spin coherence.

The new paradigm adopted a consistent theory of paramagnetic relaxation in liquids, which I proposed in 1976 when writing the book [12]. It was shown that in fact, the dipole-dipole interaction along with spin decoherence, the transfer of excitation energy between spins and spin-lattice relaxation of spins also leads to the transfer of spin coherence. One would expect that the contribution of exchange and dipole-dipole interactions to spin decoherence, spin coherence transfer, and excitation energy transfer between spins would add up.

But it turned out that the contributions of these interactions to the rate of spin coherence transfer are not added, but subtracted!

Unfortunately, for a long time this amendment of the theory [12] was not in demand for the interpretation of experimental data on spin exchange. This happened only in 2009 [13].

The existing paradigm of spin exchange was based on the wrong theory of paramagnetic relaxation caused by dipole-dipole interaction, which does not take into account the transfer of spin coherence between spins with different resonance frequencies.

The new paradigm of spin exchange takes into account the transfer of spin coherence induced by the dipole-dipole interaction.

8.2 Manifestation of Spin Exchange in the Shape of EPR Spectra

8.2.1 The Shape of Individual Lines in the EPR Spectrum

Within the existing paradigm, the EPR spectrum is represented as the sum of Lorentzian absorption lines. The integral intensity of each line is independent of the spin exchange rate. But resonance frequencies and line widths depend on the rate of spin exchange. In the region of fast spin exchange it is believed that all spins have the same resonance frequency equal to the frequency of the center of gravity of the spectrum in the absence of spin exchange.

In the new paradigm, the EPR spectrum is the sum of mixed-shape lines corresponding to the collective modes of spin motion. They are the sum of the symmetric Lorentzian absorption line and the antisymmetric Lorentzian dispersion line with the frequency and resonance width of the collective mode [7–9].

The contribution of dispersion to the resonance lines of collective modes is determined by the spin coherence transfer rate [7–9].

In the conditions of EPR experiments in the field of linear response, the microwave field excites collective modes in completely different ways, i.e. collective modes have different oscillator forces. As a result, collective modes produce resonance lines of completely different integral intensity [9].

8.2.2 The Exchange Narrowing

The new paradigm treats the phenomenon of exchange narrowing of EPR spectra in a completely different way [9].

Within the framework of the existing paradigm, the phenomenon of exchange narrowing of the spectrum is interpreted as a result of averaging the resonance frequencies of spins by frequent collisions of particles, i.e., by rapid transfer of spin coherence.

But as it turns out, this interpretation is wrong. Even at a very high rate of spin coherence transfer, the resonance frequencies of collective modes do not all become equal to the same frequency, they are different, as a rule.

The phenomenon of exchange narrowing occurs because, at a high rate of spin coherence transfer, only one of the collective modes has a large oscillator force, and the other modes have a zero-tending oscillator force, i.e. are the so-called "dark" states.

The new paradigm, in contrast to the existing paradigm of spin exchange, includes a description of the situation when the contribution of the exchange interaction to the transfer of spin coherence is smaller than the contribution of the dipole-dipole interaction.

Note that this situation is in principle impossible within the framework of the existing paradigm, as it is believed that the dipole-dipole interaction does not provide any contribution to the transfer of spin coherence. In this case, e.g., in the region of small values of the "rate" of the coherence transfer, the positions of the maxima of the resonance lines can be repelled rather than attracted, as is the case in the situation where the exchange interaction dominates in the process of the spin coherence transfer.

8.2.3 Saturation Effect

The new paradigm of spin exchange gives a completely different interpretation of the effect of saturation of EPR spectra of dilute solutions of paramagnetic particles [10]. Due to the transfer of spin coherence and the transfer of spin excitation in strong microwave fields, collective modes of spin motion are formed, in which the resonance frequency and resonance width depend on the power of the microwave field.

In the existing paradigm, it was not assumed that the frequencies of EPR lines could change in the presence of a microwave field.

In fact, due to the transfer of coherence and the transfer of spin excitation between the spins in the microwave field, coupled quasiparticles caused by the interaction of photons with the spin system are formed.

These quasiparticles can be called spin polaritons.

8.3 The Determination of the Rate of the Bimolecular Spin Exchange Processes with the Help of EPR Spectroscopy

The formation of collective modes fundamentally changes the manifestations of spin exchange in EPR spectroscopy.

In the new paradigm of spin exchange, a completely different protocol is proposed to determine the rate of spin exchange from EPR spectroscopy data. Within the framework of the new paradigm, it is recommended to determine the rate of spin coherence transfer first of all, finding the contribution of dispersion to the steady state EPR spectrum in the region of linear response, and not the rate of spin decoherence by the concentration broadening of resonance lines, as is customary in the existing paradigm of spin exchange.

The protocol for determining the rate of spin exchange processes from EPR spectroscopy data in the new paradigm is quite different from the protocol in the existing paradigm of spin exchange. Indeed, from the EPR data one determines only the total contribution of exchange and dipole-dipole interactions, e.g., in the rate of spin coherence transfer.

It is obvious that within the framework of the existing paradigm it is possible to obtain a completely erroneous rate of spin coherence transfer caused by the exchange interaction in bimolecular collisions. To illustrate, imagine the situation. Suppose that the total transfer rate of spin coherence is zero, since the contributions of the exchange and dipole-dipole interactions to the transfer of spin coherence fully compensate each other. Then, within the framework of the existing paradigm, it will be obtained from the experimental data that there is no spin coherence transfer in this system. This leads to the erroneous conclusion that bimolecular collisions do not occur at all.

Thus, there is a fundamental difference between the new and the existing paradigm of spin exchange concerning of getting of the spin exchange rate from the full rate of the process caused by the exchange and dipole-dipole interactions.

References

1. Kivelson, D.: Theory of ESR line widths of free radicals. J. Chem. Phys. **33**, 1094–1106 (1960)
2. Currin, J.D.: Theory of exchange relaxation of hyperfine structure in electron spin resonance. Phys. Rev. **126**, 1995–2001 (1962)
3. Salikhov, K.M.: The contribution from exchange interaction to the line shifts in ESR spectra of paramagnetic particles in solutions. J. Magn. Res. **63**, 271–279 (1985)
4. Bales, B.L., Peric, M.: EPR line shifts and line shape changes due to spin exchange of nitroxide free radicals in liquids. J. Phys. Chem. B. **101**, 8707–8716 (1997)
5. Bales, B.L., Meyer, M., Smith, S., Peric, M.: EPR line shifts and line shape changes due to spin exchange of nitroxide free radicals in liquids 4. Test of a method to measure re-encounter rates in liquids employing ^{15}N and ^{14}N nitroxide spin probes. J. Phys. Chem. A. **112**, 2177–2181 (2008)
6. Kurban, M.R., Peric, M., Bales, B.L.: Nitroxide spin exchange due to re-encounter collisions in a series of n-alkanes. J. Chem. Phys. **129**, 064501 (2008)
7. Zamaraev, K.I., Molin, Y.N., Salikhov, K.M.: Spin exchange. Nauka, Sibirian branch, Novosibirsk (1977)
8. Molin, Y.N., Salikhov, K.M., Zamaraev, K.I.: Spin exchange. Principles and applications in chemistry and biology. Springer, Heidelberg/Berlin (1980)
9. Salikhov, K.M.: Consistent paradigm of the spectra decomposition into independent resonance lines. Appl. Magn. Reson. **47**, 1207–1228 (2016)
10. Salikhov, K.M.: Peculiar features of the spectrum saturation effect when the spectral diffusion operates: system with two frequencies. Appl. Magn. Reson. **49**, 1417–1430 (2018)
11. Abragam, A.: The principles of nuclear magnetism. Clarendon Press, Oxford (1961)
12. Salikhov, K.M., Semenov, A.G., Tsvetkov, Y.D.: Electron spin echo and its application. Nauka, Siberian branch, Novosibirsk (1976)
13. Bales, B.L., Meyer, M., Smith, S., Peric, M.: EPR line shifts and line shape changes due to spin exchange of nitroxide free radicals in liquids 6: Separating line broadenning due to spin exchange and dipolar interactions. J. Phys. Chem. A. **113**, 4930–4940 (2009)

Afterword

Writing any book is a true adventure. It gives the opportunity to look over the entire subject matter from a bird's eye view and notice things you can never see standing on the ground.

It has been my utmost pleasure to reconnect with 50 years of research in spin exchange during the past eight months in the course of writing this book. It gave me the opportunity to revisit not only my own, but also my colleagues' previous works.

As a result, I was able to understand the value of the tremendous research done to date in this field already, and recognise that there truly was an actual paradigm shift in our understanding of the nature of spin exchange. It has been a real honour to be part of it.

This new outlook allowed me to formulate a number of interesting new research projects. Some are mentioned in the book. If you happen to spot any of these and want to explore them further, make sure to get in touch.

Most of all, I trust that this manuscript becomes a valuable resource for everyone who wants to use the spin exchange. As to paradigm shift, this is a reflection of the essence of scientific research, because everything we do is to understand the world around us better. I trust that this book becomes a staple in the library of future spin exchange researches.

Prof. Kev M. Salikhov
June, 2019, Kazan

K. M. Salikhov, *Fundamentals of Spin Exchange*,
https://doi.org/10.1007/978-3-030-26822-0

Printed in the United States
By Bookmasters